C0-BQF-818

Michail A. Zharkov

History of Paleozoic Salt Accumulation

Editor in Chief A. L. Yanshin

Translated by
R. E. Sorkina, R. V. Fursenko, and T. I. Vasilieva

With 35 Figures

Springer-Verlag
Berlin Heidelberg New York 1981

Professor Dr. M. A. ZHARKOV
Institute of Geology and Geophysics
630090, Novosibirsk-90, USSR

The original Russian edition was published by Nauka,
Siberian Department, Novosibirsk, 1978.
Academy of Sciences of the USSR, Siberian Branch,
Transactions of the Institute of Geology and Geophysics, Issue 354.
Editor in Chief A. L. YANSHIN

ISBN 3-540-10614-6 Springer-Verlag Berlin Heidelberg New York
ISBN 0-387-10614-6 Springer-Verlag New York Heidelberg Berlin

Library of Congress Cataloging in Publication Data. Zharkov, Mikhail Abramovich. History of Paleozoic salt accumulation. Translation of: Istorii͡a paleozoĭskogo solenakoplenii͡a. Includes bibliography and index. 1. Salt deposits. 2. Geology, Stratigraphic – Paleozoic. I. Title. QE471.15.S2Z4913 1981 553.6'32 81-1970 AACR2.

Offsetprinting and bookbinding: Brühlsche Universitätsdruckerei, Giessen.

2132/3130-543210

Preface

The principal aim of the present work is to understand the evolution of halogenesis in the Paleozoic. To succeed in the study it was necessary to make a general and systematic synthesis of data available on world-wide Paleozoic halogenic deposits and describe all known evaporite basins. This study succeeds the monograph *Paleozoic Salt Formations of the World* (Zharkov 1974a). The history of Paleozoic salt accumulation is based chiefly on evidence presented in the above monograph; this work should be considered as its direct continutation.

The present work mainly aims at: (1) establishment of the number of both salt and sulfate basins and salt and sulfate sequences formed therein in the Paleozoic; (2) determination of the stratigraphic position of salt and sulfate sequences in separate regions, their distant correlation and recognition of stages of evaporite sedimentation during the Paleozoic; (3) determination of the volume and areas of distribution of halite, potash, and sulfate sedimentation within basins and on continents through periods, epochs, and ages of the Paleozoic to single out epochs of the most intense evaporite sedimentation; (4) reconstruction of paleogeography of continents to recognize stages of evaporite accumulation and paleoclimatic zones of halogenic sedimentation in the Paleozoic; (5) understanding the evolution of evaporite sedimentation in the Paleozoic.

The nomenclature used in the book should be explained. The terms evaporite and halogenic are synonyms; they are used for all the deposits formed by the evaporation of waters in marine and continental basins. The terms salt-bearing and salt characterize only rock, potash, and other readily soluble salts and describe deposits, sequences, or series containing readily soluble salts. The term salt basins identifies basins wherein salt deposits have accumulated. Finally, the term sulfate refers to anhydrite and/or gypsum, and in some cases polyhalitic or other rocks, while readily soluble salt rocks of sulfate type are described as sulfate salt rocks. Sulfate sequences and sulfate basins embrace those with anhydrite or gypsum, and, in some cases, glauberite and thenardite rocks derived from halogenic rocks.

Acknowledgments. This work was carried out under the supervision of Academician A.L. Yanshin, who showed himself a scrupulous editor. The author is grateful to A.L. Yanshin for his generous contribution and constant help.

The book benefited from reviews of the manuscript by M.G. Valyashko, A.A. Ivanov, and M.P. Fiveg, who made valuable comments. The author is indebted to his colleagues G.A. Merzlyakov, T.M. Zharkova, V.V. Blagovidov, N.M. Emanova, and T.A. Danelia for constant assistance.

The translation of this book was done by R.E. Sorkina, R.V. Fursenko, and T.I. Vasilieva, to whom the author expresses his heartfelt thanks.

March 1981 M.A. ZHARKOV

Contents

CHAPTER IV Paleogeography of Continents and Paleoclimatic Zonation of Evaporite Sedimentation

CHAPTER I

Distribution and Number of Paleozoic Evaporite Sequences and Basins

Introduction

Recent more detailed studies of the geological structure of different regions and exploration for minerals, first of all for oil and gas, rock- and potash salts, gypsum, etc. have yielded new evidence specifying distribution of evaporite series in the interiors of the Earth, their number, areal distribution, and age (Lotze 1938, 1957a, 1968; Borchert and Muir 1964; Ivanov 1953; Ivanov and Levitsky 1960; Fiveg 1962; Strakhov 1962, 1963, 1971; Valyashko 1962; Stewart 1963b; Kozary et al. 1968; Lefond 1969; Meyerhoff 1970a,b; Zharkov 1971b, 1974a; Ivanov and Voronova 1972; Kalinko 1973a,b; Sozansky 1973; and a number of papers collected in *Potassium Salt Deposits of the USSR* (1973), and *Problems of Salt Accumulation* (vol. I, II, 1977).

The author used new information in an attempt to define the number of Paleozoic salt and sulfate sequences recognized on different continents of the Earth, their distribution in time and space, and the number of evaporite basins outlined in the Paleozoic. To elucidate these problems brief evidence should be cited from the previous book (Zharkov 1974a). Some salt and sulfate basins of Pelozoic age have been revealed recently and not yet described. Brief characteristics are given in the present work.

Cambrian Evaporite Deposits

Salt beds of Cambrian age have been recognized in Asia, North America, and Australia (Fig. 1).

They are widely developed in the East Siberian salt basin on the territory of the Siberian platform from the Yenisei Range on the west to the Lena River on the east and from the East Sayan and Baikal-Patom Highlands on the south to the basin of the Podkamennaya and Nizhnaya Tunguska Rivers on the north. An areal extent of rock salt there is believed to vary from 1,500,000 to 2,000,000 km^2. Cambrian halogenic deposits are better studied in the Irkutsk Amphitheater and in the Angara-Lena Basin.

The oldest salt sequence has been recognized in the lower Irkutsk horizon, in the north and central Angara-Lena Basin, in the Upper Vilyuchan area, as well as in the Murbai-Chastin Basin where it is exposed in the Chastin 2-ch borehole and Murbai No. 1 borehole. The sequence is liekly to occur in the foothills, i.e., at the most strongly subsiding margin of the Angara-Lena Basin, south and south-east of the Nepa-Botuoba Anticline. The sequence reaches 300–370 m in thickness. It consist of rock salt beds, up to 30 m thick, interlayered with anhydrite, anhydrite dolomite, and dolomite.

In other parts of the Irkutsk Amphitheater and Angara-Lena Basin the Irkutsk horizon includes chiefly carbonate-sulfate deposits usually varying in thickness from 140 to 650 m. Beds and bands of rock salt locally occur at different stratigraphic levels. They have been found in the lower horizon in the western Baikal area, in its central part, in the northern Irkutsk Amphitheater in the Sedanovo area, as well as in the upper horizon in the Cis-Sayans, in the south-western Baikal area, and in the region of Bratsk.

Another thick salt sequence is located above the Irkutsk horizon at a level of the Usolye horizon and has the same name. The sequence consists of repeatedly interbedded rock salt, anhydrite, anhydrite dolomite, dolomite, locally limestone, siltstone, and sandstone. It reaches 1000–1500 m in thickness. Salt saturation ranges from 50% to 75–80%. Rock salt beds vary from a few meters to 100–200 m in thickness. Three areas with maximum total thickness of rock salt are recognized: Kan-Taseeva, Baikal, and Berezovaya regions. In the first two regions the salt thickness exceeds 1000 m.

An overlying salt sequence is associated with the Tolbachan horizon of the Lower Cambrian. There beds and units of rock salt are assigned to the Belsk and Tolbachan Formations and their equivalents. Primarily they occur in the Upper Belsk Member. The thickness of the sequence, sometimes named the Belsk sequence, varies from 50 to 400–500 m. Salt saturation reaches 50–60%. In the Irkutsk Amphitheater the total maximum thickness of the rock salt beds of the Belsk sequence (150–200 m) is associated with the Kan-Taseeva and Angara-Lena areas. Carbonate and sulfate-carbonate deposits interbedded with minor rock salt usually occur between the Belsk and Usolye salt beds at a level of the lower Tolbachan and Elgyan horizons; salt saturation does not exceed 15–20% there.

A marker section of carbonate rocks identified as the Bulai or Olekma Formations occupies approximately the central part of the Cambrian halogenic series overlying the Belsk salt sequence. It lies at a level of the Uritsk and Olekma horizons. The sequence varies from 100 to 150 m in thickness. Thin beds of rock salt are locally present in the lower part of the sequence.

Further up the section within the Irkutsk Amphitheater and Angara-Lena basin we find another salt-bearing sequence at a level of the Chara horizon; it is known as the Angara or Chara Formation. The thickness reaches 500–600 m. The sequence consists of rock salt interbedded with anhydrite, anhydrite-dolomite, dolomite, and limestone, locally with sandstone. Salt saturation commonly ranges from 50% to 70%. The south-western Siberian platform is characterized by a higher content of potash in the upper salt units of the Angara Formation. The total maximum reaches 200–250 m in the Irkutsk Amphitheater.

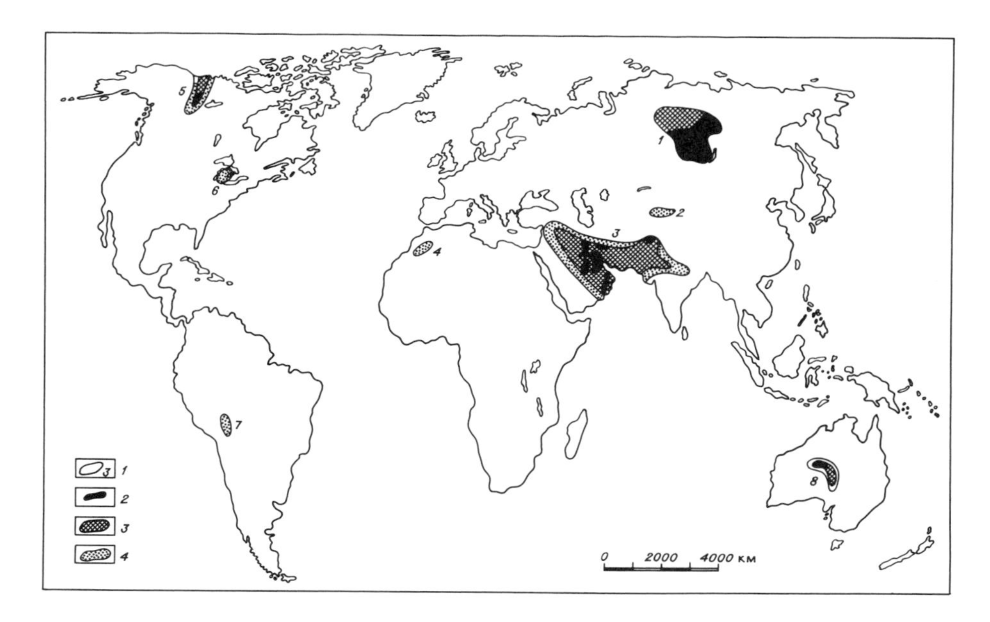

Fig. 1. Distribution of Cambrian evaporite deposits.
I Boundary and the number of evaporite basins (*1* East Siberian, *2* Tarim, *3* Iran-Pakistan, *4* Anti-Atlas, *5* Mackenzie, *6* Michigan, *7* Pre-Andean, *8* Amadeus); *II–III* areas of development of salt deposits *II* determined, *III* inferred); *IV* areas and basins of sulfate accumulation

The uppermost among the salt-bearing sequences known in the southern Siberian platform is associated with the central Litvintsevo Formation and lies at a level of the Zeledeevo horizon (Chechel 1969b). The sequence occurs in the south-eastern and central Irkutsk Amphitheater. In addition to rock salt it includes sulfate and sulfate-carbonate rocks. The thickness of the Litvintsevo salt sequence varies from 20–23 m to 165 m. Salt saturation reaches 80–85%.

In the Siberian platform, in the center of the Tungusska Syneclise, salt deposits may be present at the Mayan stage of the Middle Cambrian and in the Upper Cambrian (Zharkov 1966; Zharkov and Skripin 1971).

Thus, the information available shows that in the Cambrian deposits of East Siberia five salt sequences are recognized, namely the Irkutsk, Usolye, Belsk, Angara, and Litvintsevo sequences. Furthermore, two more sequences may be present in the central poorly known part of the Siberian platform: in the Mayan despoits of the Middle Cambrian and in the Upper Cambrian. A strict regularity can be observed in the spatial distribution of the Cambrian salt deposits. The lowermost Irkutsk sequence is developed in the marginal foothill part of the Angara-Lena Basin. The Usolye, Belsk, and Angara beds are known predominantly from the same area, but the southern limit of each overlying salt sequence gradually shifts northward toward the platform interiors. The Litvintsevo salt sequence has the least areal dimension in the Irkutsk Amphitheater but may extend into the Tunguska Syneclise. Provisionally recognized Maxan salt strata and the Upper Cambrian deposits may be situated only in the Tunguska Syneclise interiors. All the salt deposits and the carbonate and sulfate-carbonate strata separating them are closely interlinked. Such a pattern of distribution of Cambrian halogenic deposits in space and time implies the existence of a single vast evaporite basin in the Cambrian period in the Siberian platform where we can confirm a repeated change of conditions favorable for salt accumulation.

The Iran-Pakistan salt basin is situated in south-western Asia. Rock salt strata have been recognized there in three regions: (1) in southern Iran within the Persian Gulf and in the north-eastern Arabian Peninsula in Oma, (2) in central eastern Iran – in Kerman and Shirhesth, (3) in the Salt Range in Pakistan.

The presence of halogenic strata of considerable thickness in the first region is suggested by outcrops of evaporite deposits in salt domes over 200 in number (O'Brien 1957; Humphrey 1958; Kent 1958, 1970; Morton 1959; Stöcklin 1962, 1968a,b; Naquib 1963; Stöcklin et al. 1964; Mina et al. 1967; Tschopp 1967; Wolfort 1967; Ruttner et al. 1968; Ala 1974; Kent and Hedberg 1976). They are developed in Shiraz, Hormoz, South Persia, Oman-Zufar, and northern Kerman salt-bearing regions. Halogenic deposits form the Hormoz Formation consisting chiefly of rock salt, gypsum, anhydrite with fragments and blocks of other rocks. A succession of layers in the Hormoz Formation remain unknown.

Two salt sequences were reported from central Iran. One of them is developed in northern Kerman, west of the Loot Uplift, and named the Hormoz Formation (Stöcklin 1968a). It includes red saline marl and sandy shale interlayered with limestone, dolomite, gypsum, and rock salt; this salt sequence underlies the Lalun Red Sandstones. The second sequence is spread in the Shirhesth region, overlies the Lalun Red Sandstones, and occurs in the Middle/Upper Cambrian Mila Formation (Ruttner et al. 1968; Morgunov and Rudakov 1972).

The salt-bearing sequence in West Pakistan is distinguished as the Punjab Formation (Gee 1934; Bailey 1948; Schindewolf 1954; Asrarullah 1963; Krishnan 1966, 1968; Gansser 1967). It is a temporal equivalent of the Hormoz Formation. The sequence consists of alternating gypsum, anhydrite, folomite, red marl, and rock salt. A visible thickness is 879 m; an estimated thickness exceeds 1000 m.

Areal distribution of Cambrian salt beds in south-western Asia is still unknown. They may be conditionally located in the Salt Range, Pakistan, and salt-bearing regions in Iran, the Persian Gulf, and in Oman. The Cambrian evaporites appear to occur in foothills and under the Himalayas proper (Gansser 1967). An analysis of spatial distribution of Cambrian deposits of different composition suggests that salt deposits are rather widespread in south-western Asia. They have accumulated in a single vast Iran-Pakistan evaporite basin comparable in size with the East Siberian salt basin.

The Mackenzie salt basin is situated within the Mackenzie River Basin in Yukon and Northwest Territories of Canada (Williams 1923; Stewart 1945; Heywood 1955; Lefond 1969; Tassonyi 1969; Aitken et al. 1973; Aitken and Cook 1974; Meijer-Drees 1975; Norford and Macqueen 1975). A salt sequence is exposed there in deep boreholes in the Norman Wells area and to the south, near Blackwater Lake. It is identified as the Salina River Formation, rock salt mainly occurring in the lower part of the Formation in intercalations with anhydrite, dolomite, mudstone, and siltstone. West and east of the Mackenzie River Basin salt deposits wedge out and the Salina River

Formation is represented there by red to brown dolomite marl, grey anhydrite, and dolomite. A supposed age of the halogenic sequence is Late Cambrian (Aitken et al. 1973; Norford and Macqueen 1975). The areal extent of the Salina River salts is not yet known, but it may be rather considerable and embrace the Mackenzie River Basin plain and extend farther north into the Canadian Arctic Archipelago. The Cambrian Mackenzie evaporite basin is no less than 800–1000 km in length and over 600 km in width.

The Amadeus Basin in Australia is the region where Cambrian salt deposits have been recently found (Ranford et al. 1965; Wells et al. 1967, 1970; Cook, 1969; Wells 1969). They are associated with the Chandler Limestones and exposed in the Alice No. 1, Orange No. 1, and Mount Charlotte No. 1 deep boreholes. The Cambrian salt deposits appear to have rather wide distribution in the Amadeus Basin. Now they are presumed to occur over the entire eastern part of the basin. However, the Cambrian evaporite basin proper in Australia may exceed the bounds of the Amadeus Basin, and extend farther west and north-west including the Canning Basin. Cambrian halogenic deposits may continue south-eastward into the Torrens region in northern Adelaide as well, where diapiric domes are known. If the assumption concerning such a wide areal distribution of Cambrian halogenic deposits proves true, then one more large evaporite basin of the Cambrian period comparable with the East Siberian and Iran-Pakistan Basins will be outlined in Australia.

Besides the above regions of Cambrian salt deposits, the presence of evaporite rocks in sedimentary strata of the Cambrian can be recorded in other isolated regions of the Earth. Cambrian gypsum and anhydrite have been found in north-western Africa within the Anti-Atlas Basin in Morocco (Lotze 1957a), in the Cis-Andean Basin in South America (Benavides 1968), in the Tarim Basin in China (Some Aspects . . . 1965; Meyerhoff and Meyerhoff 1972). Salt deposits are suggested in the last two regions as well, but they have not been proved yet. Rodgers (1970) has suggested recently that salt deposits of Cambrian age may occur in the Central and southern Appalachians, USA. Another region of distribution of Cambrian sulfate rocks has been recently reported from North America. A 425-m thick anhydrite sequence containing interlayered anhydrite, dolomite, sandstone, and mudstone was exposed in the north-eastern central part of the Michigan Basin in the Brazos State Foster 1 deep borehole; the sequence is included in the Munising Group and compared with the Eau Claire, Galesville, and Franconia Formations of the Upper Cambrian (Catacosinos 1973). The anhydrite sequence probably occurs over a small area in the northern Michigan Basin.

Thus, only twelve salt units of Cambrian age have been defined up to now on different continents of the Earth; they were formed within four evaporite basins. Seven salt units, such as the Irkutsk, Usolye, Balsk, Angara, Litvintsevo sequences, and presumably a Mayan sequence (Middle Cambrian) and an Upper Cambrian sequence, have accumulated in the East Siberian Basin; three units, namely the Hormoz, Mila, and Punjab Formations, underlie the Iran-Pakistan Basin; the Salina River Formation is developed in the Mackenzie Basin; and the Chandler Limestones are spread in the Amadeus Basin. The number of salt units has not yet been determined finally in the basins because of inadequate information. Thus, only five units have been so far recognized in the East Siberian Basin, and two upper units are believed to occur in the central Siberian platform. Salt sequences may also occur in the Iran-Pakistan Basin and in the Amadeus Basin. The number of separate sulfate-bearing sequences can at present be estimated for Cambrian salt basins, and therefore sulfate deposits have been assigned to salt-bearing strata. Four sulfate basins, namely the Anti-Atlas, Cis-Andean, Tarim, and Michigan Basins may have existed in the Cambrian period as well. Four sulfate sequences have been recognized there so far. Rock salt sequences appear to exist in the sulfate basins, particularly in the Cis-Andean and Tarim Basins.

Ordovician Evaporite Deposits

The information available allows us to recognize ten basins where evaporite deposits of Ordovician age have been found, namely the Lena-Yenisei Basin, the Basin of Severnaya Zemlya, the Baltic Basin, the Basin of the Canadian Arctic Archipelago, and the Moose River, Williston, South Illinois, Anadarko, Canning, and Georgina Basins (Fig. 2), but apparent Ordovician salt deposits have been recently found only in the evaporite Basin of the Canadian Arctic Archipelago. The presence of rock salt in Ordovician deposits is recorded in the Williston and Canning Basins as well, but they are only tentatively considered as salt basins. The age of the salt bed in the

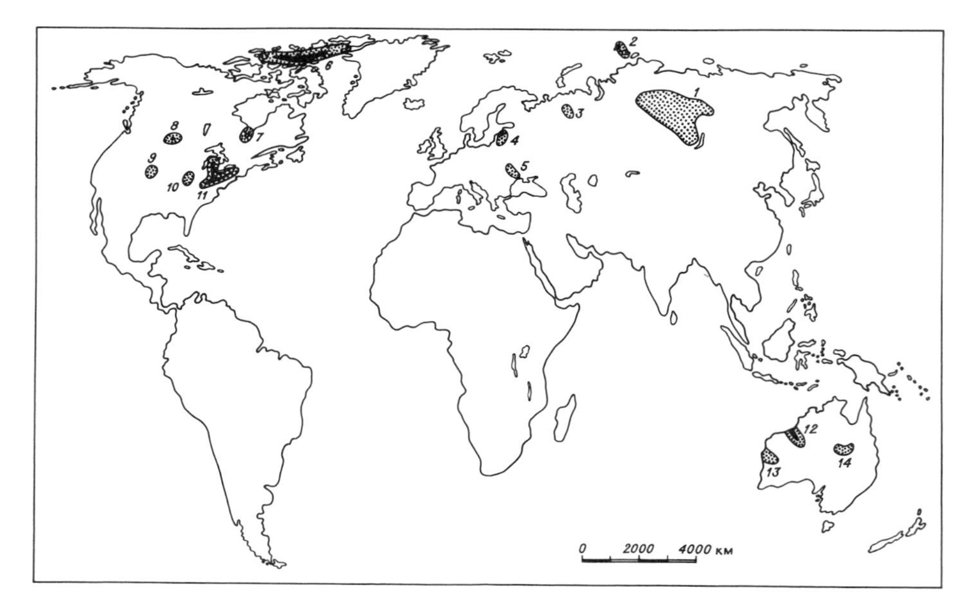

Fig. 2. Distribution of Ordovician and Silurian evaporites.

Evaporite basins: *1* Lena-Yenisei, *2* Severnaya Zemlya, *3* Pechora, *4* Baltic, *5* Dniester-Prut, *6* Canadian Arctic, *7* Moose River, *8* Williston, *9* Anadarko, *10* South Illinois, *11* Michigan-Pre-Appalachian, *12* Canning, *13* Carnarvon, *14* Georgina. For explanations see Fig. 1

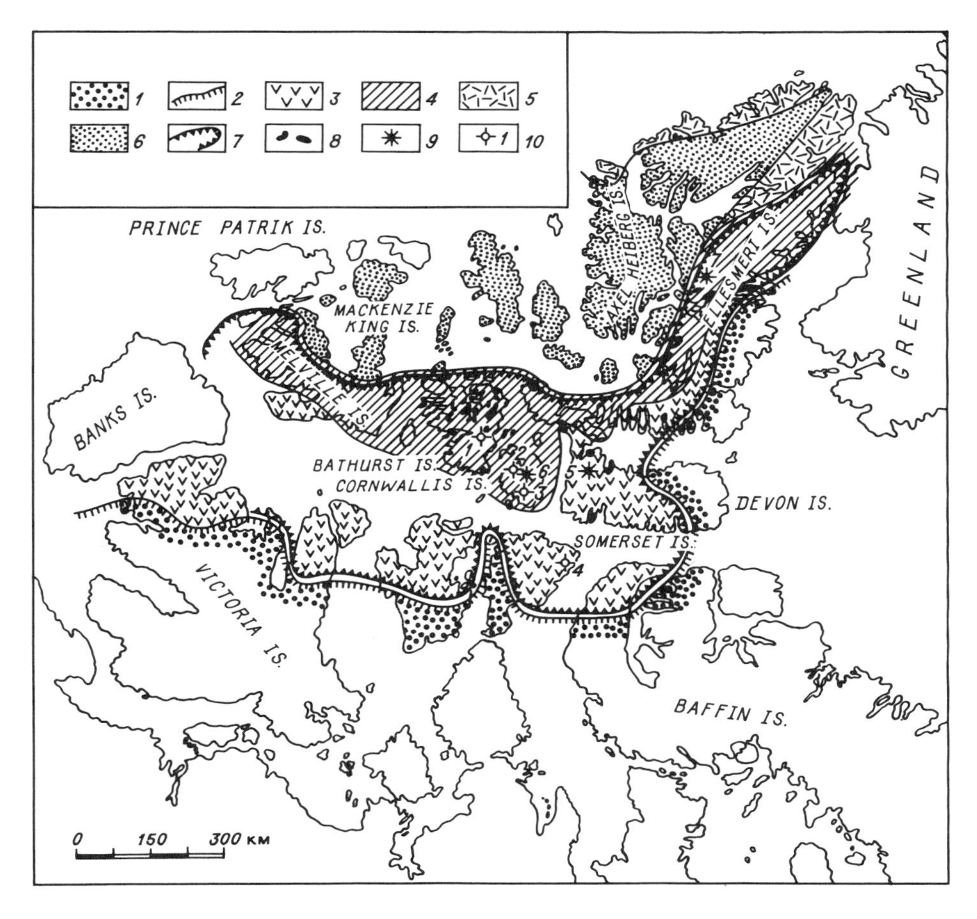

Fig. 3. Index map of the Basin of the Canadian Arctic Archipelago. Compiled from the data of Thorsteinsson and Tozer (1960), Forthier et al. (1963), Kerr (1967a), 1968, 1974, 1975), Mayr (1975), Miall (1974a,b, 1975), Mossop (1972a,b, 1973), Trettin (1969, 1971), etc.

1 outcrops of Precambrian formations, *2* limit of their distribution, *3* central stable region, *4* Franklinian miogeosyncline, *5* Franklinian eigeosyncline, *6* Sverdrup Basin, *7* boundaries of the Ordovician evaporite basin, *8* outcrops of Ordovician sulfate rocks, *9* diapiric domes, *10* boreholes which penetrate Ordovician deposits (*1* Bathurst Caledonian River J-34, *2* Panarctic Deminex Cornwall Central Dome K-40, *3* Lovitos Cornwall Resolute Bay L-41, *4* Panarctic Deminex Garner O-21)

Canning Basin has not yet been determined; the sequence is referred to as Late Ordovician/Early Devonian, but it may even be Cambrian. For the Williston Basin there is rather meager evidence for Ordovician rock salt beds exposed at great depths. Only sulfate sediment accumulation took place in all the other basins.

Within the Basin of the Canadian Arctic Archipelago Thorsteinsson and Tozer (1960, 1961) have recognized four major structural elements: Central stable region, miogeosynclinal and eugeosynclinal zones of the Franklinian geosyncline, and Sverdrup Late Paleozoic/Meso/Cenozoic basin superposed on the ancient Franklinian geosyncline (Fig. 3).

Ordovician evaporite deposits were reported from the Central stable region and from the miogeosynclinal zone. Only the south and south-west boundaries of the evaporite basin are known along outcrops of Precambrian formations. The north-western boundary of the basin is more or less precise, it runs between the miogeosynclinal and eugeosynclinal terranes of the Paleozoic on northern Ellesmere Island. As for the northern and north-eastern boundaries, they are not yet defined and coincide approximately with the southern boundary of the Sverdrup Basin. However, Lower Paleozoic, including Ordovician, deposits are buried there under sedimentary strata underlying the Sverdrup Basin; the evaporite basin may extend farther north. The eastern boundary of the basin is unknown. The basin there appears to continue on the Arctic Ocean shelf.

Ordovician deposits of the Canadian Arctic Archipelago are better known from Ellesmere, Devon, Bathurst, Cornwallis, and Somerset Islands. Furthermore, they are exposed in four deep boreholes, one of which (Bathurst Caledonian River J-34) penetrates the saliniferous bed. Data from foreign literature have been used to characterize Ordovician deposits (Fortier et al. 1963; Kerr 1967a, 1968, 1974, 1975; Trettin 1969, 1971; Mossop 1972a,b, 1973; Kerr et al. 1973; Miall 1974a,b, 1975; Mayr 1975, 1976).

At the base of the Ordovician on Ellesmere and Cornwallis Islands there was identified the Copes Bay Formation whose lower part may be placed into the Upper Cambrian. It consists chiefly of silty and argillaceous, locally gypsinate, limestone; conglomerate can be observed in the lower part of the formation. The thickness varies within a wide range from 100 to 1200 m. The formation is overlain by a bed of evaporite deposits identified as Baumann Fiord Formation (Kerr 1967a). The formation mainly crops out on Ellesmere Island along coasts of Trold Fiord, Flat Pebble and Stearfish Bays, as well as in the central part of the island and in its western regions on the Bache Peninsula. To the west the Baumann Fiord evaporites subside under younger deposits and are exposed only in the central part of Cornwallis Island in the Panarctic Deminex Cornwallis Central Dome K-40 borehole. The Baumann Fiord Formation has been discussed in detail by Kerr (1967a) and Mossop (1972a,b, 1973).

The formation was divided into three units: A, B, and C. Lower unit A consists mainly of beds of coarsely crystalline anhydrite interlayered with nodular anhydrite and thin-bedded anhydrite-carbonate rocks. Lenses of micrograined limestone 0.6–3 m thick occur locally. Silicification associated with anhydrite is rare. Unit B contains green to gray, thin to thick-platy micrograined limestone interlayered with isolated red quartz sandstone and red argillaceous limestone. Anhydrite interbedded with anhydritized limestone dominates unit C section; these rocks are thin-bedded. The thickness of the Baumann Fiord Formation varies from 200 to 500 m.

On northern Ellesmere Island the Baumann Fiord Formation continues northward to the margin of the miogeosyncline, wedges out, and is replaced by the lower Eleanor River Formation (Thorsteinsson 1974). The Eleanor River Formation overlies the Baumann Fiord evaporites elsewhere in interiors of the miogeosyncline and the Stable region. The Eleanor River Formation consists of dark to light gray limestone varying from thin-platy to massive, silicified, locally dolomitized types. The formation reaches 200 m in thickness.

Overlying Ordovician deposits have been recently united in the Cornwallis Group including Bay Fiord, Thumb Mountain, and Irene Bay Formations (Kerr 1967a, 1974; Thorsteinsson 1974; Mayr 1975). The formations can be traced along the strike of the miogeosynclinal zone from Bathurst Island to North Greenland, but in western parts of the basin they are buried under Silurian/Devonian deposits.

The lowermost Bay Fiord Formation of the Cornwallis Group contains evaporite deposits building up the second halogenic sequence in the Ordovician section of the Canadian Arctic Archipelago. Rock salt beds are exposed in this very sequence. The Bay Fiord Formation consists chiefly of thin-bedded argillaceous limestone, siltstone, anhydrite, and dolomite. Its thickness varies from 300 to 600 m.

In the Houghton Dome located on Devon Island the Bay Fiord Formation it consists of three units (Greiner 1963). Dark to light gray, thin-bedded to thin-platy gypsinate mudstone and dolomite mudstone interbedded with thin (up to 30 cm) light gray dolomite occur in the lower unit (about 120 m thick). The upper 15 m of the unit contains abundant dolomite and mudstone. The second unit includes dolomite that is thin-platy at the base and massive at the top; this unit is 30 m thick. The third unit, 65 m thick, consists of gray dolomite, mudstone, and gypsum. The gypsum sequence in the Haughton Dome is tightly folded and extremely variable in composition. Locally it contains abundant sulfate rocks.

In the Prince Alfred Bay region on Devon Island the gypsum sequence of the lower Cornwallis Group (probably, Bay Fiord Formation) includes 80-m thick thin-platy gypsum in the lower part and is interbedded with green, blue, and dark gray gypsum, calcareous siltstone, gray mudstone, and gypsinate mudstone in the upper part (about 100 m thick); gypsum dominates the upper part of the section.

On Bathurst Island the Bay Fiord Formation contains thick beds of rock salt. They are penetrated by the deep Bathurst Caledonian River J-34 borehole. A visible thickness of the Bay Fiord Formation is 1107 m. Its upper part (125 m) is represented by mudstone, anhydrite, and dolomite. Rock salt interbedded with anhydrite, dolomite, and mudstone occurs below. The saliniferous part of the section is penetrated down to 682 m. An exposed thickness of the Bay Fiord Formation reaches 807 m as a whole (Kerr 1974).

A salt sequence appears to have a considerable areal distribution within the Ordovician evaporite basin of the Canadian Arctic Archipelago. The sequence apparently extends farther north, into the Sverdrup Basin. The areal extent of rock salt is still unknown there, but it appears to include Bathurst and Cornwallis Islands, i.e., to exceed 100,000 km^2. If average thickness of rock salt in the area is at least 100 m, then, together with a visible thickness of the salt sequence, the volume of rock salt in the basin will exceed 1.10^4 km^3. Without doubt this estimate is lower than the real volume of salt, but it should be viewed as a first approximation.

The Thumb Mountain Formation forming the central part of the Cornwallis Group reaches 450–500 m in thickness. In the central miogeosyncline the formation includes thick-platy dark to brown argillaceous limestone, locally dolomitized at the top. In a number of sections on Bathurst Island the formation contains light gray dolomite and green mudstone. Locally dolomite builds up individual units up to 125 m thick. They are usually brown to chocolate-brown in color. The Irene Bay Formation is represented by thin-platy gray and green-gray limestone with lenses, bands, and beds of green mudstone. Its thickness varies from 35 to 55 m.

The section of Ordovician deposits is crowned with the Allen Bay Formation developed on Ellesmere, Cornwallis, and Devon Islands. or with its equivalent Cape Phillips Formation on Bathurst Island and in the northern regions of the miogeosyncline on Ellesmere Island. The upper members of the formations are assigned to the Silurian. The Ordovician deposits of the Cape Phllips Formation include chiefly dolomite, argillaceous limestone, mudstone, and silicified argillaceous limestone. Gray to yellow, brown, and yellow to gray dolomite dominates the Allen Bay Formation. Its thickness varies from 300 to 1200 m.

Hence, two evaporite sequences, namely the Baumann Fiord and the Bay Fiord, can be defined in the Ordovician of the Canadian Arctic Archipelago Basin. The lower sequence is assigned to the Lower Ordovician, i.e., to the Lower Arenigian. The Bay Fiord Formation is of Middle Ordovician age. Salt deposits are known only from the upper Bay Fiord evaporite sequence. The lower Baumann Fiord sequence is so far known as sulfate-bearing. All the above Ordovician deposits, together with Silurian sedimentary sequences, may have accumulated in a single sedimentary basin existing through the Early Paleozoic and probably occupying almost the entire territory of the Canadian Arctic Archipelago.

The Canning Basin is situated in north-western Australia between Precambrian basement inliers: Kimberley to the north-east and Pilbara to the south-west. The basin is filled in with deposits of Ordovician, Silurian, Devonian, Carboniferous, and Permian age; a maximum total thickness amounts to 7–8 km. Salt deposits are known to be present at the base of the section under study. Their total thickness, areal extent, and age are still uncertain. Evaporite deposits form the Carribady Formation, which reaches 1700 m in thickness. Two salt beds (120 and 500 m thick) alternating with anhydrite and dolomite can be identified within the formation. The two sequences are separated by a bed of gray dolomite interlayered with red to brown dolomitic marl and black mudstone. The Carribady evaporite series underlies Middle Devonian red beds (Veevers 1967), the lower part of which is assigned by some geologists to the Upper Devonian (Johnstone et al. 1967). Evaporites are underlain by a sequence of limestone, mudstone, and sandstone assigned to the Lower/ Middle Ordovician (Veevers 1967). This evidence allows us to assign the Carribady Formation to the Lower/Middle Ordovician. But this age is disputed by some writers (Koop 1966), who emphasize that the sequence underlying evaporites is not exposed directly in boreholes and a conclusion that the Carribady Formation overlies Lower Ordovician deposits was drawn from comparison with other salt-free beds. Therefore an older age of the evaporite series in the Canning Basin can be expected. It may be considered as an equivalent to the Cambrian Chandler salt sequence of the Amadeus Basin south-west of the Canning Basin along the same strike. This suggests that the Canning evaporite basin and salt sequences of the Carribady Formation within the basin can be tentatively referred to as Ordovician.

The Williston Basin underlies parts of North Dakota, Montana, South Dakota in the United States and parts of Manitoba, Saskatchewan, and Alberta in Canada. Evaporite deposits are widely developed within Ordovician sedimentary terranes. The Red River, Stony Mountain, and Stonewall Formations forming a section of the upper Lower, Middle, and Upper Ordovician contain six sulfate units, three in the lower, one in the middle, and two in the upper formations. Units 1–1.5 m thick consist of anhydrite interbedded with dolomite and limestone. The Stonewall Formation is believed to contain rock salt bands exposed in boreholes at depths over 2900 m along the Cedar Creek anticline in south-eastern Montana, USA (Lefond 1969), However, the evidence needs to be checked.

Within the Lena-Yenisei sulfate basin sulfate-bearing rocks have been found in the north and north-western Siberian platform, as well as in its central and south-eastern regions. They are developed mainly in the Norilsk region (north-western Siberian platform) where sulfate rocks are present throughout the Ordovician section of the Iltyk, Guragir, Angir, Amarkan, and Zagornino Formations (Ordovician Stratigraphy . . . 1975). Gypsum and anhydrite usually occur among carbonate-, variegated terrigenous-carbonate-, or marl-clayey sediments. They vary mainly from 0.2 to 0.5 m, individual beds reach 1 m in thickness. In the Nizhnaya Tunguska River Basin sulfate rocks within the Ordovician sequences are exposed in the Turin and Tutoncha boreholes. In the former borehole they occur as bands and lenticular inclusions in a Lower Ordovician limestone-dolomite sequence, while in the Tutoncha borehole anhydrite was found in the lower part of the dolomite sequence and in Lower Ordovician limestone sequence. In the northern Siberian platform, gypsum beds in the Moiero River Basin occur in the central member of the Irbukly Formation and in the Kochakan Formation. Gypsum bands, up to 0.2 m thick, have been found in the Maimecha River Basin in deposits of the Tompok and Bysyuryakh Formations as well. In the Vilyui River Basin Middle/Upper Ordovician deposits contain sulfate rocks. The Stan Formation has gypsum intercalations, several centimeters thick. Pink to white gypsum beds make up the entire Kharyalakh Formation; the thickness of upper horizons varies from 1 to 15 m. In the Markha-Morkoka district a gypsum sequence marked by the presence of gypsum beds interlayered with mudstone, marl, and dolomite is considered to be an equivalent to the Kharyalakh Formation (Ordovician Stratigraphy . . . 1975). In the Berezovaya Depression gypsum is reported from the Lower Ordovician Tochilnin Formation and from the Upper Ordovician Ilyun Series. In the Irkutsk Amphitheater gypsum bands and lenses lie among red beds of the Upper Ordovician Bratsk Formation. Hence, in the Lena-Yenisei Basin sulfate rocks form a wide belt running from the north-western and northern Siberian platform (Norilsk, the Moiero and Maimecha Rivers) south-eastward to the Berezovaya Basin.

In the Severnaya Zemlya Basin sulfate rocks dominate the dolomite-marl and gypsum-limestone members of the Komsomol Formation (Egiazarov 1970).

In the Baltic Basin gypsum lenses, inclusions, and thin layers, about 10 cm thick, occur among dolomite and terrigenous-carbonate sediments in the Iev upper horizon. Moreover, small gypsum inclusions and lenses, as well as gypsification of rocks can be observed in deposits of the Itfer, Tallin, and Porkun horizons (Selivanova 1971).

The Moose River Basin is situated in the south-west coast of the Hudson Bay, Canada. Gypsum bands and beds occur there in the Ordovician dolomite sequence (Norris and Sanford 1969).

The South Illinois Basin is recognized as a sulfate region from the presence of anhydrite bands and thin beds, 0.5 to 1 m thick, within the Middle Ordovician Ioachim Formation (Bond et al. 1968, 1974).

The Anadarko Basin embracing South Oklahoma, USA, is recognized as an evaporite basin owing to rather abundant anhydrite in the West Spring Creek Formation in which anhydrite forms individual bands and beds alternating with dolomite and dolomitic marl, mudstone, and sandstone, or occurs in the rocks as lenses and inclusions (Reedy 1968; Latham 1973).

In the Georgina Basin, West Australia, thin gypsum inclusions and lenses are known in the terrigenous-carbonate rocks of the Lower/Middle Ordovician Toko Group (Brown et al. 1968).

Thus, four salt sequences of Ordovician age can be recognized: the Bay Fiord Formation is spread in the Basin of the Canadian Arctic Archipelago; the Lower and Upper salt sequences of the Carribady Formation are developed in the Canning Basin; the Stonewall Formation underlies the Williston Basin. The last three sequences are recognized tentatively. Eighteen sulfate sequences can be identified in the Ordovician evaporite basins: the Lena-Yenisei Basin [1] contains seven sequences (the Ordovician sulfate-carbonate sequence is spread in the Norilsk district; the sulfate-dolomite sequence is developed in the central Tungusska Syneclise; Irbukla, Kochakan, Stan, Kharyalakh, Tochilnin, and Bratsk Formations are known); the Baltic Basin includes three sequences, namely Tallin, Itfer, and Ievsk horizons; the Williston Basin has two sequences, namely the Red River and Stony Mountains Formations; the Basin of Severnaya Zemlya includes the Komsomol Formation; the Moose River Basin has the dolomite sequence; the Basin of the Canadian Arctic Archipelago includes the Baumann Fiord Formation; the South Illinois Basin contains the Ioachim Formation; the Anadarko Basin includes the West Spring Creek Formation; and, finally, the Georgina Basin contains the Toko sequence. In the future salt deposits will be found in sulfate basins poorly studied as yet, particularly in those where rather thick anhydrite units occur. The promising basins appear to include the Lena-Yenisei Basin, as well as the Basins of Severnaya Zemlya and the Canadian Arctic Archipelago. It should be noted that there were three evaporite basins only in the Ordovician Epoch, i.e., the Anadarko, South Illinois, and Georgina Basins, all of them sulfate basins. The other basins, as will be shown below, probably persisted in the Silurian, and some of them as far as the Devonian, and evaporitic accumulation was discontinuous; these are Ordovician/Silurian, or even Ordovician/Silurian/Devonian basins.

Silurian Evaporite Deposits

Salt deposits of Silurian age are reliably recognized now only in the Michigan-Pre-Appalachian Basin (see Fig. 2). It occupies a vast territory of New York, Pennsylvania, West Virginia, Ohio, and Michigan, USA, as well as the south-western extremity of Ontario, Canada. Salt deposits in the Michigan-Pre-Appalachian Basin were studied in detail (Alling 1928; Martens 1943; Landes 1945; Pepper 1947; Dellwig 1955; Fettke 1955; Kreidler 1957; Alling and Briggs 1961; Pierce and Rich 1962; Sanford 1965; Dellwig and Evans 1969; Lefond 1969; Matthews 1970). According to Pierce and Rich (1962) an areal extent of rock salt amounts to about 260,000 km^2, but according to Lefond (1969), 214,000 km^2.

1 The number of sulfate sequences in the Lena-Yenisei Basin and their names are tentative because of inadequate information concerning the northern and central Siberian platform

In the basin evaporite deposits form the Upper Silurian Cayugan Series divided into the Salina and Bass Island Formations. In the Michigan Basin the Salina Formation is divided into seven units designated A through G. Three Unites (C, E, G) are essentially dolomite-argillaceous with anhydrite; three others (B, D, F) consist of rock salt with layers of anhydrite and dolomite, locally, of mudstone; the lowermost unit A is represented by dolomite, but in the central Michigan Basin beds of anhydrite and rock salt appear. The Bass Island Formation contains dolomite, but its subsiding parts are made up of anhydrite and rock salt. Bands of potash salt (sylvinite) occur, with a thickness of 28 m.

The halogenic Salina Series in the Pre-Appalachian Basin is represented by four formations, in ascending order they are: the Pittsford, Vernon, Syracuse, and Camillus Formations. All salt deposits are placed into the Syracuse Formation consisting or rock salt interbedded with gray mudstone, anhydrite, and dolomite; seven salt units with thicknesses ranging from 4.6 to 168 m are identified in the section. Moreover, anhydrite bands are present in the Vernon and Camillus Formations. The total maximum thicknesses of rock salt reach 540 m in the Michigan Basin, and 250 m in the Pre-Appalachian Basin.

Two more Silurian evaporite basins, namely, the Dniester-Prut and Lena-Yenisei Basins, which may have been salt-producting, can be recognized now.

The Dniester-Prut evaporite basin occupies the south-western margin of the East European platform within the Moldavian plate. The basin is bounded in the north-east by the Ukrainian Shield, and in the east and south probably by the uplift zone postulated for the Black Sea and extending into Rumania. In the south-west the basin was separated from the open sea of normal salinity by a shallow carbonate zone, developed within the Podolian Saddle and probably extending into northern Dobruja. The basin was first described as evaporite by Zavidonova (1956) and then outlined by P.L. Shulga and V.S. Krandievsky (Atlas of Paleogeographical . . . 1960; Atlas of Lithologo-Paleogeographical . . . 1960).

The following description of the basin is based on the data presented by the above-mentioned authors, as well as by O.G. Bobrinskaya et al. (Stratigraphy of Sedimentary . . . 1964), Vyalov (1966), Edelstein (1969), Tesakov (1971), O.I. Nikiforova et al. (Reference key . . . 1972), Grachevsky and Kalik (1976), Tsegelnyuk (1976).

Silurian deposits within the Dniester-Prut Basin and in Podolia are divided into Kitaigorod, Muksha, Ustje, Malinovtsy, and Skala horizons. The Kitaigorod horizon in Moldavia consists of dark gray argillaceous limestone, locally dolomitized, interbedded with mudstone and marl, and in some cases with siltstone and sandstone at the base. The horizon varies from 25 m to 30 m in thickness. Deposits of the Kitaigorod horizon when extending north-westward into East Podolia are identified as a formation of the same name (Tsegelnyuk 1976). In occurrences near the villages of Verkhnyakovtsy, Shidlovtsy, Yurkovtsy, Ivanovka, and Darakhov, as well as in exposures along the Dniester River, the formation is represented by dark gray cloddy limestone with marl interlayers. The formation ranges from 39 to 68 m in thickness. Farther north-west in Volyno-Podolia the Kitaigorod Formation includes, along with limestone and marl, dark to green-gray mudstone and dolomite bands. Westward (the town of Lutsk, villates of Pelcha, Romashkovka, Godomichy) dolomitized limestone, dolomite, and dolomitic marl appear in the section. Then dolomites and dolomitized carbonate rocks become subordinate and toward the Lvov Depression mudstone, siltstone, black pelitomorphic limestone dominate the Kitaigorod Formation. In the subsiding part of the Lvov Depression equivalents of the Kitaigorod Formation are known to form the Dublyany Formation (Tsegelnyuk 1976) including essentially black argillaceous siltstone with layers of black to dark gray limestone and locally sandstone; the formation is 44 to 80 m thick.

Within the Dniester-Prut Basin the overlying deposits including the Muksha, Ustje, Malinovtsy, and lower Skala horizons form a single sequence of halogenic rocks, predominantly anhydrite. In different parts of Moldavia the sequence is represented by dolomite, dolomitized limestone, limestone, marl, and mudstone interbedded with anhydrite and bentonite clay. The thickness of the sequence is 250–300m. Anhydrite occurs chiefly at a level of the Ustje, Malinovtsy, and lower Skala horizons, while dolomite limestone with rare lenses and inclusions of anhydrite occurs chiefly in the Muksha horizon. Zavidonova (1956) using V.S. Eremenko's data reported the presence of rock salt bands in the Malinovtsy horizon near the village of Sarateny-Vek, Nisporen District. Their thickness and areal extent are unknown. The same evidence reported by Zavidonova was cited by P.L. Shulga (Atlas of Paleogeographical ... 1960). However, in works published later rock salt was not recorded in the Silurian deposits of the Dniester-Prut Basin. Even in deep boreholes in Nisporen and adjacent areas only thin bands and inclusions of anhydrite were reported (Stratigraphy of Sedimentary . . . 1964; Edelstein 1969). Hence A.G. Zavidonova's information should not be relied upon completely. Though the possible presence of rock salt should be noted in the Dniester-Prut Basin, the basin itself should be so far recognized as sulfate-bearing only.

A sulfate-bearing carbonate sequence of the Muksha, Ustje, Malinovtsy, and lower Skala horizons in north-western Moldavia (the villages of Brinzeny, Morosheshty) is known as the Pigai Formation (Tsegelnyuk and Bukatchuk 1974) consisting chiefly of dolomite and dolomitic marl with bands and inclusions of anhydrite. Farther north-westward in the Dniester Basin equivalents of the sulfate-bearing carbonate sequence are divided into the Bagovitsy, Malinovtsy, and Prigorodok Formations (Tsegelnyuk 1976). The Bagovitsy Formation consists of two members: the lower Muksha member comprising dolomitized limestone, dolomitic marl, and dolomite, 13–25 m thick, and the upper Ustje member, 20–26 m thick, represented by dolomite and dolomitic marl with rare limestone layers. The Malinovtsy Formation includes cloddy and massive limestone, platy dolomite and marl. Its thickness in the Dniester region ranges from 112 to 142 m. Tsegelnyuk subdivides the Malinovtsy Formation into several dolomite and dolomitic marl members. One of them, the Perpelitsy member along the Dniester and in the Zbruch and Zhvantsik Rivers interfluve, is assigned to the lower Malinovtsy member. According to Tsegelnyuk, within the basin of the Dniester, Zbruch, Zhvantsik, Stryp, and Safet Rivers the upper Malinovtsy member includes deposits of Isakovtsy beds represented by dolomite and dolomitized limestone, interbedded with gypsum and anhydrite in the vicinity of Darakhov. The Prigorodok Formation also contains dolomite and dolomitic marl. Its thickness varies from 22 to 60 m. But near the village of Verkhniakovtsy and the town of Darakhov the formation contains gypsum and anhydrite layers as well. Hence, evaporite rocks are known to stretch from the Dniester-Prut interfluve through the Dniester Valley to the town of Darakhov and possibly farther north-westward.

West of the above region the Muksha, Ustje, Malinovtsy, and lower Skala sections are mainly limestone. Equivalents of the Bagovitsy Formation form the Stryp Formation of dark gray to black limestone, marl and argillaceous limestone begin to dominate the Malinovtsy Formation; and dark gray limestone and clay occur in the Zavadov Formation equivalent to the Prigorodok Formation. Even black argillaceous limestone, black mudstone, and siltstone forming the Kulichkov, Peremyshil, and Zavadov sections occur in the subsiding zone of the Lvov Depression and in the Cis-Carpathian Depression.

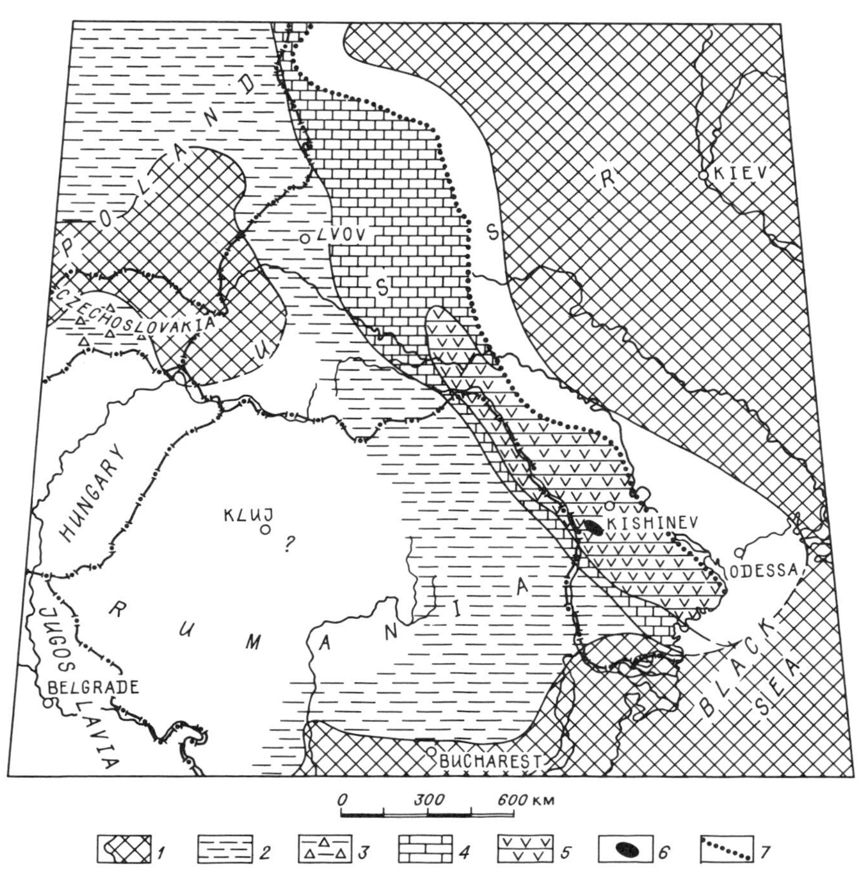

Fig. 4. Litho-paleogeographic map for the Ludlovian of south-western margin of the East-European Platform and adjacent area. Compiled from the data taken from the "Atlas of Litho-Paleogeographic maps . . ." (1960), "Atlas of Paleogeographic Maps . . . (1960), publications by Zavidonova (1956), Edelstein (1969), Tsegelnyuk (1976), Vyalov (1966), Nikiforova et al. (Reference . . . 1972).

1 land, *2–5* areas dominated by shale and graptolitic shale *(2)*, silty shale *(3)*, limestone and dolomite *(4)*, dolomite and sulfate rocks *(5)*, *6* inferred areas of rock salt development, *7* recent boundaries of deposits

The upper Skala deposits in Moldavia are represented by limestone, marl interlayered with mudstone and dolomite, up to 200 m thick. In Podolia Tsegelnyuk divided these deposits into the Rashkov and Zvenigorod Formations. The former includes limestone, dolomitic marl, dolomite, and clay; the latter consists chiefly of cloddy and platy limestone. In both formations the thickness of argillaceous rocks increases westward toward the Lvov Basin; and finally dark gray to black mudstone interbedded with dark limestone becomes dominant. This argillaceous sequence is divided by P.D. Tsegelnyuk into the Glinyan and Poltvin Formations.

Thus, within the Dniester-Prut Basin evaporite deposits occur in the center of the Silurian sedimentary sequence at a level of the Muksha, Ustje, Malinovtsy, and lower Skala horizons. Their age is mainly Wenlockian/Ludlovian.

The Dniester-Prut evaporite basin was engulfed and connected with the sea in the south-west (Fig. 4) as suggested by facial changes when a sulfate-dolomite sequence in the central basin grades into dolomite, limestone, and limestone-argillaceous sequences. Deep-water black mudstone and limestone occur in the central Lvov Depression. Recently they were penetrated by deep boreholes in southern Moldavia in Dobruja (in the Yargorin, Baurchin, and Mantov areas), in the Odessa Region (Orekhovo area), and in Romania near the towns of Constanta, Călăraşi, Borsul-Verde, Popeşti, Hirlău, Rădăuti (Grachevsky and Kalik 1976). All these regions were once covered by a deep open sea situated west and south-west of the evaporite basin. The latter was separated from the sea by a rather narrow strip of shallow water where carbonate reef structures are believed to have developed (Grachevsky and Kalik 1976). Locally in the evaporite basin conditions might be favorable for salt accumulation . However, there is no reliable evidence to support accumulation of salt sequence in the Dniester-Prut Basin. The data available allow us to regard the basin only as a sulfate one and to distinguish there only the Pigai sulfate-bearing sequence(Tsedelnyuk and Bukatchuk 1974).

In the Lena-Yenisei Basin sulfate rocks (gypsum and anhydrite) are known from the Lower and Upper Silurian deposits. According to data reported by Tesakov and Predtechensky and to the unified stratigraphic scale for the Silurian of the Siberian platform approved by the Siberian Regional Interdepartmental Stratigraphic Committee in 1976, the oldest sulfate-bearing sequence is developed in the Nyuya and Berezovaia Basins where it forms the Utokan Formation consisting of gray and variegated silty limestone interlayered with dolomite and gypsum. The Utokan Formation belongs to the Khaastyr horizon of the Upper Llandoverian. The same sulfate-bearing sequence appears to continue into the Vilyui Basin where gypsum and gypsinate dolomite were found at the same stratigraphic level in the limestone-dolomite unit of the Khaastyr horizon.

Two sulfate-bearing sequences can be recognized in the Upper Silurian. One of them, of Ludlovian age, is known in the Norilsk Region and probably stretches farther east into the Moiero Basin, and south into the Turukhan Region. In the northwestern Siberian platform the sequence is known as the Kongda Formation in which thin gypsum and gypsinate dolomite bands occur in limestone, dolomite, and marl; in the Moiero River Basin the sequence is represented by the Yangada Formation, the central part of which is characterized by gypsification of rocks, and contains lenses, inclusions, and bands of gypsum. Both formations occur at a level of the Tukal horizon.

The second uppermost sulfate-bearing sequence is rather well developed in the north-western and central Siberian platform. The sequence runs from the Norilsk Region south-eastward to the Moiero-Morkoka Rivers Basins and probably occurs in the central Tungusska Syneclise. The sequence lies at a level of the Upper Silurian Kholykhan horizon. Gypsum beds locally reach 4–5 m in thickness, and in the Moiero River Basin and in the Norilks Region their total thickness is no less than 50 m.

Salt deposits appear to occur in the lowest parts of the Tunguska Syneclise in the upper sulfate-bearing (Kholyukhan) sequence (Sokolov et al. 1977). The Turin TO-2 borehole is believed to penetrate rock salt beds, up to 4 m thick, established from the geophysical data and from the presence of rock salt crystals in dolomite-anhydrite and anhydrite. If this information is confirmed by future investigations, then the Silurian Lena-Yenisei Basin can be regarded as salt-producing, and the Kholyukhan sequence as salt-bearing; but so far the basin should be considered as sulfate-bearing.

In other Silurian evaporite basins rock salt has not yet been found: only anhydrite and/or gypsum are known to occur there. Most Silurian sulfate basins began to take shape even in the Ordovician. In essence they were unique sedimentary basins where sulfate accumulation was discontinuous during Ordovician/Silurian time. For the Silurian we can postulate the following basins: the Basin of Severnaya Zemlya, the Baltic Basin, the Basin of the Canadian Arctic Archipelago, the Moose River, Williston, and Canning Basins. Only two new sulfate basins have appeared since, namely the Pechora Basin in north-east Europe, and the Carnarvon Basin in Australia.

In the Basin of Severnaya Zemlya gypsum bands and lenses are reported from the upper part of the Silurian carbonate sequence on Oktyabrskaya Revolyutsiya Island (Egiazarov 1970). The thickness of the gypsum-bearing unit there is 75–80 m, but interbeds and lenses of gypsum and gypsinate rocks reach 15 m and can be traced along the strike for a distance of 300–400 m. The gypsum-bearing unit appears to be Pridolian in age.

In the Baltic Basin sulfate rocks were recorded throughout the entire Wenlockian (Paprenyai horizon) and in the lower part of Ludlovian deposits (the lower Pagegyai horizon) (Pashkevichus 1961; Gailite et al. 1967). Sulfate rocks occur along south and east margins of the basin. They are developed in south Estonia, east Latvia, and along eastern and southern margins of Lithuania. This belt includes dolomite, dolomitized and dolomitic marls and rare interbeds of limestone, as well as inclusions, lenticular bands of gypsum, and gypsinate rocks. Gypsum was reported from the vicinity of the towns of Prenyai and Zhezhmuryai in Lithuania, Mezhtsiems, Aluksne, Akniste, and Druvas in Latvia. Sulfate rocks are not very thick and they are apparently subordinate. The entire Paprenyai-Pagegyai dolomite sequence can be regarded as sulfate-bearing.

In the Basin of the Canadian Arctic Archipelago gypsum and anhydrite make up a rather thick unit of the lower part of the Read Bay Formation. Separate bands and lenses of gypsum reach 10 m, the total thickness of the unit exceeding 20–25 m. A sulfate-bearing unit was reported from north-eastern Ellesmere Island and Somerset Island (Fortier et al. 1963; Christie 1864). The extent of sulfate rocks is confined to the Moose River Basin. Gypsum and anhydrite occur as lenses and thin bands in the Kenogami River dolomites of Late Silurian to Early Devonian age (Norris and

Sandford 1969). In the Williston Basin anhydritized dolomites are known in the Interlake Formation (Porter et al. 1964). In the Canning Basin the Stottid sequence of the Carribady Formation appears to be referred to as the Silurian.

The Pechora sulfate-bearing basin is situated along the north-western margin of the East European platform and includes the interfluve of the upper Pechora, Lemiu, and Ilyich Rivers (Pershina 1972; Filippova 1973; Atlas . . . 1972). Sulfate rocks are found in two sequences, one of which is situated at a level of the Kosju and lower Adak horizons, the second belonging to the Lower Silurian upper Filippjelsk horizon. Both sequences consist chiefly of dolomite and dolomitic marl with interbeds, lenses, and inclusions of gypsum. In the Carnarvon Basin anhydrite bands are associated with the Dirk Hartox Formation represented mainly by dolomite, limestone, and siltstone (Brown et al. 1968).

Thus, the data available allow us to recognize fourteen Silurian evaporite sequences: one salt-bearing (the Salina Formation in the Michigan-Pre-Appalachian Basin) and thirteen sulfate-bearing ones. They formed within eleven evaporite basins. Only four basins are known to be Silurian proper, namely the Michigan-Pre-Appalachian, Dniester-Prut, Pechora, and Carnarvon Basins. Most basins were emplaced as far back as in the Ordovician, and evaporite sedimentation continued in them during Silurian time. There were five Ordovician/Silurian basins, namely the Lena-Yenisei Basin, the Basin of Severnaya Zemlya, the Baltic Basin, the Basin of the Canadian Arctic Archipelago, and the Williston and Canning Basins. The Moose River Basin existed throughout Ordovician/Silurian/Devonian time. In the Silurian only the Michigan-Pre-Appalachian Basin was a salt-producing one, The Dniester-Prut and Lena-Yenisei Basins may have been salt-producing, but reliable evidence for it is not yet available. The following sulfate-bearing sequences formed in the sulfate basins: the Utokan, Kongda, and Kholyukhan sequences formed in the Lena-Yenisei Basin; the gypsum-bearing sequence – in the Basin of Severnaya Zemlya; the Kosju-Adak and Filippjelsk sequences – in the Pechora Basin; the Paprenyai-Pagegyai sequence – in the Baltic Basin; the Pugai beds – in the Dniester-Prut Basin; the Read Bay Formation – in the Basin of the Canadian Arctic Archipelago; the Kenogami River Formation – in the Moose River Basin; the Interlake Formation – in the Williston Basin; the Dirk Hartox Formation – in the Carnarvon Basin; and the Stottid Formation – in the Canning Basin. It should be noted that accumulation of some sulfate-bearing formations continued in the Early Devonian and they are essentially of Late Silurian to Early Devonian age. The same is true of the Kenogami River Formation.

Devonian Evaporite Deposits

Devonian evaporite deposits are known in Asia, Europe, North America, and Australia (Fig. 5). Nine Devonian salt basins have been recognized up to now; they are the North Siberian, Tuva, Chu-Sarysu, Morsovo Basins, the Upper Devonian Basin of the Russian platform, the Western Canadian, Hudson, Michigan, and Adavale Basins. Ten sulfate basins can be outlined, i.e., the Minusinsk, Kuznetsk, Teniz, Turgai, Moesian-Wallachian, Tindouf Basins, the Basin of the Canadian Arctic Archipelago, the Moose River, Illinois, and Central Iowa Basins. First we shall describe in brief the salt basins.

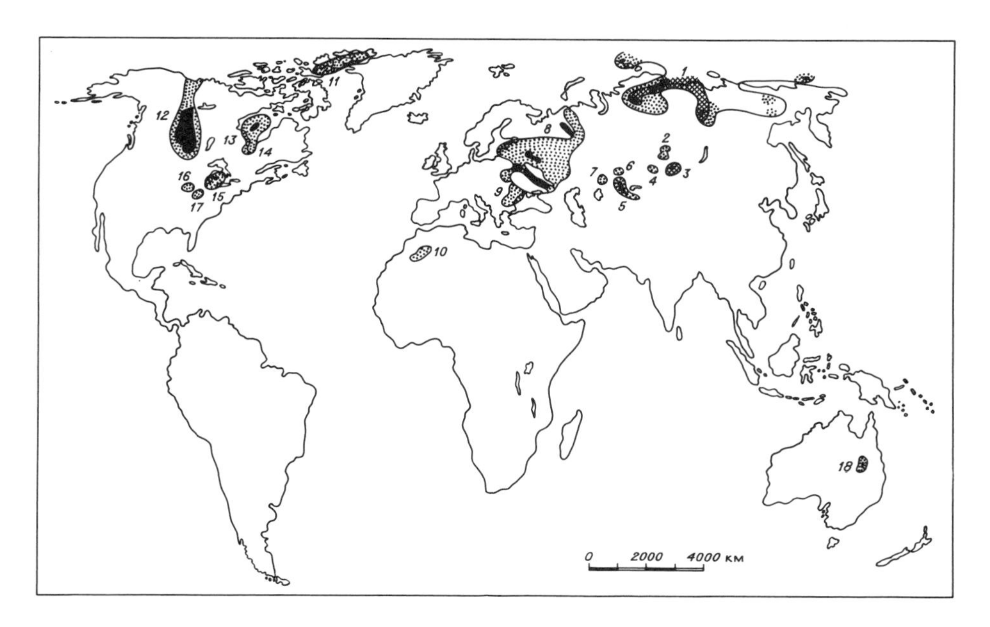

Fig. 5. Distribution of Devonian evaporites.
Evaporite basins: *1* North Siberian, *2* Minusinsk, *3* Tuva, *4* Kuznetsk, *5* Chu Sarysu, *6* Teniz, *7* Turgai, *8* Morsovo and Upper Devonian of the Russian Platform, *9* Wallachian-Moesian, *10* Tindouf, *11* Canadian Arctic Archipelago, *12* West Canadian, *13* Hudson, *14* Moose River, *15* Michigan, *16* South Iowa, *17* Illinois, *18* Adavale. For explanations see Fig. 1

The North Siberian Basin. Many Devonian evaporite strata have been recently found in the north of the East Siberia; in the Verkhoyansk-Chukotsk District, in Severnaya Zemlya, and on Vrangel Island; the strata may have formed within a single evaporite basin (Kalinko 1951, 1953, 1959; Egiazarov 1959, 1970, 1973; Bogdanov and Chugaeva 1960; Ivanov and Levitsky 1960; Menner 1962, 1965, 1967a,b; Fradkin 1964, 1967; Bgatov et al. 1967; Glusnitsky 1967; Cherkesova 1968, 1973; Salt Potential . . . 1968; Menner and Fradkin 1969, 1973; Glushnitsky and Menner 1970; Merzlyakov 1971; Menner et al. 1973a,b; Kolosov et al. 1974; Matukhin and Manner 1974, 1975; Divina and Mathkhin 1975; Kameneva 1975; Sokolov and Matukhin 1975; Fradkin et al. 1977; Kislik et al. 1977; Sokolov et al. 1977).

Salt formations are recognized in three regions of the North Siberian Basin: in the Norilsk Region, the north-western Siberian platform; and in the Khatanga and Kempendyai Depressions.

In the Norilsk Region rock salt was found in the Zubovo, Manturovo, and Fokin Formations. The presence of salt deposits in the Zubovo Formation has been reported quite recently (Sokolov et al. 1977). Rock salt beds were exposed in the GOC-4 borehole on the right bank of the Neralakh River. The beds occur in the central unit of the Zubovo Formation. The unit is represented by the alternation of red and green mudstones, gray argillaceous dolomites, light gray anhydrite and rock salt. Only two salt beds, 4.3 m and 5.8 m thick, can be observed. Single grains of sylvine occur in rock salt. Coeval salt deposits are thought to occur southward as well in the Nizhnaya Tungusska River Basin where they are associated with the Nima Formation (Sokolov et al. 1977). Drilling data of the Turin TO-2 borehole show that the lower and central parts of the formation contain rock salt beds up to 17 m thick, in variegated mudstone and siltstone. This evidence suggests that the Zubovo salt formation is developed over much of the north-western Siberian platform and western Tungusska Syneclise. Salt deposits of the Zubovo Formation are usually referred to as the Early Devonian (Downtonian/Gedinnian), but some geologists assign them to the Late Silurian (Glushnitsky 1977). The Manturovo salt sequence is exposed in the Talnakh, Mikchandin, and North Imangdin areas. Its maximum thickness is 190 m. The sequence consists of rock salt beds with thickness ranging from 2 to 47 m interbedded with anhydrite, up to 27 m thick, dolomitic marl, dolomite, mudstone, and locally siltstone. The overlying Fokin salt sequence containing rock salt alternated with anhydrite, marl, and dolomite is located on the west coast of Lake Pyasino; its thickness is 220–260 m. The Manturovo and Fokin salt sequences are assigned to the Eifelian/Lower Givetian and to the Famennian, respectively.

At Nordvik a salt sequence makes up salt domes and is exposed in several deep boreholes. Its total thickness is unknown but believed to exceed 300 m. The salt sequence appears to have considerable areal distribution in the Khatanga Depression: it may extend from Nordvik not only eastward, but westward to Norilsk as well. The information available allows us to regard the sequence as a separate member and not to unite it with the Manturovo salts developed in the north-western Siberian platform.

In the Kempendyai Depression salt deposits form the Kygyltuus Formation wherein salt units represented by rock salt with interbeds of mudstone and anhydrite alternate with siltstone, mudstone, marl, and in places with dolomite and anhydrite. There are also bands of tuff and tuffite in the section. Salt units vary in thickness from several meters to 200 m, but salt-free units are 15–20 m, locally 40 m thick. The thickness of the exposed part of the Kygyltuus salt formation is 1100 m, but evidence suggests a total thickness of 4000 m.

Sulfate deposits are extremely widespread in the North Siberian Basin. They are known almost throughout the Devonian section. The oldest sulfate-bearing sequence is associated with the Zubovo Formation in the north-western Siberian platform, at Norilsk, as well as with the coeval Koldin Formation developed in the lower parts of Devonian deposits along the north-eastern slope of the Tunguska Syneclise in the Kotui River Basin and in the Olenek-Vilyui Rivers interfluve. Beds of anhydrite and gypsum in the Zubovo Formation commonly reach 5–10 m, locally even 20 m; and lenses, inclusions, and bands of sulfate rocks occur throughout the section. In the Koldin Formation gypsum is less common. In the western and central Tunguska Syneclise, in the Nizhnay and Podkamennaya Tunguska Rivers Basins, bands and lenses of gypsum and anhydrite occur in deposits of the Nima Formation at the same stratigraphic level. Hence, in the west, north-west, and central Siberian platform rather thick (up to 200 m) sulfate-bearing deposits appear to occur ubiquitously at the base of the Devonian. They seem to form a single sulfate-bearing sequence known as the Zubovo Formation. The same sequence is likely to include the lower Kureika Formation at Norilsk, i.e., its lower sulfate-bearing unit.

Additionally, the Lower Devonian gypsum-bearing deposits are known in two regions of the North Siberian Basin: the Taimyr Peninsula, in the lower Taryan Basin; and in Severnaya Zemlya. In the former region a gypsum-bearing unit of banded gypsum, argillaceous dolomite, and dolomitic clay, over 60 m thick, is recognized at the base of the Devonian section. In Severnaya Zemlya, on Oktyabrskaya Revolyutsiya Island, interbeds and lenses of gypsum, up to 35–40 m thick, occur in dolomite of the carbonate sequence assigned to the upper Lower Devonian, corresponding probably to the Dezhnev Formation. Both sulfate-bearing sequences are regarded as separate units.

Some sulfate-bearing sequences can be observed in Middle Devonian deposits of the North Siberian Basin. One of them includes the Sidin Formation containing variegated mudstone, marl, dolomite, and siltstone with gypsum in the Kotui River Basin and in the Olenek-Vilyui interfluve. The Sidin sulfate-bearing sequence is regarded as an equivalent to the Manturovo sequence at Norilsk, where anhydrite and gypsum are very common as well. The Sidin Formation appears to be developed also in the central Tunguska Syneclise, where inclusions of anhydrite occur in the Tynepa Formation at the same stratigraphic level. The second sulfate-bearing sequence can be recognized as the upper part of the Yukta (or Makusov) Formation in the Norilsk Region where beds of gypsum and/or anhydrite occur in limestone and dolomite. The third sequence is known in Severnaya Zemlya. It includes the upper Rusanov Formation consisting of dolomite, dolomitized limestone, marl, and gypsum with a total thickness of 135 m; gypsum locally reaches 8–10 m. The fourth sulfate-bearing sequence including deposits of the Atyrkan Formation, where beds and units of gypsum and anhydrite occur, is known in the northern Verkhoyansk Range, in the Uel-Sikryakh River Basin. The fifth sequence is widespread in the Sette-Daban Range, southern Verkhoyansk Range. Sulfate rocks are associated there with the upper sequence of the Burkhala Formation. The sequence has been recently identified as the separate Zagadochnino Formation (Khaiznikova 1970; Sedimentary and Volcanogenic-Sedimentary Formations . . . 1976). The sixth Middle Devonian sulfate-bearing sequence can be recognized as the upper Sebechan Formation known in the basin of the Sebechan, Khobochalo, and Danynie Rivers in the Tas-Khayakhtakh Range. Finally, the seventh Middle Devonian sulfate-bearing sequence includes rocks of the Vayakh Formation on the eastern slope of the Omulevskie Mountains where gypsum and anhydrite are very common.

Upper Devonian sulfate-bearing sequences are known in the north-western and central Siberian platform, in Severnaya Zemlya, and on Vrangel Island. In the Siberian platform sulfate rocks are most widespread in the Norilsk Region, where they make up the essential part of the Nakokhoz, Kalargon, and Fokin Formations. The Nakokhoz sulfate formation is most extensive. It extends from the Norilsk Region farther east into the Kotui River Basin. A sulfate-bearing type of the Kalargon section is recognized north and west of Norilsk. Rock salt beds are present in the Fokin Formation. However, they occur in the lower part and only there can a salt-bearing sequence be distinguished. As for the upper part of the Fokin Formation consisting chiefly of gray and red to brown marl with gypsum and anhydrite, it can be regarded as a separate upper Fokin sulfate-bearing sequence.

In the Kempendyai Depression the upper part of the Devonian section refers to the Namdyr Formation represented by red siltstone and sandstone interbedded with dolomite, limestone, tuff, and tuffite with anhydrite bands. In the Ygyattin Basin the Vilyuchan Formation contains sulfate rocks of the same age. Probably, the Namdyr sulfate-bearing sequence is widespread in the Vilyui Syneclise. In Svernaya Zemlya the sulfate-bearing sequence includes the Matusevich Formation where lenses of gray to dirty gray gypsum are present in dolomite and marl units among variegated sandstone and siltstone. A separate Upper Devonian sulfate-bearing sequence can be recognized on Vrangel Island. It is developed in the upper reaches of the Khishchnikov River and made up of sandstone, limestone, and shale interbedded with gypsum. Until recently these deposits were referred to as Early Carboniferous (Tilman et al. 1970), but later the Late Devonian age was confirmed (Kameneva 1975).

Thus, Devonian evaporite deposits are recognized over the vast territory above 2,500,000 km^2 of northern East Siberia and the north-western USSR. Up to now there are five known salt-bearing sequences (the Zubovo, Manturovo, and Fokin Formations are developed around Norilsk, the Kygyltuus and Nordvik Formations are spread in the Kempendyai and Khatanga Depressions respectively). The real number of sulfate-bearing sequences developed there is still uncertain because of inadequate study of the territory, but sixteen sequences can be tentatively outlined: Zubovo Formation, gypsum-bearing sequence (Taimyr Peninsula); Dezhnev, Sidin, Yikta, Rusanov, Atyrkan, Burkhala, Sebechan, Vayakh, Nakokhoz, Kalargon, Upper Fokin, Namdyr, Matusevich Formations, and gypsum-bearing sequence (Vrangel Island). The number of sequences may be reduced in future studies.

In the Tuva Basin the Devonian salt-bearing sequence has long been known in foothills of the western Tannu-Ola Range in southern Tuva (Matrosov 1954; Levenko 1955, 1956, 1960; Lepeshkov et al. 1958; Ivanov and Levitsky 1960; Pastukhova 1960, 1965; Zaikov et al. 1967; Minko 1972; Meleshchenko et al. 1973; Kolosov et al. 1977). It is placed into the Ikhedushiingol Formation and assigned to the Upper Eifelian/Lower Givetian. An exposed thickness of the sequence exceeds 600 m. It may contain some salt units with a total thickness over 900 m (Minko 1972; Kolosov et al. 1977). In addition to rock salt the section has clay, siltstone, and anhydrite. Thin interbeds of sylvine and sylvine-halite rocks; rare occurrences of potash minerals and rinneite are found in upper horizons. The rock salt sequence is known so far over a rather small area. However, it is believed to occupy a vast territory stretching farther south and south-west into Mongolia and fill a rather big evaporite basin.

The Chu Sarysu salt basin is situated in western Central Kazakhstan. Salt-bearing deposits build up there numerous domes in the Bestube and Sarysu diapir areas. They are exposed in a number of deep boreholes (Petrushevsky 1938a,b; Zaitsev and Pokrovskaya 1948; Aleksandrova and Borsuk 1955; Ditmar 1961, 1963, 1965, 1966; Varentsov et al. 1963; Shakhov 1965a,b, 1968; Gulyaeva et al. 1968; Kunin 1968; Visloguzova et al. 1968; Orlov et al. 1969; Blagovidov 1970; Li and Mailibaev 1971; Mikhailov et al. 1973; Sinitsyn et al. 1977). Salt deposits are developed in the northern and western Chu Sarysu Synecline in the Dzhezkazgan, Tesbulak, and Kokpansor Basins. The salt formation exceeds 1000 m in thickness. The section contains two salt sequences separated by terrigenous-carbonate beds, 50 to 100 m thick. The lower unit, 500 m thick, consists of rock salt with interbeds of dolomite and anhydrite. It was assigned to the Upper Devonian (Sinitsyn et al. 1977). The upper unit, built up of rock salt alternated with mudstone, is of Early Carboniferous age. Its thickness is

350 m. Hence, the Chu Sarysu evaporite basin is Late Devonian to Early Carboniferous in age. Only the lower salt unit may have formed there in Late Devonian time.

The Morsovo salt basin embraces the central, eastern, and south-western parts of the East European platform (Tolstikhina 1952; Sarkisyan and Teodorovich 1955; Devonian . . . 1958; Ivanov and Levitsky 1960; Strakhov 1962; Polyanovskaya 1964; Tikhy 1964, 1967, 1973; Khizhnyakov and Polyanovskaya 1967; Lyashenko and Lyashenko 1967; The Devonian . . . 1967; Tikhomirov 1967; On the Potassium . . . 1970; Utekhin 1971; Blagovidov et al. 1972; Filippova and Krylova 1973; Golubtsov 1973; Gurevich et al. 1973; Kirikov 1973; Sammet 1973; Kislik et al. 1976a, 1977). Salt deposits of the Morsovo (Narva) horizon (Eifelian) are known in two regions; in the Moscovian Syneclise and in the Pripyat Depression. In the Moscovian Syneclise rock salt forms a belt, about 500 km long and 170–200 km wide, extending from south-east to north-west. This salt unit reaches 50–60 m in thickness. Rock salt interbeds enriched in potash minerals have been found near the town of Yartsevo, Smolensk District (On the Potassium . . . 1970). In the Pripyat Depression salts have been found in the Narva horizon in the Vishan, Marmovichi, Davydov, Chernin, Kormyan, Sosnovka, and Eastern Pervomaysk areas. The unit is believed to occupy 1500 km^2. Its thickness does not exceed 30–35 m. Both salt units form isolated occurrences in the Morsovo Basin within the most subsiding zones of the Moscovian Syneclise and the Pripyat Depression. Salt sequences wedge out rapidly and are replaced by sulfate and sulfate-carbonate deposits toward the periphery of the structures. It allows us to recognize, apart from salt-bearing, some sulfate-bearing units in the Morsovo Basin. One of them is spread in the Moscovian Syneclise. It overlies and flanks this salt sequence on all sides and can be traced both south-east toward the Riazan-Saratov Trough, Don-Medveditsa Swell, and Volga Monocline, and west into the Baltic area. In the Pripyat Depression sulfate deposits are so far included in the salt sequence because of their low thickness and little area extent. Another sulfate-bearing sequence is known in the Lvov Basin and in south-western Poland. Interbeds and units of gypsum, anhydrite, and anhydrite-dolomite are found there in the middle Lopushan member which is an age equivalent of the Narva horizon of the Pripyat Depression and of the Morsovo horizon of the Moscovian Syneclise. The sulfate-bearing middle Lopushan sequence, defined distinctly as a separate member, possibly marks an isolated evaporite basin. It is tentatively considered to have formed in the single Morsovo Basin. If new information becomes available it could be placed into a marginal zone of the Moesian-Wallachian evaporite basin described below.

The Upper Devonian evaporite basin of the East European platform occupies most of the European part of the USSR. Evaporite deposits accumulated there in the Baltic and Moscovian Syneclises, Volga-Ural Region, Timan Trough, Pechora Syneclise, Pripyat and Dnieper-Donets Depression. Thick salt deposits formed in the Pripyat and Dnieper-Donets Depression, and in the Timan Trough. Sulfate sediments accumulated elsewhere.

In the Pripyat Depression Upper Devonian deposits include two salt sequences: the lower sequence associated with the Evlanovo and Liven horizons of the Frasnian, and the upper sequence occurring at a level of the Lebedyan and Dankov horizons of the Famennian (Bruns 1956; Fursenko 1957; Kirikov 1959, 1963a,b, 1973; Shcherbina 1959, 1960a,b, 1961, 1962; Golubtsov and Makhnach 1961; Bayazitov 1963;

Gorkun 1964; Kislik and Lupinovich 1964, 1968; Makhnach and Kurochka 1964; Kislik 1966; Kurochka 1966a,b; Lithology . . . 1966; Eroshina 1968; Lupinovich et al. 1970; Eroshina and Vysotsky 1972; Eroshina et al. 1976; Kislik et al. 1976b; Bordon et al. 1977; Garetsky and Konishchev 1977; Kislik et al. 1977; Korzun 1977); The lower sequence consists of rock salt units, up to 50–70 m thick, alternated with salt-free rocks, i.e., clay, marl, limestone, dolomite, and anhydrite ranging from 2–3 to 15–30 m. The thickness of the salt ranges from 60–200 to 1158 m.

In the southern Vishan area the sequence contains three horizons of potash salt with 4–6 cm thick intercalations of sylvine rocks. The upper salt-bearing sequence reaches 3260 m in thickness. It is usually divided into two members: the lower halite or potash-free and the upper clayey-halite or potash-bearing. The halite member is respresented by rock salts (from 7–30 to 300 m thick) and by salt-free rocks (from 1–2 to 30 m). The potash-bearing member consists of rock salt (from 3–7 to 33–45 m thick) with interbeds and horizons of potash salt, as well as with beds of carbonate clay, marl, and argillaceous dolomite. The upper member contains about 60 potash horizons, the thickness of which varies from 0.5 to 40.0 m; the horizons are made up of sylvine-, carnallite-sylvine-, sylvine-carnallite-, and carnallite rocks interbedded with rock salt and halopelite (Kislik et al. 1976b).

Apart from salt-bearing, three sulfate-bearing sequences can be recognized in the Upper Devonian of the Pripyat Depression. One of them includes rocks belonging to the lower part of the carbonate sequence and resting at a level of the Shchigrov horizon where anhydrite and anhydrite-dolomite interbedded with argillaceous dolomite, dolomitic marl, and argillaceous limestone occur. The thickness of the sequence reaches 20–22 m in the west of the basin. The second sequence includes deposits of the Voronezh and lower Evlanovo horizons and underlies the upper salt sequence. Anhydrite and anhydrite-dolomite alternate there with marl, argillaceous dolomite, and limestone as well. The thickness of the Voronezh-Evlanovo sulfate-bearing sequence varies from 12 to 48 m. The third sequence is known as the Mezhsolevaya [2] in the Pripyat Depression. It consists chiefly of clay, marl, limestone, and dolomite, interbedded with siltstone and sandstone. Anhydrite occurs in places as thin beds in the upper and lower parts of the sequence. The Mezhsolevaya sequence usually varies from 100 to 500 m in thickness.

In the Dnieper-Donets Depression salt deposits are distributed over an area of at least 400,000 km^2. Three salt sequences are tentatively recognized in the basin. The lowermost sequence lies approximately at a level of the Shchigrov horizon of the Frasnian (Novik 1952, 1954; Kutsyba 1954, 1959; Kirikov 1962, 1963a,b). It is thought to build up salt stocks. An initial thickness of the sequence may have been about 1000 m (Halogenic Formations . . . 1968). The second salt sequence is named lower by most investigators (Britchenko et al. 1968, 1977; Kityk, 1970; Makhnach et al. 1970; Kityk and Galaburda 1977). The salt-bearing part of the sequence belongs to the Evlanovo and Liven horizons. However, the underlying sulfate-bearing part of the sequence occurs within the Voronezh horizon and therefore this halogenic series is regarded as Voronezh/Evlanovo/Liven in age (Kityk and Galaburda 1977).

2 The Mezhsolevaya sequence is Russian for a sequence (in this case sulfate sequence) between two salt sequences

The normal maximum thickness of the entire salt-bearing series ranges from 600 to 1200 m, but decreases to 400–500 m in uplifts. Two beds of sylvine-halite rocks 6 to 80 m thick are known in the salt sequence in the Romny area (Brichenko et al. 1977). Some investigators (Makhnach et al. 1970; Katyk and Galaburda 1977) assign the upper salt sequence of the Dnieper-Donets Depression to the Elets horizon, while others (Britchenko et al. 1977) assign it to the Elets and Lebedyan horizons. It is divided into three parts: the lower sulfate-carbonate, the middle thickest salt-bearing, and the upper sulfate-carbonate again. The total thickness of the sequence varies from 300 to 1000 m. Wide variations in thickness are chiefly caused by omission of beds or units of rock salt from the section.

Salt deposits of the Timan Trough are poorly known. They were reported from some salt domes (Seregov, Chusovaya domes) within the Vychegda Basin, and their areal distribution is outlined approximately from geophysical and geological evidence (Kalberg 1948; Lyutkevich 1955, 1963; Pakhtusova 1957, 1963a,b; Ivanov and Levitsky 1960; Zoricheva 1966, 1973; Privalova et al. 1968; Dedeev and Raznitsyn 1969). The salt sequence consists of rock salt irregularly interlayered with terrigenous and carbonate rocks of which 1200 m crops out, and the total thickness reaches about 2500–3000 m. The most probable age of the sequence is Upper Devonian (Frasnian).

One cannot be sure of the real number of sulfate-bearing bedas developed in the central and northern Upper Devonian basin of the East European platform because of extremely wide distribution of sulfate rocks (gypsum and anhydrite) in Upper Devonian deposits of the Baltic and Moscovian Syneclises and in those of the Volga-Ural area. In fact, they are common through the section of Frasnian and Famennian sediments (Tolstikhina 1952; Narbutas 1959, 1961, 1964; Ivanov and Levitsky 1960; Zhaiba et al. 1961; Strakhov 1962; Gailite 1963; Liepinsh 1963a,b, 1973; Ulst 1963; Makhlaev 1964; Berzin and Ozolin 1967; Tikhomirov 1967; Vodzinskas and Kadunas 1969; Sammet 1971, 1973; Utekhin and Sorskaya 1971; Filippova and Krylova 1973; Tikihy 1973; Zoricheva 1973). Therefore, all the Upper Devonian deposits of the Baltic and Moscovian Syneclises and those of the Volga-Ural area containing beds and units, as well as lenses and inclusions of gypsum and anhydrite, have to be united in a single Upper Devonian sulfate-bearing sequence. The Ustukhta and Idzhid-Kamensk sulfate-bearing sequences can be tentatively distinguished within the Pechoar Syneclise. The Ustukhta sequence is developed in the Ukhta and probably upper Izhma districts (Ivanov and Levitsky 1960; Tszu and Kossovoy 1973), and the Idzhid-Kemensk sequence is spread in the upper Izhma district; it seems to stretch farther into the Pechora Range and into the lower reaches of the Pechora River where gypsum interbeds are known to occur in deposits tentatively assigned to the Lebedyan and Dankov horizons (Nalivkin 1963; Tszu and Kossovoy 1973).

On the whole, six salt sequences are known in the Upper Devonian evaporite basin of the East European platform, namely the Evlanovo-Liven and Dankov-Lebedyan sequences are developed in the Pripyat Depression; the tentatively recognized Shchigrov sequence, as well as the lower and upper salt-bearing sequences, are spread in the Dnieper-Donets Basin; and the salt-bearing sequence is developed in the Timan Trough; additionally there are six sulfate sequences, namely the Shchigrov, Voronezh-Evlanovo, and Mezhsolevaya sequences underlie the Pripyat Depression; the Upper

Devonian sequence extends within the Baltic and Moscovian Syneclises and Volga-Ural area; the Ustukhta and Idzhid-Kemensk sequences are known in the Pechora Syneclise.

The West Canadian evaporite basin is situated in North America, flanked by the Rocky Mountains on the west and by the Canadian Shield on the east. From south to north it stretches from North Dakota and Montana, USA, through Saskatchewan, Alberta, and Northwest Territories, Canada, to the Arctic Ocean. Much information on halogenic deposits of this basin is presented elsewhere (Law 1955; Andrichuk 1958, 1960; Belyea 1960, 1964; Pearson 1960, 1963; Sandberg 1962; Grayston et al. 1964; Lane 1974; Schwerdtner 1964; Norris 1965, 1967; Carlson and Anderson 1966; Keyes and Wright 1966; Klingspor 1966, 1969; Bassett and Stout 1967; Danner 1967; Griffin 1967; Harding and Gorrell 1967; Jordan 1967, 1968; Kent 1967, 1968; Sandberg and Mapel 1967; Langton and Chin 1968; McCamis and Griffith 1968; Prather and McCourt 1968, Wardlaw 1968; Fuller and Porter 1969; Holter 1969; Price and Ball 1971, etc.). A composite Devonian section of the basin contains eight salt-bearing sequences, in ascending order Lotsberg, Gold Lake, Prairie (or Muskeg), Black Creek, Hubbard, Davidson, Dinsmore, and Stettler Formations. The Lotsberg, Gold Lake, and Prairie Formations are most widespread.

The occurrence of Lotsberg salts is restricted to the North Alberta Basin. It consists chiefly of rock salt. Its maximum thickness is about 300 m. The Gold Lake salts fill in the subsiding zones of the Central and North Alberta Basins. Rock salt predominates. Its thickness varies from 45 to 80 m. Both sequences belong to the Lower Elk Point and are referred to as the Middle Devonian (Eifelian).

The Prairie salt sequence belongs to the Upper Elk Point and is regarded as the Givetian. The thickness ranges from 75 to 300 m. There are two zones of increased thickness: the northern one situated in the North and Central Alberta Basins, and the southern zone situated in the Saskatchewan subbasin. In Saskatchewan the upper part of the sequence contains three horizons of potash salt, namely Esterhazy, Belle Plaine, and Patience Lake. The maximum thicknesses of potash beds are 18–21 m, 15–18 m, and 10–15 m for the Esterhazy, Belle Plaine and Patience Lake sequences, respectively. They consist of sylvinite and carnallite rocks interbedded with rock salt. They are separated by rock salt units. In Saskatchewan the lower part of the Prairie sequence is known as the lower salt unit, wherein two units of anhydrite can be identified. In Alberta the equivalents of the Prairie salt sequence are named the Muskeg Formation, represented by the alternation of rock salt, anhydrite, and in places mudstone and dolomite. It includes seven units: Telegraph, Mikwa, Wabaska, Chipevian, Wolverine, Mink, and Bear. North-westward the salt deposits of the Muskeg Formation are replaced by sulfate rocks, and then by carbonates building up reefs. Near the Barrier Reef in the Zema area between reefs of the Keg River Limestones the Black Creek salt is recognized in the lower Muskeg Formation. Its thickness does not exceed 80 m.

The Hubbard salt sequence of the upper Dawson Bay Formation (Manitoba Group) is known only in the southern West Canadian Basin. The sequence is assigned to the Givetian. It consists of rock salt interlayered with anhydrite and mudstone. Its thickness usually does not exceed 20 m. The Davidson salt sequence, regarded as Frasnian, is known only from southern Saskatchewan. It is represented by rock salt with lenticular interbeds of mudstone and brown anhydrite. The thickness ranges from 10–15 m to 60 m. The overlying salt-bearing beds, known as the Dinsmore sequence, occur in the upper Wymark member of the Duperow Formation. It is exposed in single boreholes south of Saskatoon. Its thickness reaches 60–65 m. The sequence consists of rock salt interbedded with anhydrite and mudstone. In the West Canadian Basin the uppermost salt sequence has been found in the lower Stettler Formation in south Alberta. The thickness of rock salt beds there ranges from 10 to 30 m, locally reaching 165–170 m. The section contains anhydrite as well.

Sulfate and sulfate-carbonate rocks making up several separate sulfate-bearing sequences are known from the Devonian of the West Canadian Basin at different stratigraphic levels. The oldest sequence is associated with the upper Ernestian Lake Formation where a rather thin (up to 10 m) anhydrite unit is identified. It underlies the Gold Lake salt-bearing sequence. The next sulfate-bearing sequence is known as the Chinchaga Formation. It is developed in the northern part of the North Alberta Basin and equivalent to the Contact Rapids Formation. It consists chiefly of anhydrite alternated with argillaceous dolomite and dolomitic marl. The sequence reaches 60 m in thickness. Another sulfate-bearing sequence can be outlined in the sulfate deposits of the Muskeg Formation of the North Alberta Basin. The Prairie (or Muskeg) salt sequence wedges out there, and at this level a thick (up to 250 m) series consisting of anhydrite with small interbeds of carbonate rocks appears in northern Alberta. The Beaverhill Lake sulfate-bearing sequence can be also regarded as a separate unit in which anhydrite bands are found in south-eastern Alberta, in southern Saskatchewan, and in northern Montana. It surrounds the Davidson salt sequence on all sides and is regarded as the Lower Frasnian.

Two isolated, though probably coeval sulfate-bearing sequences, are outlined in the West Canadian Basin among the Upper Frasnian deposits. One of them is recognized in central Alberta within the Woodbend Group, and the other sequence associated with the Duperow Formation is situated in south-eastern Alberta, in southern Saskatchewan, and in south-western Manitoba. The Duperow sulfate sequence flanks the Dinsmore salt-bearing sequence on the south-west, south, and south-east. There are many isolated occurrences of sulfate rocks among the Upper Frasnian/Lower Famennian deposits belonging to the Birdbear and Winterburn Formations in the West Canadian Basin, However, all of them can be united in two sulfate-bearing sequences, such as the Birdbear and Nisku Formations; one of them includes sulfate deposits of the Saskatchewan sub-basin, and the other contains sulfate rocks of the Nisku Formation in the Central Alberta Basin. The sulfate-bearing Stettler sequence, developed ubiquitously in southern and south-eastern Alberta and in south-western Saskatchewan, is clearly identified in the Upper Famennian.

Thus the evidence available suggests the presence of eight salt-bearing and nine sulfate-bearing strata in the West Canadian evaporite basin.

The Hudson and Moose River evaporite basins. Until recently sulfate deposits have been reported only from the Moose River Basin. As for the Hudson Basin, salt-bearing beds were found there only in deep boreholes in 1969–1972. The results obtained show that Devonian deposits fill in two isolated basins in the Hudson Bay water area that may have formed a single sedimentary structure in Devonian time, but now they are known as separate basins, i.e., the Moose River Basin in the south and the Hudson Basin in the north. The details will be given below using data recently reported by Sanford (1974) and Sanford and Norris (1975).

Within the Hudson Bay and adjacent regions Devonian deposits, together with Silurian and Ordovician ones, fill in a large syneclise entirely situated within the Canadian Shield surrounded by Precambrian outcrops. This Paleozoic terrane belongs to the Hudson platform (Fig. 6). Devonian deposits formed there within two epi-continental basins, nemaly the Moose River and Hudson Basins separated by the Henriette-Maria Cape stretching from south-west to north-east. The Moose River Basin embraces the southern Hudson Bay and adjacent coastal regions. It is separated from the Michigan and Allegheny Basins to the south and south-west by the Fraserdale Arch. The Hudson Basin occupies the central Hudson Bay and the coastal part of the Tatnam Peninsula in boundary regions of north-eastern Manitoba and north-western Ontario. The basin is flanked by the Severn Arch and Bell Arch on the south-west and north-east, respectively.

In the Moose River and Michigan Basins Devonian deposits can be distinguished as separate formations showing good correlation and similar age; in ascending order they are: the Kenogami River, Stooping River, Kwataboahegan, Moose River, Murray Island, Williams Island, and Long Rapids Formations.

On the whole, the Kenogamy River Formation is of Late Silurian/Early Devonian age. Only the upper part of the formation is assigned to the Early Devonian. In the Moose River Basin it is represented by light calcareous dolomite, oolitic dolomite, argillaceous dolomite, as well as by brecciform dolomitized limestone and silicified dolomite. Gypsum inclusions and lenses are common. The thickness of this part of the formation reaches 50–52 m in the Moose River Basin. In the Hudson Basin the upper Kenogami River Formation was penetrated by three wells drilled on the Hudson Bay coast, namely the Kaskattama No. 1, and the Penn No. 1 and 2 wells (Fig. 7). In the Kaskattama No. 1 well the section includes creamy to light gray fine-grained and aphanitic limestone with thin bands of light gray, argillaceous, silty and sandy dolomite up to 10 m thick. The Penn No. 1 well includes, in ascending order:

		Thickness (m)
1.	Light creamy, gray, and light greenish gray argillaceous dolomite	5.2
2.	Light and red to brown micrograined limestone dolomitized to a variable degree with inclusions and bands of light gray anhydrite	15.0
3.	Light gray, micro- and fine-grained oolitic dolomite limestone	5.2

The total thickness is 25.4 m.

The Penn No. 2 borehole penetrates the following section of the upper Kenogami River Formation, in ascending order:

		Thickness (m)
1.	Light gray, strongly argillaceous dolomite	about 1.0
2.	Brown thin-grained limestone	4.5
3.	Light to brown dolomite with bands of bituminous dolomite with bands of anhydrite and crystals of pyrite	7.0
4.	Pink dolomitized limestone	9.0
5.	Light microgranular dolomitic limestone	6.0

The total thickness is 27.5 m.

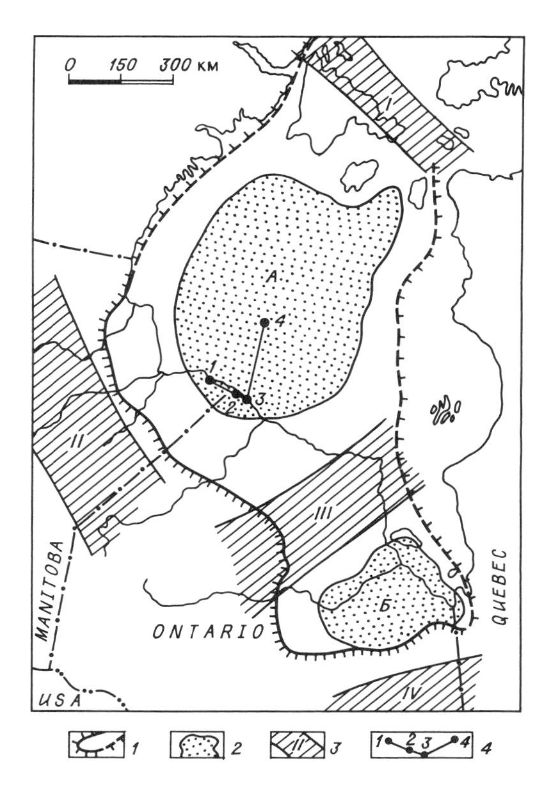

Fig. 6. Distribution of Devonian deposits in the Hudson Bay. Compiled from data of Sanford and Norris (1975).

1 determined and inferred boundaries of Paleozoic deposits, *2* distribution of Devonian sediments (*A* Hudson Bay, *B* Moose River Basin), *3* arched uplifts and their numbers (*I* Bell, *II* Severn, *III* Henriette Maria, *IV* Freserdale); *4* boreholes and their numbers (*1* Kascatama No. 1, *2* Penn No. 2, *3* Penn No. 1, *4* Hudson Walrus A-71), and the profile shown in Fig. 7

The Penn No. 2 borehole penetrates the following section of the upper Kenogami River Formation, in ascending order:

		Thickness (m)
1.	Light gray, strongly argillaceous dolomite	about 1.0
2.	Brown thin-grained limestone	4.5
3.	Light to brown dolomite with bands of bituminous limestone, with inclusions and lenses of anhydrite and crystals of pyrite	7.0
4.	Pink dolomitized limestone	9.0
5.	Light microgranular dolomitic limestone	6.0

The total thickness is 27.5 m.

The upper Kenogami River Formation wedges out toward the central Hudson Basin. This part of the Kenogami River Formation is Late Gedinnian/Early Siegenian.

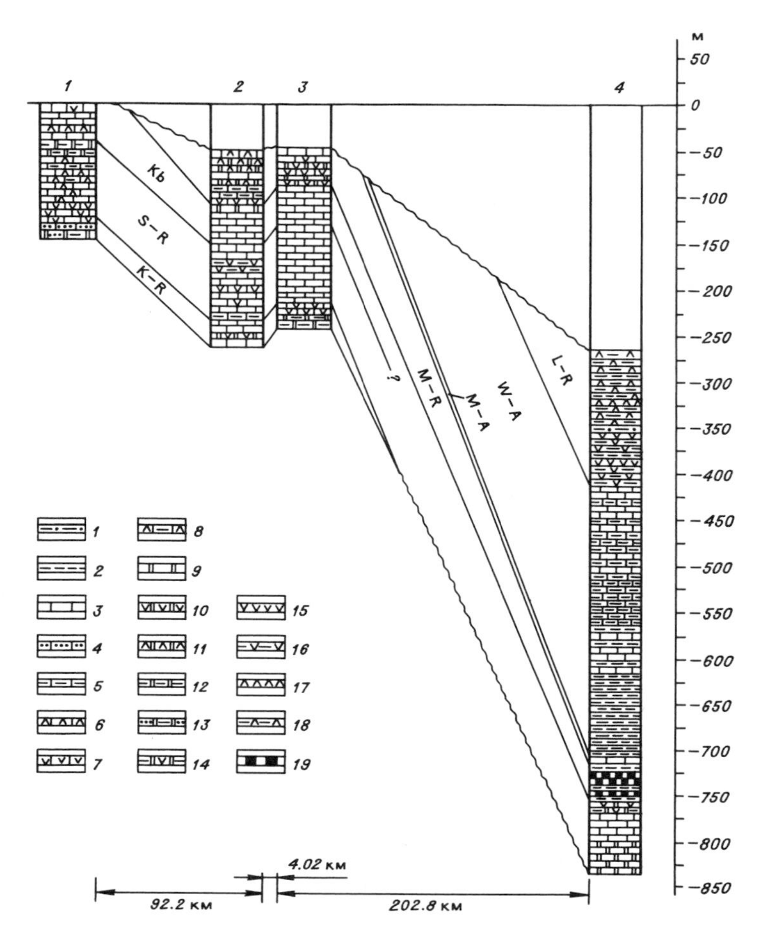

Fig. 7. Correlation chart of Devonian deposits along profile *1–4* (see Fig. 6).
1 siltstone, *2* mudstone, *3* limestone, *4* arenaceous limestone, *5* clayey limestone, *6* limestone with inclusions, lenses, and interbeds of gypsum, *7* limestone with inclusions, lenses, and interbeds of anhydrite, *8* clayey limestone with interbeds and lenses of gypsum, *9* dolomite, *10* dolomite with interbeds, lenses, and inclusions of anhydrite, *11* dolomite with interbeds, lenses, and inclusions of gypsum, *12* clayey dolomite, *13* clayey dolomite, locally arenaceous dolomite, *14* clayey dolomite with interbeds and lenses of anhydrite, *15* anhydrite, *16* anhydrite with mudstone interbeds, and mudstone with inclusions and interbeds of anhydrite, *17* gypsum, *18* clayey gypsum and gypsinate mudstone, *19* rock salt. ***Figures above columns*** show numbers of boreholes, for names see Fig. 6. Formations: *K-R* Kenogami River, *S-R* Stooping River, *Kb* Quatobahegan, *M-R* Moose River, *M-A* Murray Island, *W-A* Williams Island, *L-R* Long Rapids

In the Moose River Basin, in the Albany River delta area, and along the lower reaches of the Stooping River at the northern margin of the basin the Stooping River Formation contains limestone, dolomitic limestone, and dolomite. Limestone is usually calcarenitic, brown to grayish-brown, sometimes orange. Dolomitic limestone is brown, fine- and micro-grained. Dolomite is dark to light brown, thin-crystalline. There are inclusions of gray and black chert in these rocks. Along the northern and north-western margins of the basin limestone dominates the Stooping River Formation, while the amount of dolomite decreases. In the center of the basin the lower part of the formation is dominated by dolomite and dolomitic limestone usually containing lenses of chert, as well as inclusions and veins of anhydrite. Limestone, locally silicified, is common in the upper part of the formation. Along the eastern margin the Stooping River Formation grades into limestone alternated with red sandy mudstone and green sandstone forming a separate Sextant Formation.

In the Hudson Basin the Stooping River Formation was penetrated by four deep wells. In the section of the Kaskattama No. 1 borehole the formation contains the following units, in ascending order:

		Thickness (m)
1.	Light gray creamy thin-grained and aphanitic limestone	8.0
2.	Light to dark gray aphatinitic limestone with inclusions and veins of anhydrite	34.5
3.	Light brown, blue to light gray medium-grained dense limestone with interbeds of gray argillaceous limestone	19.0
4.	Creamy aphanitic dolomite	8.2
5.	Light to dark brown argillaceous, calcareous dolomite	4.5

The total thickness of the formation is 74.2 m.

In the Penn Nos. 1 and 2 wells limestone and dolomitic limestone dominate the lower part of the formation. In the upper part of the Penn No. 1 gray chert occurs in limestone, and creamy dolomite, argillaceous limestone, and gypsinate mudstone appears in the Penn No. 2 well. The thickness of the formation ranges there from 81 m to 85 m. Equivalents of the Stooping River Formation failed to be traced in the Hudson Walrus A-71 borehole due to their similarity to the overlying Kwataboahegan Formation; they form a single sequence.

Organic remains suggest Middle to Late Emsian age of the Stooping River Formation.

In the type locality of the Moose River Basin, the Kwataboahegan Formation consist of dark brown, calcarinate, bituminous, massive, and thick-platy limestones more or less dolomitic. Carbonate banks and bioherms are common. Locally limestone contains gypsum inclusions and selenite veins. The maximum thickness of the deposits is 75 m. In the Hudson Basin the Kwataboahegan Formation also consist chiefly of light gray and light creamy limestone, commonly argillaceous and silicified, locally gypsinate (Kaskattama No. 1 borehole). In more subsiding areas (Penn No. 1 and 2 wells) bituminous limestone and calcareous dolomite are widespread. In the central zone of the basin (the Hudson Walrus A-71 borehole) dolomite dominates the lower part of the section, and lenses of anhydrite occur up the section. The Kwataboahegan Formation is difficult to distinguish there from the underlying Stooping River Formation. The thickness of the undifferentiated section is 76.5 m.

The Moose River Formation is represented by thin-grained and argillaceous limestone, dolomite, carbonate breccia, gypsum, and anhydrite. The maximum content of gypsum in outcrops was found in the belt extending from south-east to north-west for a distance of 72 km, the width being about 16 km, i.e., from the Wakwaywkastic

River to the Cheepash River. Peculiar land forms, building up gypsum mountains, are known there. There are many gypsum quarries in the Cheepash Basin. In the Moose River Basin the formation reaches 60 m in thickness. The thickness of gypsum in one of deep wells drilled in the Moose River Basin on the coast of Mike Island is 44 m. Brecciform carbonate rocks and breccias dominating the formation may have formed due to leaching of gypsum beds. The data obtained from three deep borholes suggest highly variable composition of the Moose River Formation in the Hudson Basin. In two boreholes (Penn Nos. 1 and 2) the formation consist chiefly of thin- and medium-grained dolomitic and argillaceous limestone interbedded with dolomite. Rocks usually contain inclusions, lenses, and bands of gypsum and anhydrite, light gray flint and minor red mudstone. The Hudson Walrus A-71 borehole drilled in the central Hudson Basin penetrated the following section, in descending order:

		Thickness (m)
1.	White thin-grained calcarinate limestone	4.9
2.	Light gray anhydrite	3.0
3.	Light brown to gray anhydrite with inclusions of red mudstone	6.1
4.	Pale yellow to brown thin-grained calcarinate limestone with bands of red argillaceous limestone	3.0
5.	Bright red mudstone	1.5
6.	Interval poorly represented by a core consisting chiefly of rock salt	10.5
7.	Orange to bright red mudstone and shale	10.0

The thickness exposed in the borehole is 39 m. The formation is of Middle Eifelian age.

In the Hudson Basin salt deposits occupy the central part of the Hudson Bay proper. An areal extent of evaporite rocks reaches 80,000 km^2, and that of rock salt is about 50,000 km^2. The average thickness of rock salt beds is about 10 m. Thus the salt content may amount to 5.10^2 km^3. Only sulfate rocks (gypsum and anhydrite) have been so far found in the Moose River Basin, and rock salt has not yet been known even in deep boreholes which are very few in number there. Therefore rock salt may be probably developed in subsiding parts of the Moose River Basin where sulfate thickness increases (Fig. 8).

The Murray Island Formation is easily recognized in the Devonian section of the Hudson and Moose River Basins. It is represented mainly by limestone and argillaceous and bituminous dolomite. The thickness of the formation varies from 6 m to 20 m in the Moose River basin and does not exceed 6–8 m in the Hudson Basin. The age of the formation is Upper Eifelian.

In the Moose River Basin the Williams Island Formation including Givetian (Middle Devonian) and Frasnian (Upper Devonian) deposits is divided into the lower and upper units. The lower unit consists of greenish to gray shale and mudstone interbedded with yellow to brown sandstone, brown and gray gypsinate sandy mudstone, siltstone, and sandstone. Minor limestone and carbonate breccia are present. The lower unit reaches 45–50 m in thickness. The upper consists of thin- to medium-platy argillaceous limestone and light blue to gray calcareous mudstone. Locally gypsum and anhydrite occur up the section. Its thickness reaches 40–45 m. The total thickness of the formation is about 90 m.

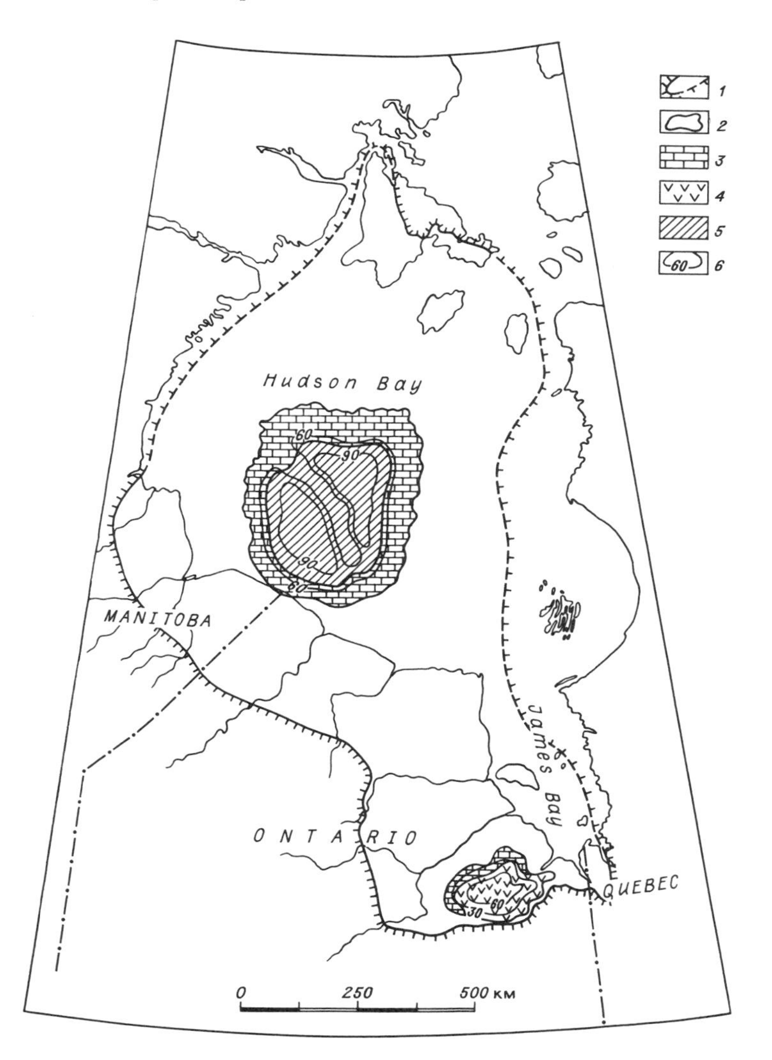

Fig. 8. Lithologic map and thicknesses of the Moose River Formation in the Hudson and Moose River Basins. After Sanford and Norris (1975).

1 determined and inferred boundaries of Paleozoic deposit distribution, *2* the Moose River Formation, *3–5* of sediments, *3* limestone, clayey limestone and dolomitized limestone, *4* gypsum, carbonate rocks and carbonate breccia, *5* rock salt, gypsum, anhydrite, mudstone, and carbonate (salt zones), *6* lines of equal thicknesses (m)

In the Hudson Basin the Willimas Island Formation reaches 270–290 m in thickness. The formation is represented there by gray and pale yellow argillaceous and silty limestone, mudstone with inclusions and bands of gypsum and quartz sandstone. Red gypsinate mudstone and siltstone are common.

In both basins the Devonian section is crowned by the Long Rapids Formation of Late Frasnian/Early Famennian age. It includes dark gray to black bituminous shale and mudstone, in places gray shale interlayered with dolomite and limestone. Purple red sandstone, brown to red mudstone interbedded with white gypsum is also present in the Hudson Basin. The rocks are usually saline and gypsinate there. In the Moose River Basin the thickness of the formation does not exceed 85 m, and in the Hudson Basin it reaches 150 m.

The foregoing suggests that the Hudson Basin includes a salt-bearing unit (the Moose River) and three sulfate-bearing units (namely, the lower unit comprising the Kenogami River and Stooping River Formations and two upper units corresponding to the Williams Island and Long Rapids Formations). Three sulfate-bearing units: the Kengami River, Moose River, and Williams Island Formations are known in the Moose River Basin. It should be noted that in the Moose River Basin the Devonian sulfate part of the Kenogami River Formation makes a single whole with the underlying Silurian part of the same formation and forms a sulfate-bearing unit ranging in age from the Late Silurian to Early Devonian.

The Michigan Basin is situated in the state of Michigan, USA, and partly in Ontario, Canada. The basin lies between the Michigan and Huron Lakes; on the north the basin is bounded by Precambrian rocks exposed within the Canadian Shield, on the west and east it is flanked by the Wisconsin Arch and the Algonquin Arch, respectively. On the south-west and south-east it is separated from the Illinois Basin by the Kankakee Arch, and from the Appalachian Basin by the Findlay Uplift, respectively. In the Michigan Basin all the Devonian evaporite deposits are placed in the Lucas Formation of the Eifelian (Middle Devonian). The formation varies from 30 m to 250–300 m in thickness. In the center of the basin it includes eight rock salt beds totaling 30 m (Landes, 1951). They are interbedded with units of anhydrite and, in places, dolomite. The thickest anhydrite units occur at the top and base of the formation. The rock salt beds thin out toward the margins of the Michigan Basin, and sulfate and sulfate-carbonate, and then carbonate appear in the Lucas Formation.

The Adavale Basin is situated in southern Queensland, East Australia. It is a relatively narrow submeridionally striking trough. In the south and north the Adavale Basin is connected with the Cobar and Drummond Basins, respectively. In the east the basin is bordered by the Condobolin Uplift. The evaporite Boree sequence containing rock salt with thin interbeds of anhydrite is recognized in subsiding parts of the basin; at the top anhydrite makes up a thin unit. The salt Boree sequence reaches 300 m in thickness. It is placed in the Etonval Formation tentatively assigned to the Eifelian (Hill 1967; Tanner 1967).

One, or in places two sulfate-bearing units usually occur in each Devonian sulfate basin. Thus, within the Minusinsk Basin sulfate rocks were found in the Middle Devonian deposits of the Abakan, Askyz, and Beysk Formations. In the former two formations gypsinate rocks with lenses and thin interbeds of gypsum and anhydrite occur among red sandstone and siltstone (the Abakan Formation), or among pale yellow siltstone and marl (the Askyz Formation). The Abakan-Askyz sulfate-bearing sequence was recognized tentatively. The Beysk Formation contains rather thick beds of gypsum in limestone, marl, siltstone, and sandstone. There are some thick deposits of gypsum (e.g., Dadonkov and Khamzas deposits).

A sulfate-bearing sequence is known also in the Kuznetsk Basin. Gypsum bands were found there along eastern and southern margins among variegated and red chiefly terrigenous deposits of the Podonin horizon of the Late Devonian. The Podonin sulfate is exposed in a number of deep boreholes, such as the Berdov, Ermakov, etc. Within the Teniz Basin there is a sulfate-bearing sequence tentatively regarded as Late Devonian in age. The sequence was penetrated by the Teniz test well down to 1430 m. Dolomite interbedded with anhydrite, up to 50 m thick, is exposed there (Ditmar 1966). In the Turgai Basin sulfate rocks were found in the terrigenous-carbonate sequence of Famennian age (Litvin 1973). Gypsum bands were recorded in the central zone of the Borovsk Anticline. A thick Middle Devonian sulfate-bearing sequence has been recently reported from the Moesian-Wallachian Basin [3] (Patrulius et al. 1967; Polster et al. 1976). The sequence occupies the vast areas in Bulgaria and Rumania, filling in the subsiding parts of the Varna and Wallachian Basins and the Tutrakan Trough. Its total thickness exceeds 1500 m. The thickest beds of anhydrite are associated with Givetian deposits where they are interbedded with clayey-carbonate and carbonate rocks, chiefly with dolomite; sulfate rocks also occur in the Upper Eifelian deposits. Another sulfate-bearing variegated sandy-clayey sequence was reported from the Tindouf Basin (Hollard 1967). Gypsum bands were recorded there in Middle to Upper Devonian deposits. The sequence is tentatively recognized as a separate unit. There are maybe several sulfate-bearing series in the Tindouf Basin. The Bird Fiord sulfate-bearing sequence was reported from the Basin of the Canadian Arctic Archipelago (McLaren 1963a,b; Kerr 1967a,b, 1974). Gypsum interbedded with gray dolomite and dolomitic marl makes up much of the sequence. Some gypsum beds are 2 m thick, and the total thickness is 100–120 m. Gypsinate rocks are also present in the underlying Eids Formation (Kerr 1974).

In the Illinois Basin the Jeffersonville Formation of Eifelian age is sulphate-bearing (Collinson et al. 1967).

Two sulfate-bearing sequences, namely the Kenwood and Cedar Valley, are known in the Central Iowa Basin. The Kenwood sequence, including the formation of the same name, contains several meters thick bands and units of anhydrite, interbedded with dolomite, limestone, and clay. The Cedar Valley sequence includes the Salon Formation and the lower Rapid Formation that also contains sulfate rocks, such as gypsum and anhydrite (Collinson et al. 1967).

Hence, 75 halogenic units can be recognized within the Devonian evaporite basins. Among them 26 salt-bearing units, namely the Zubovo, Manturovo, Fokin, Kygyltuus, and Nordvik sequences are known in the North Siberian Basin; the Ikhedushiingol sequence underlies the Tuva Basin; the lower salt-bearing sequence is reported from the Chu Sarysu Basin; the Morsovo sequence of the Moscovian Syneclise and the Narva sequence of the Pripyat Depression are spread in the Morsovo Basin; the Evlanovo-Liven and the Dankov-Lebedyan sequences of the Pripyat Depression, the Shchigrov sequence, as well as the lower and upper salt-bearing sequences of the Dnieper-Donets Depression, and the salt-bearing sequence of the Timan Trough are known in the Upper Devonian Basin of the Russian platform; the Lotsberg, Gold Lake, Prairie, Black Creek, Hubbard, Davidson, Dinsmore, and Stettler sequences are

3 In earlier work by Zharkov (1974a) it was named the Wallachian Basin

situated in the West Canadian Basin; the Moose River sequence occupies the Hudson Basin; the Lucas sequence is reported from the Michigan Basin; the Boree sequence underlies the Adavale Basin. Then, 40 sulfate-bearing sequences show the following distribution: sixteen sulfate-bearing sequences, such as the Zubovo sequence, the gypsum-bearing sequence of Taimyr, the Dezhnev, Sidin, Ukhta, Rusanov, Atyrkan, Burkhala, Sebechan, Vayakh, Nakokhoz, Kalargon, Upper Fokin, Namdyr, Matusevich, and the gypsum-bearing sequences of Vrangel Island are developed in the North Siberian Basin; two sequences, i.e., the Morsovo and middle Lopushan are spread in the Morsovo Basin; six sequences, namely the Shchigrov, Voronezh-Evlanovo and Mezhsolevaya sequences of the Pripyat Depression, the Upper Devonian sequence of the Baltic area, Moscovian Syneclise, and Volga-Ural area, as well as the Ustukhta and Idzhid-Kamensk sequences are known in the Upper Devonian Basin of the Russian platform; nine sequences, the Ernestina Lake, Chinchaga, Muskeg, Beaverhill Lake, Woodbend, Duperow, Birdbear, Nisku, and Stettler occur in the West Canadian Bain; three sequences, namely the Kenogami River, Williams Island, and Long Rapids are developed in the Hudson Basin; three sequences, the Kenogami River, Moose River, and Williams Island are known in the Moose River Basin; two sequences, the Abakan-Askyz and Beysk are spread in the Minusinsk Basin; the Kuznetsk Basin contains the Podonin sequence; the Teniz Basin includes the sulfate sequence; the Moesian-Wallachian Basin is underlain by the carbonate-sulfate sequence; the Tindouf Basin includes the sulfate-bearing variegated sequence; the Basin of the Canadian Arctic Archipelago contains the Bird Fiord sequence; the Illinois Basin includes the Jeffersonville sequence; the Central Iowa Basin includes two sequences: the Kenwood and Cedar Valley.

Carboniferous Evaporite Deposits

Evaporite series of Carboniferous age are known in Eurasia, North and South America, Africa, and Australia (Fig. 9).

Nine Carboniferous salt basins, namely the Chu Sarysu, Mid-Tien Shan, Sverdrup, Williston, Maritime, Paradox, Eagle, Saltville, and Amazon Basins are first to be described in brief, and then we shall try to determine the number of salt- and sulfate-bearing sequences within them.

In Carboniferous time the Chu Sarysu Basin repeatedly became a site of evaporite sedimentation. Salt deposits, developed there within the Dzhezkazgan, Tesbulak, and Kokpansor Depression, are Late Devonian to Early Carboniferous in age. However, some investigators (Sinitsyn et al. 1977) recognize in a salt-bearing series two salt units: the lower Late Devonian unit, and the upper Early Carboniferous one. They place the upper unit at the level of the Tournaisian. It consists of rock salt alternated with mudstone. The upper unit is 350 m thick. It is separated from the lower unit by variegated mudstone interbedded with limestone and dolomite. In the regions of the Chu Sarysu Basin where salt deposits are absent, their equivalents are represented by variegated mudstone, brecciated limestone, dolomite, marl, gypsum, and anhydrite.

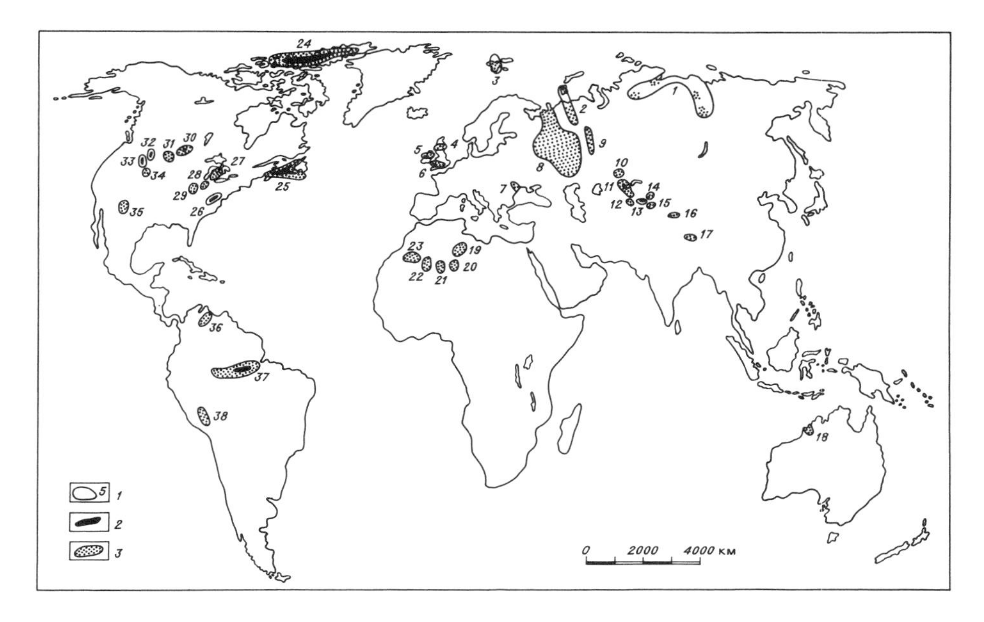

Fig. 9. Distribution of Carboniferous evaporites. Evaporite basins:
1 North Siberian, *2* Pechora-Novaya Zemlya, *3* Spitsbergen, *4* Northumberland, *5* North Ireland, *6* Central England, *7* Dobruja, *8* East European, *9* East Uralian, *10* Teniz, *11* Chu Sarysu, *12* Chimkent, *13* Mid-Tien Shan, *14* Tyup, *15* Aksu, *16* Achikkul, *17* Lhasa, *18* Fitzroy, *19* Rhadames, *20* Illisie, *21* Ahnet, *22* Reggane, *23* Tinfouf, *24* Sverdrup, *25* Maritime, *26* Saltville, *27* Michigan, *28* Illinois, *29* South Iowa, *30* Williston, *31* East Wyoming, *32* Eagle, *33* Paradox, *34* San Juan, *35* Orogrande, *36* Venezuela, *37* Amazon, *38* South Peruvian. For explanations see Fig. 1.

The higher level of halogenic rocks in the Chu Sarysu Basin is associated with Visean and Namurian deposits. Thus, the halogenic-carbonate sequence of Early/Middle Visean age with bands of anhydrite and gypsum up to 1 m thick is recognized in the Akkol Anticline, in the Talas Uplift, within the East Karakol Trough, in the Bestube Dome (Ditmar 1966; Aleksandrova 1971; Bakirov et al. 1971). Sulfate rocks in Upper Visean/Namurian deposits are present in the central and southern Chu Depression, in the Talas Uplift (Ditmar 1966; Bakirov et al. 1971). the Lower to Middle Visean sulfate rocks can be tentatively regarded as a single sulfate-bearing sequence extremely widespread in the Chu Sarysu Basin. It is separated from the Upper Visean/Lower Namurian sulfate-bearing sequence by a red sandstone, but in places the two sequences are interlinked and become undistinguishable. The Upper Visean/Namurian sequence has the same extent as the underlying one. Both sequences appear to reach the north-western part of the Kirghiz Range where bands and rather thick gypsum and anhydrite beds are developed in Visean/Early Namurian sedimentary sequences of the Ulkunburul, Kishiburul, and Tekturmas Mountains (Shcherbina 1945; Ivanov and Levitsky 1960; Bakirov et al. 1971). However, bearing in mind that sulfate rocks do not stretch continuously from the Chu Sarysu Basin into the Kirghiz

Range, it is more reliable to recognize a separate Visean/Namurian sulfate-bearing sequence in the Kirghiz Range. The two sequences must have formed in a single sedimentary basin. In the Chu Sarysu Basin the uppermost sulfate-bearing sequence is known as the Kyzylkanat sequence. It includes Middle/Upper Carboniferous terrigenous red beds and variegated deposits with lenses, inclusions, and veins of gypsum, and gypsinate rocks as well; in places thin interbeds of anhydrite and/or gypsum occur (Bakirov et al. 1971). Hence, the data available allow us to recognize, as a first approximation, one salt-bearing and four sulfate-bearing sequences within the Chu Sarysu Basin, including the north-western Kirghiz Range.

The Mid-Tien Shan Basin is tentatively regarded as salt-bearing. The basin was previously known as the Kokomeren-Tekes Basin (Zharkov 1974a) Moreover, the evidence for the presence of rock salt in Carboniferous deposits in some inland regions of Tien Shan was doubtful, as saline rocks may prove to be not Carboniferous, but Neogene in many intermontane depressions of Tien Shan. Recent data suggest that Carboniferous evaporite series, not only sulfate-bearing, but probably salt-bearing, are developed in Mid-Tien Shan [4]. However, these data are rather discrepant, and, in some cases, debatable; therefore we can only suggest the presence of the Carboniferous salt basin in Tien Shan. The Mid-Tien Shan Basin is described in detail below.

The data available suggest that the Mid-Tien Shan Basin, presumably salt-producing, is associated chiefly with the Naryn Depression (Fig. 10). It forms part of a single Carboniferous evaporite belt stretching sublatitudinally from the Chimkent Depression in the west to the Turuk Trough on the south slope of the Terskey-Alatau Range in the east. Within this belt halogenic rocks were found at different stratigraphic levels in Carboniferous deposits. In the Naryn Trough and adjacent mountain ranges evaporite deposits are generally Early/Middle Carboniferous; however, Late Carboniferous deposits occur as well.

The oldest sulfate rocks, mainly gypsum, occur as interbeds and lenses among shale and calcareous conglomerates and were found in Upper Tournaisian and Lower Visean deposits of the Dzhamandavan and Naryntau Ranges, i.e., in the south-western and eastern framing of the Naryn Depression. The thickness of the sulfate-bearing deposits ranges from 100 m to 400 m in the Dzhamandavan Range and reaches 1000 m in the Naryntau Range (Poyarkov 1972). As early as 1958 A. Ya. Galitskaya

4 Mid-Tien Shan includes the Chatkal-Narym zone (after Nikolaev 1933; Knauf and Rezvoi 1972)

Fig. 10. Generalized lithological map of Visean and Namurian (Lower Carboniferous) and Bashkirian (Middle Carboniferous) deposits of Tien Shan and adjacent areas. ▶

1 land, *2* land areas showing development of volcanic rocks, *3* episodically existed islands, *4–14* areas of development of different sediments, *4* terrigenous, *5* terrigenous and carbonate, *6* terrigenous, carbonate, and sulfate, *7* terrigenous, mostly red sulfate, *8* sulfate and carbonate, locally terrigenous, *9, 10* salt (*9* inferred areas of rock salt development, *10* inferred areas of rock salt distribution), *11* dolomitic, *12* calcareous, probably reef, *13* siliceous-calcareous, *14* argillaceous and arenaceous, mostly gray-colored, *15* a zone of evaporite distribution in the Tien Shan, *16* southern boundary of the zone of development of island areas and reef carbonate buildups. Evaporite basins: *I* Chu Sarysu, *II* Chimkent, *III* Mid-Tien Shan, *IV* Tyup

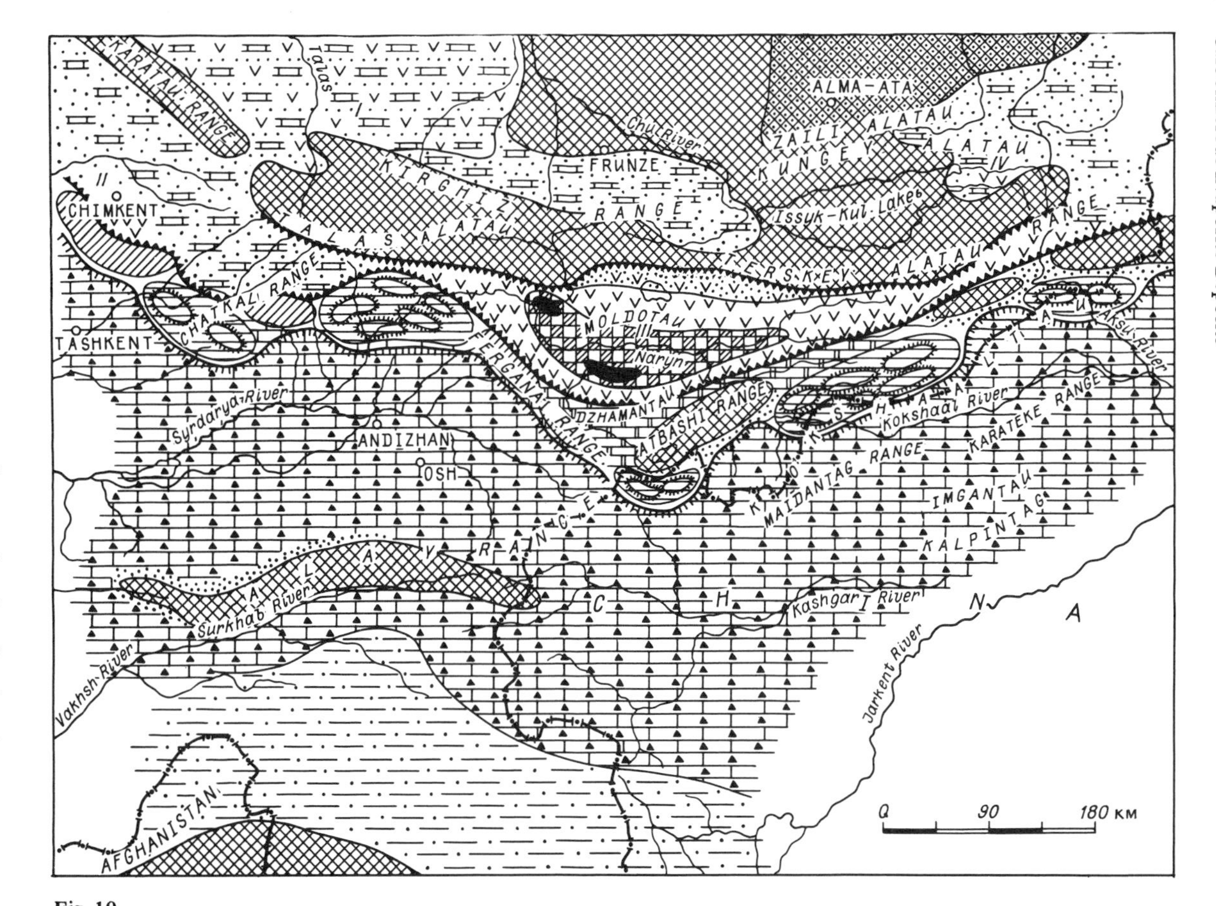
1
2
3
4
5
6
7
8
9
10
11
12
13
14
15
16
KARATAU RANGE
Talas
ALMA-ATA
ZAILIY ALATAU
KUNGEY ALATAU
Chu River
FRUNZE
KIRGHIZ RANGE
Issyk-Kul Lake
CHIMKENT
TALAS ALATAU
TERSKEY ALATAU RANGE
CHATKAL RANGE
TASHKENT
MOLDOTAU
Naryn
FERGANA RANGE
DZHAMANTAU
ATBASHI RANGE
Aksu River
KOKSHAAL
Kokshaal River
KARATEKE RANGE
Syrdarya River
ANDIZHAN
OSH
MAIDANTAG RANGE
IMGANTAU
KALPINTAG
ALAY RANGE
Surkhab River
Kashgar River
CHINA
Vakhsh River
Jarkent River
AFGHANISTAN
0 90 180 KM

Fig. 10

described a 700–800 m thick Middle Visean sequence within the western framing of the Naryn Depression in the Kokiyrim Range; the deposits are gray marly limestone with bands and lenses of gypsum (Poyarkov 1972). North of the Naryn Depression, at different localities in the Moldotau Range, gypsum-bearing sediments are known among Visean and Namurian (Lower Carboniferous) deposits, as well as in the Bashkirian (Middle Carboniferous). Thus, in the Mount Dyudyunbel area, A.G. Lasovsky and L.N. Mozylev described a Namurian/Bashkirian section with a unit of alternated gray to yellow thin-bedded silicified limestone, black clay shale interbedded with gypsum, sandstone, and siltstone at the base. Some investigators reported the presence of bands and units of rock salt in addition to gypsum in the lower reaches of the Kokomeren River. T.M. Borisova and E. Yu. Egoshin have recorded rock salt lenses as large as 40 by 200 m among Carboniferous gypsum-bearing deposits on the left bank of the Kokomeren River at the mouth of its right tributary Chon-Dobe and a salt bed on the right bank of the Karachauli River, 8 km higher up the mouth. A.A. Luyk and P.A. Kovalev reported thin lenticules of halite in green to gray sandy-clay shales on the right bank of the Kokomeren River near its junction with the Kavyuk-Su River, 5 km south of the village of Kaa-Bulak.

Recently O.I. Karas, N.P. Koroleva, V.A. Romanov, A.A. Churin, A.A. Zharikov, and V.V. Ermikova studied the Upper Paloezoic, Carboniferous deposits including those in the Naryn Depression and those of its framing. They reported the presence of sulfate and salt rocks in Bashkirian sedimentary sequences in the Bokaly and Karalarga Rivers Basins. The section along the Karalarga River composed by A.A. Zharikov is of particular interest. It includes, in ascending order: (a) saline sandstone – 118 m; (b) sandstone and mudstone with seams and inclusions of rock salt in the upper and lower parts – 100 m; (c) mudstone and sandstone with inclusions of rock salt – 30 m; (d) saline sandstone with limestone interbeds at the top – 90 m. In the Bokaly River Basin sulfate and possibly salt rocks were placed into the Chemanda Formation of the Bashkirian, though some investigators (Poyarkov 1972) assign it to the Visean and Namurian (Lower Carboniferous).

North of the Naryn Depression gypsum is present in Visean to Namurian and Bashkirian deposits almost over the entire area embracing the Balykta and Sonkul Troughs. However, they occur there mainly in terrigenous red beds. For example, in the Kavak-Tau Mountains, on the north slope of the Moldotau Range, variegating gypsinate siltstone is present in the upper Karachauli Formation of Visean to Namurian age (Poyarkov 1972). Gypsiferous sediments occur also in the Aktaylyak Formation assigned to the Lower Bashkirian (Ektova and Belgovsky 1972). Additionally, along the south slope of the Dzhamandavan Range gypsum interbeds were reported from the upper Kodzhagul Formation assigned to the Upper Carboniferous (Belgovsky 1972). Carboniferous sulfate rocks are also known east of the Naryn Depression in its continuation within the Turuk Trough (Poyarkov 1972).

Thus, Carboniferous evaporites are developed around the mountains framing the Naryn Depression. To the north-west, in the Kokomeren, Bokaly, and Karalarga Rivers Basins rock salt is present. Recently the presence of the Carboniferous rock salt was reported from the Naryn Depression as well. According to O.I. Karas et al., the sequence is exposed in the Chelokkoin 6 borehole at depths of 2906 m to 3150 m in Bashkirian deposits. The presence of rock salt was determined from geophysical data.

New data and core samples described by G.A. Glazatova allow us to outline rather approximately six rock salt beds there. Four upper beds probably alternate with limestone and anhydrite, but the two lower, thickest beds contain, aside from anhydrite, numerous terrigenous bands, such as siltstone, sandstone, and gritstone. The upper part of the salt-bearing sequence is essentially sulfate-carbonate, and it is capped with limestone, overlain by effusive rocks at depths of 2308 m to 2906 m. Probably this sequence goes as far as the north-western framing of the Naryn Depression, where it crops out in the Kokomeren River Basin and in adjacent areas. At present, it is impossible to determine the extent of salt deposits in subsiding zones of the Naryn Depression, but they may be developed in its interiors. The age of the salt deposits is unclear. It is probably Visean, but some workers (O.I. Karas et al.) assign them to the Bashkirian (Middle Carboniferous).

The Mid-Tien Shan evaporite basin, presumably recognized as salt-producing, is rather extensive. It stretches from west to east for a distance of more than 400 km, the maximum width exceeding 100 km. Salt deposits can be developed in the basin over an area about 3000 km^2. If further investigations confirm the size of the basin, and if an average thickness of the salt sequence is at least 200 m and salt saturation amounts to 30%, then the volume of rock salt in the basin sill account for about 150 km^3. It is impossible to determine at present the number of evaporite sequences in the basin. One salt-bearing sequence is suggested in the basin's interior. Toward the framing the sequence wedges out and is replaced by sulfate-bearing terrigenous-carbonate rocks, which to the north can be singled out into the Chamanda sequence, and to the south and south-east they form an unnamed separate sulfate-bearing terrigenous-carbonate sequence. Another sulfate-bearing sequence is probably present in the northern part of the basin where the Aktaylyak gypsum-bearing formation of Early Visean age is developed. Finally, the third terrigenous-carbonate sulfate-bearing sequence associated with the upper part of the Upper Carboniferous Kodzhagul Formation may be present south of the Naryn Depression, along the south slopes of the Dzhamandavan Range. Inadequate knowledge of the stratigraphy of Carboniferous deposits in the Mid-Tien Shan regions and uncertain correlation do not allow an estimation of the number of evaporite sequences within the basin; at the present time five sequences are known, i.e., one salt and four sulfate sequences.

The zone of evaporite sedimentation including the Mid-Tien Shan evaporite basin occupied a certain paleogeographic position. On the south it was bounded by a zone of variable width consisting of an island chain; some islands existed throughout the Early to Middle Carboniferous, but others repeatedly subsided. The Mid-Tien Shan evaporite basin proper was bounded on the south by an island chain situated within the Atbashi Range, and some other islands in the central Fergana Range. To the east some islands can be recognized within the Sarydzhaz Block, and to the west gently sloping islands can be outlined within the Chatkal Range. During Early to Middle Carboniferous time lengthy reef and other carbonate bioherms formed between the islands. Thus, the zone including the islands and reef carbonate structures separated the belt of evaporite sedimentation from the open sea of normal salinity to the south. Calarsiliceous sediments accumulated chiefly in the open sea. It should be noted that a reef-insular zone was built up as a series of island arcs with the convex side facing south. One of these island arcs flanks the Mid-Tien Shan evaporite basin, and the other the Chimkent Basin that also contained evaporites in Early Carboniferous time. The land was situated north of an evaporite sedimentation zone; terrigenous red beds or volcano-terrigenous deposits accumulated in intermontane basins developed on the mainland where sea came from time to time. In some of these basins evaporite strata were accumulated.

Thick Carboniferous salt-bearing deposits have been recently recognized within the Basin of the Canadian Arctic Archipelago. Some years ago only sulfate-bearing sequences building up most diapir domes on Axel Heiberg, Amund Ringnes, Ellef Ringness, Melville, and Ellesmere Islands were known there (Blackadar 1963; Greiner 1963; McLaren 1963b,c; Norris 1963; Roots 1963a,b,c; Souther 1963; Thorsteinsson 1963; Tozer 1963; Christie 1964; Tozer and Thorsteinsson 1964). A wide distribution of sulfate-bearing sequences and their considerable thickness made many investigators believe that rock salt may also be present within the Canadian Arctic Archipelago in subsiding parts of the Sverdrup Basin. This suggestion was confirmed in 1972 when the deep Hoodoo L-41 borehole was drilled in southern Ellef Ringnes Island on the vault of the Hoodoo salt dome. Data obtained were first reported by Davies (1974a,b, 1975a,b). Evidence available suggests that evaporite, including salt-bearing Carboniferous sediments, are present over much of the Sverdrup Basin. Thus, a new Paleozoic salt basin was recognized there and named the Sverdrup Basin. It will be briefly discussed below using data reported by Davies (1974a,b, 1975a,b), Thorsteinsson (1974) and Mayr (1975).

The Sverdrup Basin is situated within the Canadian Arctic Archipelago embracing north extremities of Prince Patrik and Melville Islands on the south-west, and stretching farther on the north-east up to north-western Ellesmere Island. Aside from these islands, the basin includes such large islands as Borden, Mackenzie King, Ellef Ringnes, Amund Ringnes, Cornwallis, Axel Heiberg, and many small islands (Fig. 11). The north-western boundary of the basin is defined only at two localities (in the extreme north-west of Ellesmere Island and on northern Axel Heiberg Island). In other areas the boundary vanished in the Arctic Ocean and is still unknown.

Carboniferous and Permian deposits are developed all over the Sverdrup Basin. Their thickness exceeds 2500 m. A great number of formations are recognized within the section (Fig. 12). The oldest sediments form the Emma Fiord Formation of the Visean (Lower Carboniferous) age. It consists mainly of sandstone and siltstone interbedded with gray and black calcareous mudstone. Conglomerate bands and coal lenses can be observed as well. The formation reaches 100–130 m in thickness. The above lying Borup Fiord Formation rests on the Emma Fiord Formation, and even Lower Ordovician and Cambrian rocks. The base of the formation is represented by red conglomerate; red and green-gray conglomerate and sandstone interbedded with siltstone, mudstone, and limestone, predominantly gray, appear higher in the section. Some sections contain dolomite. The thickness varies from 120 m to 350 m.

Mayr (1975) studied in detail the Borup Fiord Formation in the Yelverton area on northern Ellesmere Island and recognized three units: A, B, and C. Unit A consists of white sandstone and conglomerate with rare interbeds of red mudstone. Unit B contains dark gray to black crystalline dolomite, black limestone, and calcareous mudstone. Gypsum and anhydrite interbedded with dark gray marl occur in the lower part of unit C in the Yelverton area. They are overlain by carbonate clay,

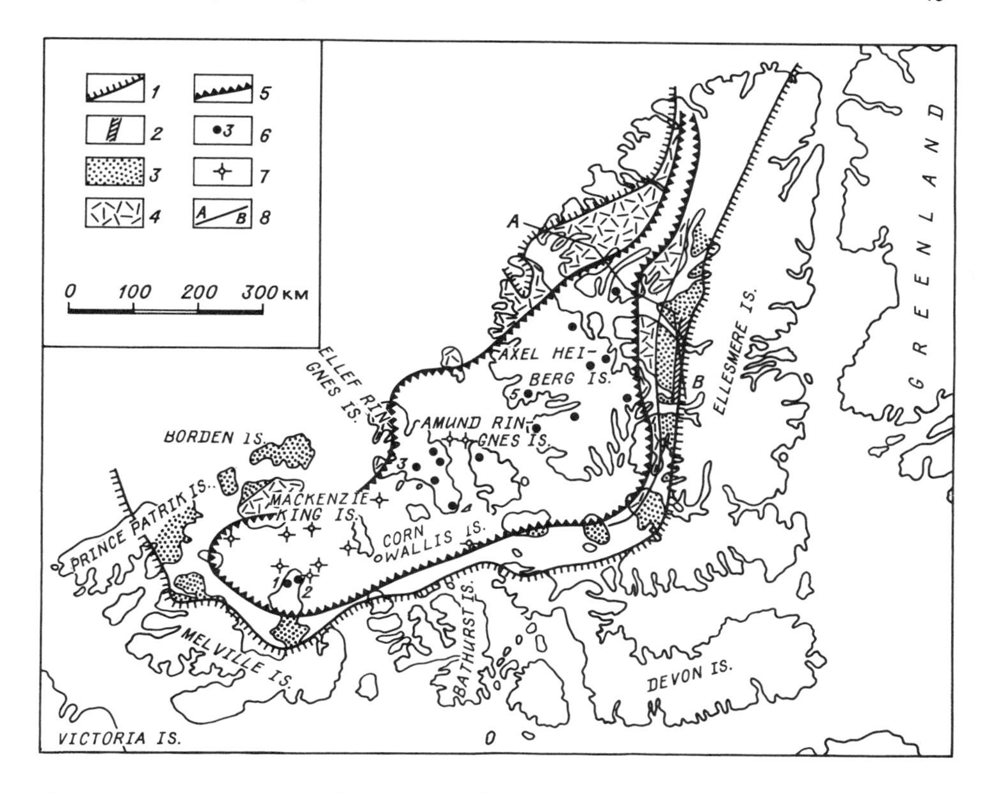

Fig. 11. Distribution pattern of Permo-Carboniferous evaporite deposits in the Sverdrup Basin. Compiled from Thorsteinsson (1974) and Davies (1975a,b).

1 boundaries of the Sverdrup Basin, *2* near-shore terrigenous deposits, *3* near-shore terrigenous, sulfate-carbonate and carbonate deposits, *4* predominantly sulfate and carbonate deposits, *5* distribution of salt deposits, *6* diapiric domes with outcrops of gypsum and anhydrite (*1* Barrow, *2* Cape Colqukoun, *3* Isachsen, *4* Hoodoo, *5* Southern Fiord), *7* inferred diapiric domes, *8* facial profile shown in Fig. 12

massive dolomite, dark gray shale, and siltstone. U. Mayr suggested that unit C and probably unit B are to be compared with the Otto Fiord evaporite formation developed chiefly in the Sverdrup Basin interiors. The Audhild Formation is known only in the Kleybolt Peninsula of Ellesmere Island. Between the Borup Fiord and Nansen Formations it includes volcanogenic basic rocks.

The Otto Fiord Formation contains most of the Carboniferous evaporite deposits in the Sverdrup Basin. Recently the formation was believed to consist chiefly of medium- and thick-platy to massive anhydrite turning into gypsum at the surface. Anhydrite may amount to 80% of the total thickness of the formation, and dark gray limestone with minor mudstone together account for the remainder 20%. Anhydrite of the Otto Fiord Formation builds up many salt domes (see Fig. 11). The thickness of the formation was estimated to be about 300–330 m. Its stratigraphic position remained uncertain. It was assigned to the Bashkirian (Middle Carboniferous) (Thorsteinsson 1974). These notions have been clarified in the last few years (Davies 1974a, Nassichuk 1975).

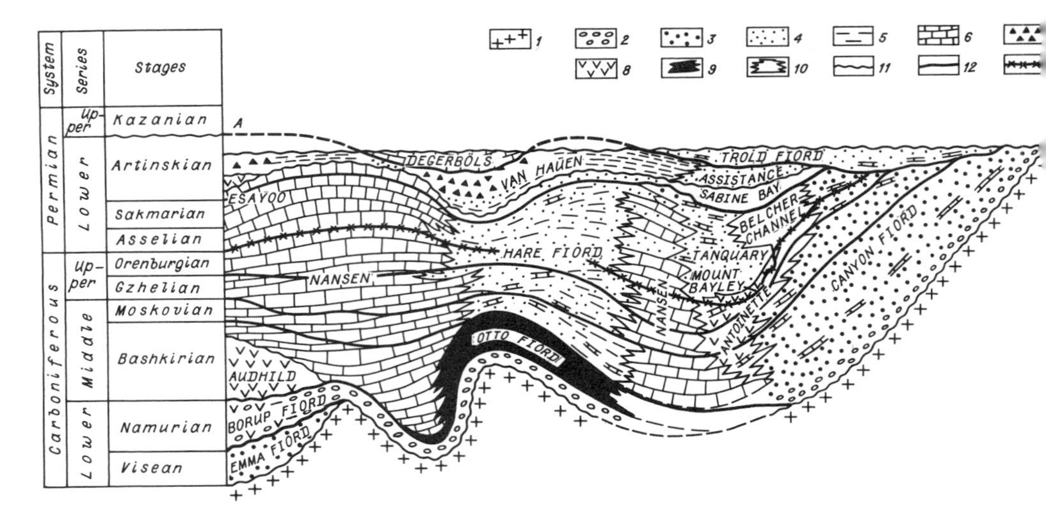

Fig. 12. Facial profile of Permo-Carboniferous deposits in the Severdrup Basin showing formations versus stratigraphic position (after Thorsteinsson 1974).

1 Pre-Carboniferous deposits, *2* conglomerate, *3* sandstone, *4* siltstone, *5* mudstone and shale, *6* limestone and dolomite, *7* chert, *8* anhydrite and gypsum, *9* rock salt and anhydrite, *10* distribution of formations, *11* unconformable occurrence of deposits, *12* stratigraphic boundaries, *13* Carboniferous/Permian boundary

After the drilling of the first Hoodoo L-41 deep borehole in the Hodoo diapir on south-eastern Ellef Ringnes Island thick salt beds were recognized in the deep parts of the Otto Fiord Formation. The borehole penetrated the following rocks, in descending order:

	Thickness (m)
The top of gypsum and anhydrite was penetrated at a depth of 331.5 m	
1. Sulfate cap rock	12.0
2. Gypsum and anhydrite	23.0
3. Recrystallized limestone with stylolites	44.4
4. Anhydrite interbedded with limestone, dolomite, sandstone, and siltstone	201.0
5. Pure, essentially recrystallized rock salt. Inclusions of halopelites and seams of anhydrite were encountered in the rock salt core at a depth of 1200 m. Two beds of intrusive rocks of gabbroid composition were penetrated at a depth of 1890 m and 3210 m. A core of deformed anhydrite and carbonate rocks with abundant inclusions of rock salt was recorded from a depth of 3780 m. The anhydrite is thin-bedded. Such fossils as gastropod, mollusk, brachiopod, and ostracod were found in carbonate interbeds	3882.0

The well was abandoned at a depth of 4213.5 m in rock salt.

On the basis of these data and fossil occurrences, Davies (1975a,b) determined the age of the Otto Fiord Formation as Late Namurian/Early Moscovian. Rock salt is believed to be present at depth in other diapir domes of the Sverdrup Basin. The present depth of the salt base in the most downwarped parts of the basin reaches 7500 m. It was concluded that salt deposits of the Otto Fiord Formation may occupy the entire interior zone of the basin (see Fig. 11).

Age equivalents of the Otto Fiord Formation in the marginal zones of the Sverdrup Basin are the upper Borup Fiord Formation and the lower Nansen Formation. The Otto Fiord Formation is overlain by the Hare Fiord Formation consisting chiefly of quartz siltstone, mudstone, limestone with various amounts of chert and quartz sandstone. The formation shows abrupt changes in composition over the entire area. The thickness varies widely from 300 m to 1230 m. The age of the Hare Fiord Formation is Carboniferous to Permian ranging from the close of the Bashkirian (Middle Carboniferous) to the beginning of the Artinskian (Lower Permian).

The Nansen Formation contains limestone, in places with minor quartz. Limestone yields abundant fossils. Rocks are light to yellow-gray, in places green-gray in color. Chert lenses are encountered. The deposits are comparable with the Otto Fiord and Hare Fiord Formations. The age of the Nansen Formation varies from the upper Namurian (Lower Carboniferous) to the lower Artinskian (Lower Permian). The thickness ranges from 1200 m to 2340 m.

The Cañon Fiord, Antoinette, Mount Bayley, Tanquary, and Belcher Channel Formations are age equivalents to the Nansen Formation. The Cañon Fiord Formation consists of variegated (gray, yellow, green, brown, and red) sandstone and siltstone, the amount of limestone, marl, and mudstone varies in different sections. Conglomerate occurs at the base of the formation. The formation reaches 1650 m in thickness. It ranges in age from the beginning of the Bashkirian (Middle Carboniferous) to the Sakmarian (Lower Permian). The Antoinette Formation consists mainly of dark gray limestone with interbeds of siltstone, sandstone, and mudstone. Anhydrite and gypsum, as well as gypsinate limestone occur throughout the section. The thickness varies from 465 m to 800 m. The age of the formation varies from the upper Middle Carboniferous (the close of the Moscovian) to the Late Carboniferous (Orenburgian). The Mount Bayley Formation is dominated by anhydrite and in this respect it is similar to the Otto Fiord Formation. But in the Mount Bayley Formation, unlike the Otto Fiord Formation, anhydrite is interbedded with quartz siltstone and sandstone. The formation reaches 200–250 m in thickness. It is assigned to the Asselian (Lower Permian). The Tanquary Formation, up to 650 m thick, is represented by sandstone and limestone with interlayers of siltstone. The deposits are assigned to the Asselian, Sakmarian, and Lower Artinskian (Lower Permian). The Belcher Channel Formation consists of limestone with bands and lenses of sandstone and siltstone. Its thickness reaches 120 m. The age of the formation is Late Carboniferous/Early Permian (Gzhelian of the Upper Carboniferous to the Lower Artinskian of the Lower Permian).

Deposits of the Sabine Bay, Assistance, and Van Hauen Formations assigned to the Upper Artinskian are of continental origin. The Sabine Bay Formation, up to 200 m thick, is represented by light gray, yellow to orange, and red to brown quartz sandstone and conglomerate. The Assistance Formation consists of gray and yellow to brown sandstone and siltstone, up to 400 m thick. The Van Hauen Formation includes dark gray to black mudstone, siltstone, gray quartz sandstone, as well as dark gray to black cherts interbedded with silty schists. The thickness varies from 400 m to 670 m.

The Lower Permian deposits are unconformably overlain by the Upper Permian sediments assigned to the Kazanian stage. They embrace the Trold Fiord and Degerböls Formations. The Trold Fiord Formation consists of gray, green, and brown sandstone interbedded with limestone. Its thickness varies from 30 m to 300 m. The Degerböls Formation, up to 380 m thick, consists mainly of gray and yellow limestone.

Thus, in the Sverdrup Basin the evaporite deposits occur throughout the Carboniferous/Lower Permian section (from the Namurian, Lower Carboniferous, to Asselian, Lower Permian). Salt deposits range from the late Namurian to the Moscovian, i.e., completely within the Early/Middle Carboniferous. Salt deposits form the Otto Fiord Formation. This rock salt unit appears to occupy 200,000 km^2. Its exposed thickness is 3882 m, and the total thickness exceeds 5000 m. However, great thicknesses do not exceed 1500–2000 m. If an average thickness of rock salt over the area is 500 m, then rock salt in the Sverdrup Basin will amount to about 1.10^5 km^3. But in fact the value should exceed this approximate estimate. In addition to the Otto Fiord salt sequence, three other sulfate sequences: the Borup Fiord, Antoinette, and Mount Bayley Formations are known in the Sverdrup Basin. The latter formation is of Early Permian age.

The Williston salt basin underlies parts of the North Dakota and Montana in the United States and parts of Saskatchewan and Alberta in Canada. The basin is rather well known (Anderseon and Hansen 1957; Edi 1958; Illing 1959; Willis 1959; Pierce and Rich 1962; Macauley et al. 1964; Carlson and Anderson 1965; Haun and Kent 1965; Norwood 1965; McMannis 1965; Proctor and Macauley 1968; Lefond 1969). In the Williston Basin evaporite deposits form the Madison Group of the Mississipian age. It contains, in ascending order, the Lodgepole, Mission Canyon, and Charles Formations. Halogenic rocks made their appearance in the lower Mission Canyon Formation where anhydrite bands occur among limestone and dolomite. The lowermost rock salt bed is known from the upper part of the formation. The Charles Formation consists mainly of rock salt, anhydrite, dolomite, and limestone. In the central parts of the basin the Charles Formation consists of six rock salt horizons wedging out toward the margins; carbonate-sulfate grades into carbonate. The Madison Group salt deposits cover an area of 80,000 km^2. The total maximum thickness of rock salt does not exceed 100 m. The Charles salt is overlain by the Kibbey Formation represented chiefly by red mudstone interbedded with sandstone, siltstone, and limestone with lenses and seams of anhydrite. Thus, there are one salt-bearing and two sulfate-bearing sequences in the Williston Basin. The salt sequences may be named the Charles Formation; it also includes lower beds of rock salt known in the Mission Canyon Formation. The lower Mission Canyon Formation forms a separate sulfate sequence of the same name. The upper sulfate sequence includes deposits of the Kibbey Formation.

The Maritime Basin is located in the New Brunswick and Nova Scotia Provinces of Canada and stretches into the Gulf of St. Lawrence and the Antlantic Ocean.

There is a long list of publications on Carboniferous evaporite deposits (Gussow 1953; Bancroft 1957; Bell 1960; Hamilton 1961; Greiner 1962; Crosby 1962; Boyle 1963; Howie and Cumming 1963; Collins 1964; Belt 1965; Evans 1965, 1967; Withington 1965; Cooper 1966; Benson 1967, 1970 a,b; Kelley 1967; Shaw and Blanchard 1968; Lefond 1969; Taylor 1969). Evaporite deposits are completely associated with the Mississipian formations. They are present in the Albert Formation of the Horton Group where they form the Gautreau sequence, as well as in the Windsor Group in a subzone B. The first salt sequence is recognized only in the Moncton Basin. The area covered by the sequence does not exceed 100 km^2. The second salt sequence is more widespread. Recent data (Howie and Barss 1975) on evaporite deposits of the Windsor Group suggest their distribution not only in western New Brunswick, in northern Nova Scotia, and on Cape Breton Island, but also on Prince Edward Island, in areas adjacent to the Gulf of St. Lawrence, and in a shelf zone south of Newfoundland. The greatest thickness of the Windsor deposits is known in the southern Gulf of St. Lawrence near Cape Breton Island where it reaches 5000 m. The Windsor salt generally builds up numerous domes most widespread in the Cumberland sub-basin. An areal extent of the Windsor Group evaporites exceeds 200,000 km^2. According to the improved data, in the Maritime Basin the volume of rock salt of both sequences can exceed 4.10^3 km^3.

The Saltville Basin is located in eastern North America, Virginia, USA. Halogenic rocks are associated there with the Mississippian deposits and developed in a narrow belt following the Saltville Fault from south-west to north-east. Salt deposits belong to the Maccrady Formation, the thickness of which varies from 20 m to 30 m, and increases to 250 m, or even to 600 m in areas where gypsum, anhydrite, and rock salt are distributed. These evaporites may be grouped into the Maccrady salt sequence. It is overlain by Cambrian formations. The Saltville Basin has not yet been outlined (Withington 1965; Cooper 1966).

The Paradox and Eagle salt basins are situated in the western regions of the USA in the Rocky Mountains within the Colorado Plateau. The Paradox Basin encompassing the frontier territories of four states, namely Utah, Colorado, Arizona, and New Mexico, is bounded by the Defiance Uplift on the south and by the Circle Cliffs Uplift and the Uncompahgre Uplift on the east. The Eagle Basin is located exclusively in Colorado. It is flanked by the Uncompahgre Uplift and the Front Range Uplift to south-west and north-east, respectively. In the Paradox Basin all halogenic deposits are grouped into a single formation of the same name. It was studied in detail (Herman and Sharps 1956; Herman and Barkell 1957; Hite 1960, 1961, 1968; Elston et al. 1963; Ohlen and McIntyre 1965; Baars et al. 1967; Mattox 1968; Peterson 1968; Peterson and Hite 1969). Its thickness exceeds 1200 m in most localities; the thickness rises to 2500–3000 m and locally exceeds 4000 m in the crest of salt anticline. The formation consists of rock salt interbedded with potash salt, anhydrite, dolomite, and black shale. The formation is represented by five cycles, in ascending order: the Alkali Gulch, Barker Creek, Akah, Desert Creek, and Ismay cacles. The areal extent of potash salt exceeds 15,000 km^2. The total thickness of rock salt ranges from 150–200 m to 900 m and more. The areal extent reaches 30,000 km^2. To the margins of the basin salt deposits thin out, and the Paradox Formation becomes exclusively sulfate-carbonate, and then carbonate. The boundaries of the formation are essentially facial, it has a very intricate outline; the margins are represented by sulfate and sulfate-carbonate rocks, and the interior part contains sulfate rocks, shale, rock and potash salt.

In the Eagle Basin evaporites form a single sequence known as the Eagle Valley Formation (Lovering and Mallory 1962; Quigley 1965; Mallory 1966, 1971; Peterson and Hite 1969). It is represented by rock salt, anyhdrite (or gypsum), marl, mudstone, siltstone, and sandstone. Beds of potash salt occur locally. The formation varies from 300 m to 2700 m in thickness. Salt deposits underlie only the center of the basin within the Alkeli Trough. The data available suggest with some certainty that the Paradox and Eagle Valley sequences have accumulated in the individual evaporite basins separated by the Uncompahgre Uplift.

The Amazon Basin is one of the recently recognized Carboniferous salt basins situated in South America within the Amazon River Basin. The Nova Olinda salt series of Pennsylvanian age was defined there (Morales 1959, 1961; Oliveira 1959). The formation varies from 650–800 m to 1200 m in thickness. It consists of sandstone, siltstone, mudstone, dolomite, limestone, anhydrite, and rock salt. Salt deposits dominate the upper part of the Nova Olinda Formation. They were found over an area of 100,000 km^2. The total maximum thickness of rock salt is about 400 m. The formation is underlain by the Itaituba Formation containing interbeds and beds of anhydrite, of a thickness of 200–300 m.

A great number of Carboniferous evaporite basins containing only sulfate rocks such as gypsum and/or anhydrite have been recently recognized. There are 20 sulfate basins, namely the North Siberian, Pechora-Novaya Zemlya, Spitsbergen, Northumberland, North Ireland, Central England, Dobruja, East European, East Uralian, Teniz, Chimkent, Tyup, Aksu, Achikkul, Lhasa, Fitzroy, Radames, Illisie, Ahnet, Reggan, Tindouf, East Wyoming, San Juan, Orogrande, South Iowa, Illinois, Michigan, Venezuela, and South Peruvian basins. We will try to determine the number of sulfate-bearing sequences within these basins.

Within the North Siberian Basin in which evaporites persisted not only in the Devonian, but also during the Early Carboniferous, there are halogenic rocks (exclusively gypsum and anhydrite) of Carboniferous age in the north-western Siberian platform at Norilsk, in the Khatanga, Kyutingda, and Kempendyai Troughs. West of Norilsk sulfate rocks are exposed in the Tundrin Formation section containing several anhydrite beds associated with two lower beds. Anhydrite is interbedded with gray dolomite, dolomitic marl, and dolomitic limestone in the first bed, while in the second overlying bed anhydrite occurs among red marl (Matukhin and Menner 1974). The areal extent of the Tundrin sulfate-bearing sequence is still uncertain. In the Khatanga Trough interbeds of gypsum and anhydrite occur in the carbonate sequence. Menner (1965) places them in the central part, nevertheless their position in the section is still uncertain. In the Kyutingda Trough situated in the north-eastern Siberian platform in the Kyutingda River valley and bounded by the Olenek and Daldyn Uplifts in the north-east and south-west, respectively, sulfate rocks are associated with the Tournaisian deposits and assigned to the gypsum-bearing sequence recognized there (Kuteinikov and Syagaev 1962; Fradkin 1964, 1967; Menner 1965; Biterman et al. 1970). Its lower part is composed of interbedded dolomitic limestone, dolomite, and gypsum, and the upper part of dolomitic marl, clay, and gypsum.

East of the trough gypsum bands are rather persistent; their thickness varies from 1 m to 10 m. The thickness of the entire gypsum-bearing sequence ranges from 50 m to 90 m. In the Kempendyai Trough halogenic rocks (gypsum and anhydrite) occur only in the lowermost Carboniferous, in the lower Kurunguryakh Formation (Fradkin 1964, 1967; Nakhabtsev and Fradkin 1970). They are interbedded there with dolomite and tuffaceous siltstone. The thickness of the sulfate-bearing part of the Kurunguryakh Formation is 90 m. The above regions of the East Siberian Basin where Carboniferous sulfate-bearing deposits were recognized are widely spaced. Therefore separate sulfate sequences have been recognized in each region: the Tundrin sequence is known in the Norilsk region, the sulfate-carbonate sequence in the Khatanga Trough; the gypsum-bearing sequence in the Kyutingda Trough; and the Kurungupyakh sequence in the Kempendyai Trough. If their continuation is proved, the number of sulfate-bearing sequences will be reduced.

The Pechora-Novaya Zemlya evaporite basin is recognized from the presence of two sulfate sequences; one sequence is known in the Pechora Basin, it runs from the Usa River mouth to the Ilyich River Mouth, and the other is known on Novaya Zemlya along the western coast of the Makarov Bay and on Alebastrovyi Island. Early Carboniferous age is proposed for both sequences. They are usually assigned to the Visean, and partly to the Namurian (Atlas . . . 1969). In the Pechora Basin anhydrite and gypsum are interbedded with marl, dolomitic limestone, and dolomite; the thickness of the sulfate-carbonate sequence varies from 100 m to 150 m (Chermnykh 1962, 1966). On Novaya Zemlya the thickness of the gypsum-bearing sequence is about 30 m (Rogozov 1970).

The Spitsbergen Basin was the site of sulfate accumulation during Carboniferous and Permian time. Two gypsum-bearing sequences have been recognized there; the lower sequence is assigned by Harland (1964) to the Bashkirian (Middle Carboniferous). Some geologists (Sosipatrova 1967) consider it to be the Visean. The lower gypsum-bearing sequence is confined to western Spitsbergen.

Within the East European Basin gypsum and anhydrite are known from the Carboniferous deposits in many regions. However, the distribution pattern is inadequately known, though the Carboniferous deposits are sufficiently well-studied; this does not allow us to determine with some certainty the number of sulfate sequences developed in the basin; for the time being they are outlined approximately. The oldest sulfate unit can be identified as the Ozersk sequence. It is developed in the Moscovian Syneclise, as well as in adjacent regions of the Volga-Uralian area (Tokmov, Tatar, and Zhiguli-Pugachev Arches, Riazan-Saratov Trough). A high portion of sulfate rocks in the sequence is reported from the southern Moscovian Syneclise in the Tula District. It is represented there by gypsinate banded dolomite, sulfate-dolomitic rocks, and gypsum. In the Novomoskovsk, Venev, Domnin, and Zaraisk Districts the number of gypsum beds in the section increases, with a continuous gypsum unit, 15–23 m thick in its upper part. The maximum sulfate saturation is reported from the area near the town of Laptev. The thickness of the Ozersk sequence reaches there 65–70 m (The Carboniferous . . . 1971). In the north the sequence is essentially sulfate-dolomite. These sections can be observed near Podolsk, Monino, Rybinsk, Lyubim, and Soligalich. The thickness varies there from 20 m to 62 m. In the north-western Moscovian Syneclise the Ozersk sequence contains rather thick (about 10 m) gypsum deposit.

Within the Tatar Arch the sequence becomes essentially carbonate, but anhydrite inclusions locally occur in its central part (Rauzer-Chernousova et al. 1967). Lenses and seams of gypsum and anhydrite are encountered in some sections in the Zhiguli-Pugachev Arch (the Krasnaya Polyana-1 borehole). The Ozersk sulfate sequence seems to be rather widespread in the East European Basin. Many investigators assign it to the Zavolzhsk horizon of the Tournaisian.

The overlying sulfate-bearing sequence developed almost throughout the East European Carboniferous evaporite basin occurs at a level of the Oka and Serpukhov superhorizon of the Visean, and at a level of the Protvin horizon of the Namurian (Lower Carboniferous). The sequence is tentatively named the Oka-Serpukhov sequence. It is present in the central and western parts of the Moscovian Syneclise (districts of Nepeitsevo, Tutaev, Sharya, Lyubim, Soligalich, Kashin, Krasny Kholm), where it is represented mainly by dolomite and dolomitized limestone with gypsum inclusions, lenses, and bands (The Carboniferous . . . 1971); the sequence extends into the Gorky-Volga Region (districts of Oparino, Kotelnich) and into the Upper Kama Basin (the Glazov key borehole) as well, where bands and inclusions of thin anhydrite occur in terrigenous-carbonate variegated deposits; the sequence is reported from the northern and southern parts of the Tatar Arch (Golyshurma and Baitugan areas) and from the Melekes Depression where sulfate rocks occur in the form of inclusions and veinlets of anhydrite throughout the carbonate section of the Oka and Serpukhov superhorizons; the sequence is known in the Zhiguli-Pugachev Arch where numerous inclusions of gypsum and anhydrite occur in dolomite and dolomitized limestone (Rauzer-Chernousova et al. 1967). The same Oka-Serpukhov sulfate sequence seems to continue in the northern East European Basin and farther into the Basin of the middle reaches of the Severnaya Dvina River where Barkhatova (1963) reported the presence of anhydrite bands in dolomites tentatively assigned to the Devyatino Formation. The upper horizons of this sulfate sequence locally include deposits of the Namurian Protvin horizon; this can be observed in the zone of Vyatka folding, in the Upper Kama Basin, in the Tokmov and Tatar Arches, and in the Sergiev-Abdulin Depression (Rauzer-Chernousova et al. 1967). The thickness of the Oka-Serpukhov sequence varies over a wide range of 50–60 m to 350 m.

The third sulfate sequence is recognized at a level of the Kashira, Podolsk, and Myachkov horizons of the Moscovian (Middle Carboniferous). It can be named the Kashira-Myachkov sequence. It lies within the Moscovian Syneclise, in the zone of Vyatka folding, in the upper Kama Basin, in the Cherdyn region of the Cis-Uralian Depression, and in the Riazan-Saratov Trough; the sequence probably remains sulfate-bearing in some areas of the Tatar and Tokmov Arches. The sequence contains mainly carbonate rocks with bands, lenses, and inclusions of gypsum and anhydrite. Sulfate rocks become more extensive in the Sharya, Soligalich, Chukhloma, Lyubim areas (Barkhatova 1963; Rauzer-Chernousova et al. 1967; Nalivkin and Sultanaev 1969; Shik 1971; The Carboniferous System . . . 1971). The thickness of the Kashira-Myachkov sequence is about 200 m in the most sulfate-bearing sections.

The fourth sulfate sequence is associated with the Kasimov superhorizon of the Gzhelian (Upper Carboniferous). It is clearly defined in the Moscovian Syneclise where its thickness ranges from 40 m to 70 m. South-western and western boundaries of sulfate rocks in the Kasimov sequence stretch from the town of Kasimov roughly to Moscow and the town of Krasnyi Kholm. North-east and east of this boundary minor gypsum bands, 0.3–0.5 m thick, appear in the sequence consisting chiefly of limestone and dolomite. Then, north-west of the boundary running through the towns of Kovrov, Gavrilov Yam, and Tutaev, the sequence becomes highly gypsinate, and the thickness of separate gypsum beds reaches 0.5–1 m (Goffenshefer 1971). Rather reliable data suggestive of the Kasimov sulfate sequence continuing eastward and north-eastward into the zone of Vyatka folding and into the upper Kama Basin, as well as toward the Tokmov and Tatar Arches, have been reported (Rauzer-Chernousova et al. 1967). The sequence possibly stretches northward into the Timan area and into the Vychegda River Basin (Barkhatova 1963).

The overlying fifth sulfate sequence includes deposits of the Klyazma horizon belonging to the Gzhelian (Upper Carboniferous). It is known all over the East European Carboniferous evaporite basin. In the Moscovian Syneclise its thickness reaches 60–80 m. The sequence is represented there chiefly by dolomitized limestone and dolomite with gypsum bands, the thickness of which varies from 1 m to 3–4 m (Goffenshefer 1971). Within the Vyatka folding zone, in the Upper Kama Basin and in the Cherdyn region of Cis-Urals the Klyazma sequence remains sulfate-bearing carbonate sequence; however, the portion of sulfate rocks there decreases and they form rare bands, lenses, and inclusions. The thickness of the sequence ranges from 70 m to 100 m. In the area of the Tokmov and Tatar Arches, the Melekes and Sergiev-Abdulin Depressions lenses and bands of gypsum and/or anhydrite are common in the Klyazma sequence among dolomitized limestone, calcareous dolomite, and dolomite. The section is not altered within the Zhiguli-Pugachev and Orenburg Arches. Two anhydrite units, up to 50 m thick each, were reported (Orekhovo, Kuleshovka, and Neklyudovo Districts) (Grachevsky et al. 1969). The Klyazma sulfate-bearing sequence probably reaches the eastern margin of the Pre-Caspian Depression where sulfate rocks were found in the Upper Gzhelian (Dalyan and Posadskaya 1972). It should be noted that the Kasimov and Klyazma sulfate sequences recognized as separate members can be differentiated with some certainty only within the Moscovian Syneclise. In other regions of the East European Basin, particularly in the Volga-Uralian area, they are hardly distinguishable. Only one sulfate-bearing sequence may be present in this region at a level of the Gzhelian.

The uppermost sixth sulfate-bearing sequence can be arbitrarily recognized as the Noginsk sequence among the Carboniferous deposits of the East European Basin. It is situated at a level of the Orenburgian stage of the Upper Carboniferous and becomes most distinct in the Moscovian Syneclise where north of the boundary running through the town of Murom, Sudogda, Kolchigino, and Zagorsk inclusions, lenses, and bands of gypsum and anhydrite, 2–4 m thick, appear among carbonate rocks (Goffenshefer 1971). The same sequence may have been reported from the Timan area, and from the Vychegda and Pinega Rivers Basin where bands of gypsum and anhydrite, 0.7–1 m or even 5.7 m thick, are known locally in the Orenburgian deposits (Barkhatova, 1963a). The Noginsk sulfate sequence may stretch into the Upper Kama Basin and the Tokmov Arch where anhydrite bands were recognized among the Orenburgian deposits (Rauzer-Chernousova et al. 1967).

Within the Northumberland Basin in Scotland sulfate rocks are present in Visean calcareous sandstone. Inclusions and bands of gypsum occur among dolomite, mudstone, and sandstone over a small area (George 1958, 1963).

The distribution of sulfate-bearing deposits in the Northern Ireland Basin is restricted. They are associated with the Roscunish argillaceous sequence assigned to the Upper Visean (Padget 1953; Oswald 1955; Caldwell 1959; George 1963; Dixon 1972). West et al. (1968) have recognized the sulfate-bearing sequence in the Northern Ireland Basin as a separate Aghagrania Formation subdivided into units. Gypsum beds are associated there with the lower Meenymore member and with the uppermost Corry member. Large gypsum lenses are locally 2.2 m in size, but usually gypsum is present as intercalations, inclusions, and veinlets.

The sulfate basin of Central England has a limited areal extent as well. The Visean anhydrite series is exposed there in boreholes near Hathern, Leicestershire (Llewelling and Stabbins 1968).

The Dobruja Basin is recognized from the presence of sulfate rocks (anhydrite) in the Tournaisian and Visean deposits exposed within the Dobruja Depression in Moldavia (Kaptsan et al. 1963; Kaptsan and Safarov 1965b, 1969a). The deposits were penetrated there by three (R-30, R-31, and R-32) boreholes. Anhydrite beds occur among limestone and calcareous dolomite. The thickness of the carbonate sulfate-bearing sequence is 176 m. It is believed to continue southward.

In the East Urals evaporite basin the number of sulfate-bearing sequences is unknown as the basin has not yet been outlined and it is determined roughly as a single region of halogenic sedimentation. Along the eastern slope of the Urals individual outcrops of the Carboniferous sedimentary successions may have accumulated in isolated basins, and several basins of evaporitic sedimentation may have existed there in different times. Three separate widely spaced outcrops of sulfate rocks are presently known there. The largest gyprock outcrop is located near Magnitogorsk (Ivanov and Levitsky 1960). A sulfate sequence consisting of marl, calcareous sandstone, and conglomerate with two gypsum units was outlined there as early as 1932. The thickness of the lower unit varies from 27 m to 99 m, and that of the upper one from 17 m to 20 m. The second terrane of Carboniferous gypsum-bearing deposits is located along the eastern slope of the South Urals in the Bagaryak district (Ivanov and Levitsky 1960). Gypsum beds, the thickness of which varies from 5–10 m to 15–37 m, are associated with variegated sequence represented by conglomerate, sandstone, limestone, and marl. Their total thickness is 300 m. The third terrane of sulfate rocks is known north of the Magnitogorsk terrane on the eastern slope of the Central Urals (Pronin 1969). Red beds with thin layers and lenses of gypsum occur here. Thus, three isolated sulfate sequences can be recognized within the tentatively outlined East Uralian Basin.

As stated above, the Chimkent evaporite basin is a part of the Carboniferous evaporite belt stretching into the Mid-Tien Shan Basin. The basin, situated along the western margin of the zone (see Fig. 10), is poorly known. Sulfate rocks were reported from the Tournaisian deposits. The thickness of the gypsum-bearing sequence ranges from 25 m to 60 m. It occurs within a carbonate series including tuff and tuff porphyrite bands (Shcherbina 1945; Ivanov and Levitsky 1960).

In the Teniz Basin sulfate rocks show a limited distribution. They are known in the upper Kirey Formation where limestone bands with large crystals of gypsum, as well as gypsinate sandstone and siltstone occur among red conglomerate, gravelstone, and sandstone. A proposed age of this part of the Kirey Formation is the Bashkirian (Litvinovich 1972).

The Tyup sulfate basin including the Tyup and in part Tekes Troughs of the North Tien Shan is situated east of Lake Issyk Kul between Kirghizia and Kazakhstan. Sulfate rocks occur as gypsum bands in the Tyup and Chaarkuduk Formations of the Bashkirian (Middle Carboniferous) (Chabdarov and Sevostianov 1971; Ektova and Belgovsky 1972).

The Aksu sulfate basin of Carboniferous age is situated along the north-western margin of the Tarim Depression in Sinkiang (China). Sulfate rocks (gypsum) are associated there with the Middle Carboniferous Kurukusum Group developed in the Aksu River Basin in the Asgan-Bulag-Taga area. They are observed among variegated sandstone and marl. The thickness of the sulfate-bearing sequence exceeds 300 m (Regional Straigraphy . . . 1960).

The Achikkul Basin is recognized tentatively. It embraces the northern foothills of the Przhevalsky Range between Ayagkumkul and Achikkul Lakes. The basin includes the Arktag sequence of red coarse-grained sandstone with limestone interbeds and gypsum lenses. The age of the sequence is uncertain and regarded as Permo/Carboniferous. Sulfate deposits in the Achikkul Basin and in other basins of Central Asia are probably Middle Carboniferous (Regional Stratigraphy . . . 1960).

The Lhas Basin was named arbitrarily. Sulfate rocks are known from the Pando Formation consisting of dark gray and green shales, quartzites, and limestones with gypsum lenses. The formation is 2000–2500 m thick. It is rather widespread south of the Nyenchhen-Thanglha. The location of gypsum in the Pando section is not yet clear. The entire formation is assigned to the Carboniferous. But as to stratigraphic level of sulfate rocks, it is similar to that of northern evaporite basins of Central Asia (Regional Stratigraphy . . . 1960).

The Fitzroy Basin situated in western Australia includes the Anderson sulfate-bearing formation. It is represented by sandstone, siltstone, and mudstone interbedded with dolomite and anhydrite. The formation is 1500–2000 m thick. The age is Late Carboniferous (Westphalian/Lower Stephanian) (Thomas 1959; Veevers and Wells 1961; Brown et al. 1970).

Small Rhadames, Illisie, Ahnet, Reggane, and Tindouf Basins are situated in north-western Africa, mainly in the Algerian Sahara (Aliev et al. 1971). In the Rhadames Basin sulfate was recorded in the Upper Namurian where unit E is represented chiefly by alternation of clay and dolomite with rare intercalations of white anhydrite and some siltstone and fine-grained sandstone. Within the Illisie Basin two sulfate-bearing sequences are known: the lower El Adeb Larach and the upper Tigentourin sequences. In both sequences gypsum bands occur among terigenous-carbonate variegated deposits. In the Ahnet Basin gypsum and anhydrite beds occur along the western slope of Djebel Berg Uplift in the uppermost horizons of the Visean deposits represented mainly by limestone and mudstone. In the Reggane Basin a gypsum-bearing unit is conspicuous at the Visean D level; it consists of white and pink gypsum and anhydrite interbedded with clay, marl, dolomite, and limestone; the thickness of the complex ranges from 120 m to 207 m. Another sulfate-bearing sequence is determined in the same basin at the level of the Upper Namurian. It is represented by variegated clay, sandstone, and limestone with lenses and thin layers of gypsum. The Late Visean Quarkziz sulfate sequence is widespread in the Tindouf Basin. It is divided into three units: the lower limestone unit consists of limestone and dolomite interbedded with variegated (black, red, green, and creamy) mudstone, sandstone, and white anhydrite, up to 356 m thick; the central clayey-anhydrite unit is represented by the alternation of anhydrite and mudstone; anhydrite locally is 45 m thick, the total thickness amounts to 111.5 m; the upper limestone unit, 102 m thick, contains thin layers of anhydrite. In the Tindouf Basin anhydrite is known not only in the central part, but in the southern part as well, where horizons of sulfate rocks among the upper Visean Deposits are tens of meters thick. They belong to the Ain El Barka Formation.

Six evaporite basins in addition to the salt basins described above are known in the USA. They are: Illinois, South Iowa, Michigan, Orogrande, San Juan, and East Wyoming Basins. In the Illinois Basin beds of gypsum and anhydrite build up the St. Louis Formation (Bond et al. 1968; Bond et al. 1974); in the South Iowa Basin sulfate rocks are associated with the St. Louis and Pella Formations (Johnson and Vondra 1969; Parker 1974); in the Michigan Basin sulfate rocks are known in the Michigan and Saginaw Formations (Vary et al. 1968; Ells 1974; Wanless and Shideler 1975). In the Orogrande Basin, situated in south New Mexico and in westernmost Texas west of the Pedernal Uplift, the Late Carboniferous (Missourian/Virgilian) upper part of the Magdalena Formation is sulfate-bearing. Gypsum beds among carbonate rocks and marls are known in the El Paso region (Crosby and Mapel 1975), but thickest gypsum beds (3 to 12 m) were reported from the uppermost sequence in the south of the basin in New Mexico. They are also interbedded there with carbonate rocks (Bachman 1975). The San Juan Basin is situated south-east of the Paradox salt basin. They are separated by a zone of carbonate biogenic buildups, where the Paradox halogenic sequence thins out. However, outside this zone gypsum-bearing mudstone interbedded with dense dolomite and limestone appears at the same level, but this time in the San Juan Basin. They form the Paradox Formation. The Basin of East Wyoming is recognized tentatively because it is not confined to a certain sedimentary basin, but traceable over much of a shelf zone which in Pennsylvanian time occupied eastern and north-eastern Wyoming and adjacent areas of South Dakota and Nebraska (Meughan 1975; Prichard 1975; Paleotectonic . . . 1975). Sulfate rocks (gypsum and anhydrite) occur there among terrigenous-carbonate deposits forming the Pennsylvanian Minnelusa Formation. They are developed in eastern Wyoming and stretch into central South Dakota.

The Venezuela and South Peruvian sulfate basins of Carboniferous age are known in South America. In both basins sulfate-bearing sequences are of Pennsylvanian age. In the Venezuela Basin gypsum is associated with the Palmarito Formation, and in the South Peruvian Basin gypsum is assigned to the Tarma Group (Benavides 1968).

Thus, ten salt-bearing sequences can be recognized within the Carboniferous evaporite basins. They have accumulated within nine salt basins and are distributed as follows: the Upper salt sequence of the Chu Sarysu Basin; the tentatively recognized salt-bearing sequence of the Mid-Tien Shan; the Otto Fiord sequence in the Sverdrup Basin; the Charles sequence of the Williston Basin; the Gautreau and Windsor sequences in the Maritime Basin; the Maccrady sequence in the Saltville Basin; the Paradox sequence in the basin in the Saltville Basin; the Pradox sequence in the basin of the same name; the Eagle Valley sequence in the Eagle Basin; and the Nova Olinda sequence in the Amazon Basin. There are 58 sulfate sequences, 13 of them have been deposited within salt basins, and the remaining sequences in the sulfate basins.

Permian Evaporite Deposits

Permian evaporite deposits are found in North and South America, Western and Eastern Europe, and in some regions of Asia (Fig. 13).

The Midcontinent Basin is one of the largest Permian evaporite basins. It lies in the western United States. Halogenic rocks (gypsum, anhydrite, and rock salt) were recorded throughout the entire Permian sequence over a vast area running from the southern to the northern borders of the United States through Texas, New Mexico, Oklahoma, Kansas, Colorado, Nebraska, Wyoming, Montana, South and North Dakota. Permian sediments have been studied in great detail.

Salt rocks within the Midcontinent Basin are recorded at five stratigraphic levels: (1) in the upper Wolfcampian, (2) in the Leonardian, below the San Andreas and Blaine Formations and their equivalents, (3) in beds intermediate between the Leonardian and the Guadalupian, within the San Andres and Blaine Formations and their equivalents, (4) in the Guadalupian above the Andreas and Blaine Formations and their equivalents, (5) in Ochoan.

The salt-bearing deposits in the Wolfcampian are rather scarce. They are found in the Denver Depression where the rock salt composes two units of the upper Ingleside Formation, and in the Yulesburg Depression among the Minnelusa beds known also

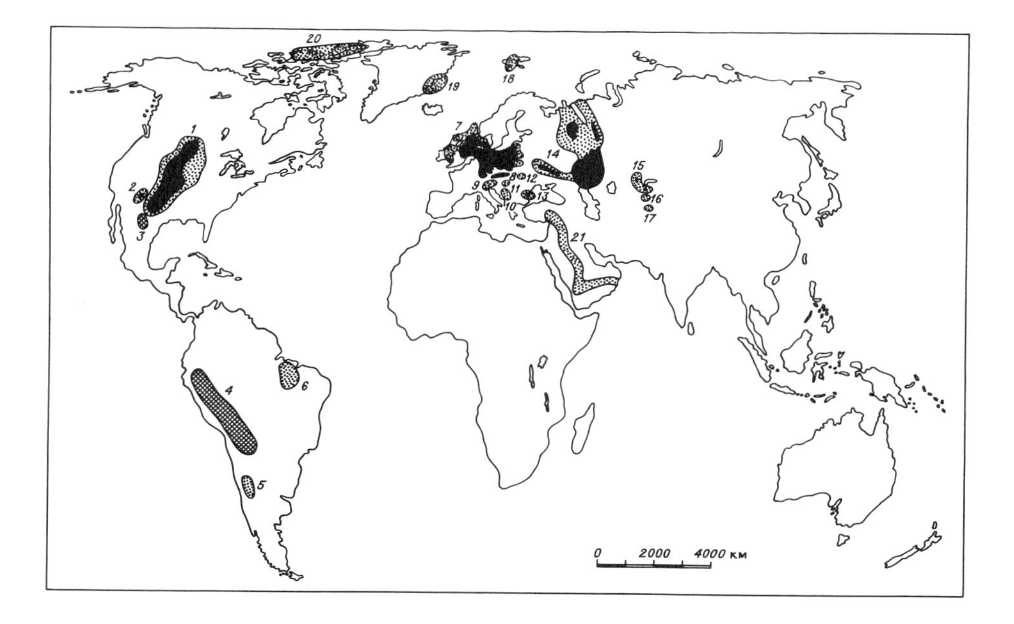

Fig. 13. Distribution of Permian evaporite deposits. Evaporite basins:
1 Midcontinent, *2* Supai, *3* North Mexican, *4* Peru-Bolivian, *5* Rio Blanco, *6* Parnaiba, *7* Central European, *8* Alpine, *9* North Italian, *10* Dinarids, *11* Mecsek, *12* Rahov, *13* Moesian, *14* East European, *15* Chu Sarysu, *16* Karasu Ishsay, *17* Darvaza, *18* Spitsbergen, *19* East Greenland, *20* Sverdrup, *21* Arabian. For explanation see Fig. 1

as the Broom Creek Strata. These salt strata are assumed to have been formed in isolated small water basins. An area of salt accumulation within the first and second basins was 4,500–5,000 km², and 11,000 km², respectively. The thickness of these salts does not exceed 25–30 m.

It is impossible to determine the actual number of individual salt units among the Leonardian sediments, firstly, because they are poorly studied and, secondly, due to the uncertain stratigraphic position of individual salt units found in different regions of the Midcontinent. Three levels of occurrence of salt deposits can be roughly proposed for the Leonardian. The Wellington salt series is confined to the lower stratigraphic level. It is extensive in the Anadarko Basin where its thickness varies from 150–200 m to 400 m. Obviously, the same salt sequence extends into the Kansas Basin; it is placed there in the Hutchinson Formation; its thickness is 150–180 m. The isolated salt units, probably occurring at the level of the Wellington Series, are found in north-eastern Colorado, in the Denver Basin where they are confined to the lower Satanka Formation. Another area of salt deposits also belonging to the Lower Leonardian is confined to the Yulesburg Basin. They occur within the Owl Canyon Formation. Two stratigraphically lower salt sequences of the Leonardian developed within two isolated regions of the Midcontinent can be outlined: (1) a salt sequence underlying the interior of the Anadarko and Kansas Basins and distinguished under the names of the Wellington or Hutchinson Formations, (2) a salt sequence confined to the Yulesburg Basin in north-eastern Colorado and in eastern Nebraska related to the Owl Canyon Formation.

The overlying salt accumulation level of the Leonardian comprises the deposits found within the Cimarron Formation in north-eastern Oklahoma, within the Tubb and Cimarron Formations in the northern regions of the West Texas Basin, the Ninnescah Formation of the Kansas Basin, and the Middle Satanka Formation of the Denver Basin. All the above deposits appear to compose a single salt unit that may be named Cimarron and is developed in Texas, Oklahoma, and Kansas. Additionally, salt sediments of the same level are developed in the uppermost Owl Canyon Formation, within the Yulesburg Basin. These strata are placed in the earlier distinguished Owl Canyon salt sequence. It is most likely that at the stratigraphic level discussed the number of the individual salt units may be greater.

The next, uppermost, level of the Leonardian comprises salts of the Flowerport, Blaine, Yelton, and Opeche Formations. Three spatially isolated salt sequences probably occur at this level: one, comprising salt rocks of the Flowerport, Blaine, and Yelton Formations, is developed in Texas, Oklahoma, and Kansas, the other two, confined to the Opeche Formation, form separate units within the Denver, Yulesburg, and Williston Basins.

The salt-bearing Guadalupian deposits mostly occur within two isolated areas of the Midcontinent: in the south, in the West Texas Basin, and in the north, in the Williston Basin. In the West Texas Basin rock salt is developed in the San Andres, Grayburg, Queen, Seven Rivers, Yates, and Tansill Formations. It is most likely that all these salt formations form a single salt sequence over the area not less than 130,000 km². The thickness of the strata attains 600–800 m, the total thickness of the rock salt often exceeding 300–400 m. In the Williston Basin the Guadalupian is represented by the Pine salt sequence, more than 100 m thick, and of an area not less than 60,000 km². Besides these major salt units, the Guadalupian is also represented by isolated salt units less thick and less extensive, recorded in the Glendo Formation in north-western Nebraska and in the Whitehorse Formation in Colorado. However, it is not yet clear whether we can regard them as separate salt units, for they might quite well be united with the Pine sequence, or might represent noneroded parts of these units.

The Ochoan salt-bearing deposits are found only in the southern Midcontinent and in the West Texas Basin. Their total area is about 100,000 km². Three salt-bearing formations are distinguished there: the lower one, Castile, the thickness of which varies from 450 m to 640 m, the total thickness of rock salt being 250–300 m; the middle one, Salado, in which alongside with the rock salt the potash salts are developed; the upper one, Rustler, having the total thickness of 150–170 m, of which about 110–120 m fall within the rock salt. All these formations should, probably, be placed into a single salt sequence, the Castile sequence.

Thus, within the Midcontinent Basin we can distinguish at least 11 salt units among the Permian deposits, which either have a different stratigraphic position, or occur at the same stratigraphic level in different parts of the Midcontinent.

It is very difficult to determine the number of the salt units developed within the Midcontinent Basin, and only an approximate estimate can be given, and it is twice as difficult to distinguish sulfate-bearing strata there. Twelve sulfate-bearing units can be tentatively distinguished. The lower sequence is known as the Upper Minelusa. These beds are found within the Yulesburg and Williston Basins in Nebraska and South Dakota, they are represented by the alternation of dolomite and anhydrite, surrounding the Broom Creek salt sequence. It is of the Late Wolfcampian age. The second, the Lower Wellington, sulfate-bearing sequence comprises the lower anhydrite unit of the Wellington Formation developed in the Anadarko Basin. Its thickness varies from 100 m to 200 m. The sequence is composed of anhydrite alternating with mudstone. The third sequence comprises the upper anhydrite unit of the Wellington Formation and is known as the Upper Wellington. These beds also occur within the Anadarko Basin. Their thickness usually varies from 30 m to 54 m, increasing to 75 m in the subsiding areas. The sequence is composed of anhydrite and dolomite with mudstone and shale interbeds. The anhydrite attains 20–30%. Both sequences are Early Leonardian in age. The fourth sequence is apparently developed in the northern West Texas Basin in New Mexico and Texas, where the lowermost Leonardian, marked by dolomite with anhxdrite and green marl interbeds, compose the Wichita Formation. The fifth, the Stone Corral sequence, is found in the Leonardian in Kansas and in south-eastern Colorado, and is traced as a marker bed. Usually, the exposed part of the sequence is composed of dolomite with marl interbeds, but at small depths, in case of subsidence, it is enriched in anhydrite which starts to dominate the section. In south Kansas the Stone Corral sequence is chiefly composed of anhydrite, alternating with dolomite marl. In the south-west two anhydrite units separated by red marl become very conspicuous in the section. The thickness of the strata attains 30 m. The sixth sulfate unit is called the Blaine sequence. These beds are rather widespread in the central regions of the Midcontinent Basin, mostly in the Anadarko and Kansas Basins. The sequence is composed of anhydrite (or gypsum), dolomite, and mudstone. Its thickness is 25–75 m. The age is the Late Leonardian/

Early Guadalupian. The seventh unit, distinguished under the name of the Owl Canyon, comprises deposits confined to a formation of the same name and developed in the Yulesburg Basin. Anhydrite alternates there with the red marl and mudstone; the thickness of the strata is about 100 m. The boundaries of this sequence are isochronous. The eighth sequence, extremely widespread in the northern regions of the Michigan Basin within the Denver, Yulesburg, and Williston Basins, encloses deposits of the Minnekahta Formation, of an age intermediate between the Leonardian and the Guadalupian. It is regarded as an equivalent to the Blaine Formation and is also composed of anhydrite, dolomite, and mudstone. Both sulfate-bearing sequences should, probably, be united, but for the lack of reliable evidence they are considered conventionally as separate units.

The younger sulfate-bearing sequences in the Midcontinent Basin occur at the level of the Guadalupian. One of these sequences, Artesia, is developed in the West Texas Basin. It is mostly salt-bearing within the Midland Basin, only its lower part (Grayburg Formation) being representing by dolomite, anhydritized dolomite, and anhydrite. However, in surrounding regions, especially in the Central Plateau Basin, i.e., south and south-west of the salt accumulation area, the Artesia sequence is sulfate-carbonate in composition, anhydrite predominates upon wedging-out, and is later replaced by the alternation of anhydrite and dolomite. The second sulfate-bearing Guadalupian sequence is recorded in the northern and north-western Midcontinent Basin. There, the Glendo red beds are developed and represented by mudstone, marl, and siltstone among which gypsum is common. The thickness of these sulfate-bearing red beds is 25–30 m. The third sequence occurs in the uppermost Guadalupian and represents the marker beds over much of Kansas and Colorado. This sequence comprises the Day Creek Formation composed of dolomite and anhydrite with red marl interbeds. The Dewey Lake Formation, crowning the Permian section in the southern Midcontinent with the West Texas Basin, can be distinguished as the last sulfate-bearing sequence. It is assigned to the Ochoan Series and is composed of red marl, siltstone, and, more rarely, sandstone with gypsum interbeds, veinlets, and lenses. The thickness of this unit slightly exceeds 1000 m.

To conclude this brief characterization of the evaporite deposits of the Midcontinent Basin we should once again emphasize that the number and range of the salt- and sulfate-bearing sequences distinguished are only approximate. The basin and the stratigraphy of the halogenic strata are as yet inadequately studied for drawing any final conclusions, and, thus, the proposed distinguishing of salt- and sulfate-bearing sequences might be regarded as the first attempt taken in this direction.

The Supai Basin lies in the west of the United States, at the southern margin of the Colorado Plateau (Pierce and Rich 1962; Gerrard 1966; Pierce and Gerrard 1966; McKee 1967; McKee et al. 1967). The salt-bearing deposits occur there at two stratigraphic levels in the Supai Formation at the upper Middle Supai sequence assigned to the Wolfcampian, and the Upper Supai sequence belonging to the Leonardian. The Middle Supai salt sequence is developed over an area of 5000 km^2, their thickness reaching 200–300 m. Within the Upper Supai salt sequence rock salt occurs in the form of separate lenses, the thickness of which varies from 60 m to 150 m. These strata cover an area of 6000 km^2. In north-western Arizona and in adjacent regions of southern Utah there is a small area, characterized by the presence of an isolated terrigenous-carbonate sulfate-bearing unit within the Leonardian formations. Its thickness attains 350–400 m; it is represented by sandstone and siltstone, alternating with limestone and dolomite, as well as anhydrite and gypsum. This unit may be conventionally placed within the Supai Basin.

The North Mexican Basin is tentatively dated as Permian. The salt deposits, penetrated by deep boreholes near Cuchillo Prado, are arbitrarily placed in the Permian (Salas 1968a,b; Lefond 1969). They probably occur at the level of the Plamozas Formation which comprises the entire Wolfcampian and Leonardian succession in the north-western part of the Chihuahua Basin. A drilled thickness of the salt-bearing unit is about 400 m, but one of the boreholes penetrated to 2585 m, without reaching the base. The area underlain by the proposed Permian evaporites within this basin is not known.

The Peru-Bolivian Basin. At present Permian evaporite deposits are known to be developed not only in Peru, but to extend farther south into Bolivia. The Peru-Bolivian Basin seems to be rather large. It extends for about 2500 km along the Central and East Andes (South American Cordillera) from the frontier regions of Columbia, Ecuador, and Peru in the north to Argentina in the south. The width of the basin locally exceeds 800 km. Its total area can attain 1,250,000 km^2.

In the Peruvian part of the basin the Permian is subdivided inot two formations: Capacabana and Mitu. They belong to the Wolfcampian and Leonardian, respectively. Sulfate rocks are found in both formations. The Capacabana Formation is mostly composed of limestone and shale with dolomite, siltstone, and sandstone interbeds. In its upper part gypsum and gypsified rock interbeds were recorded near Cuzco (Newell et al. 1953). The Mitu Formation lying above is represented by terrigenous red beds among which gypsinate rocks as well as gypsum layers, up to 8 m thick, occur. They were reported both from northern Peru, in the Marañon River Basin and in the south, near Cuzco (Newell et al. 1953; Benavides 1968). The Mitu Formation comprises, presumably, the salt-bearing deposits, composing most of the diapiric and salt structures in Peru. The salt domes are most abundant in the Huallaga River Basin and its left tributary, the Mayo River. Gypsum, anhydrite, and rock salt alternating with pink, red, and purple mudstone and siltstone crop out along the crest of the dome. The age of the salt-bearing sediments of the domes is unclear, probably Permian, though some authors assign them to the Triassic and even Jurassic (Newell et al. 1953; Benavides 1968).

In Bolivia the Permian succession is often placed into the Capacabana Group (Helwig 1972), but in some regions all the Permian sedimentary series are either included in the Yarichambi Formation (Schlatter and Nederlof 1969), or in the Titicaca Group. In the Titicaca Lake area they are subdivided into two parts: the lower Capacabana Group and the upper Chucuchambi (Lohmann 1970). Helwig (1972) tentatively assigns all the evaporites occurring in this region to Permian red beds.

In general, the Permian succession in Bolivia is quite similar to that of Peru. Two levels of the evaporite rocks can also be distinguished: the lower one, confined to the Capacabana Group, and the upper one, enclosing the salt-bearing Chucuchambi formations which might possibly be correlated with the presumably salt-bearing Mitu Series. The Chucuchambi salts in Bolivia make up several salt domes in the area between Titicaca and Poopo Lakes. In fact, salt domes of Peru and Bolivia fall into a single tectonic zone within which a thick salt sequence might be developed. Another viewpoint expressed by Lohmann (1970) seems quite probable; it presupposed that in Bolivia and in adjacent regions of Chile there might be three salt basins of different age. The Peruvian Basin might also be a separate basin. In view of the uncertain age and spatial distribution of salt-bearing units the Peru-Bolivian Basin is conventionally distinguished there as a Permian salt basin. At present one sulfate-bearing sequence (Capacabana) and two salt-bearing sequences (Mitu in the north, in Chile, and Chucuchambi in the south, in Bolivia) can be tentatively outlined there.

The Rio Blanco Basin is located in north-western Argentina. It is recognized as a sulfate-bearing basin. It is highly probable that this basin was a part of the Peru-Bolivian Basin situated farther north. Gypsum and gypsinate interbeds are recorded there from the upper Patqua Formation, mostly composed of red siltstone and sandstone. Gypsum is most abundant in the La Rioja and San Juan Provinces (Teruggi et al. 1969; Volkheimer 1969).

The Parnaiba Basin lies in western Brazil. Gypsum and anhydrite are found among the Pedra-do-Fogo red beds, the thickness of the formation being about 300 m. Sometimes gypsum layers are up to 18 m thick.

The Central European Basin extends for more than 1600 km from east to west, with a width from 300 km to 600 km and occupying the territory of the Lithuanian Republic, Poland, DDR, West Germany, Holland, Denmark, Great Britain, and most of the North Sea. A comprehensive study of salt-bearing deposits of this basin was reported previously (Zharkov 1974a). Additionally, many new publications have appeared lately which have clarified the structural features and distribution pattern of the Permian evaporite deposits both in separate regions and within the basin as a whole (Hoyningen-Huene 1967; Geology of Poland 1970; Molewicz 1970a,b; Milewicz and Pawłowska 1970; Pawłowska 1970a,b,c; Pawłowska and Poborski 1970; Poborski 1970; Poborski and Pawłowska 1970; Siedlecka 1970; Sokołowski 1970; Wyźykowski 1970a,b; Glennie 1972; Hemman 1972; Seifert 1972; Stołarczyk 1972; Bush et al. 1973; Langbein 1973; Falke 1974; Glushko et al. 1974, 1976; Charysz 1975; Dickenshtein et al. 1975; Döhner and Elert 1975; Flügel 1975; Gurary 1975; Katzung 1975; Lyutkevich 1975; Meier 1975; Neumann and Schön 1975; Permian . . . 1975; Watson and Swanson 1975; Ziegler 1975; Dickenshtein et al. 1976; Döhner 1976; Stolle and Döhner 1976). This evidence allows definite determination of the number of salt and sulfate sequences developed within the Central European evaporite basin.

The oldest salt sequence was recorded from the Rotliegende in the Lower Permian sediments. They occur in the western part of the Mid-European Basin where they fill in the subsiding zones of the Lower Elba and East England Basins, as well as the adjacent North Sea areas. The thickness of this sequence often exceeds 1000–1500 m. The determined area under these salt deposits is 80,000 km^2.

In the Zechstein succession developed in the Central European Basin over an area of more than 700,000 km^2 four salt units might be distinguished, corresponding to the Werra, Stassfurt, Leine, and Aller series. The uppermost Zechstein Series known as the Ore sequence is not thick (about 10 m); it is incommensurate with other series and, thus, not being regarded as separate, it should be placed in the Aller salt sequence.

As a rule, the Werra sequence is confined to the marginal zone where it is rather thick with numerous potash-salt layers. In fact, the Werra salt deposits occur in isolated areas separated by extensive anhydrite banks from the center. Probably, it would be much more reasonable to assume that each of these areas encloses separate salt sequences which might be given names of their own. The data available suggest that rock salt beds may occur between sites of rather thick Werra salts and in interior parts of the basin as well. Though their thickness is only several meters, we can distinguish a single salt sequence, i.e., the Werra series within the basin. In regions with highest salt content, thickness is as follows: in the Werra Fulda Basin – 350 m, in the Thüringia Basin – 300 m, in the Weser and Lower Rhine Basins – 200 m, in the south-east of the North East German Basin – 340 m, in the south of the Polish-Lithuanian Lowland – 150 m.

In all these marginal basins the Werra sequence encloses potash salts, which sometimes make up two horizons, as in the Werra Fulda Depression where the Thüringia and Hessen potash salt layers are known. Probably, the sulfate rocks of the Werra Series should in most cases be regarded as a part of the salt sequence, since their thickness is not great, and they constitute together with salts a single cycle of evaporite sedimentation. However, this pattern is not always observed. One of the peculiar features of the lower Zechstein sedimentary cycle was the formation of anhydrite banks framing areas of thick salt accumulation to the north in the Lower Rhine and Werra Fulda Basins, and to the north-west in the Thüringia Basin. The anhydrite banks are traced along the northern margin of the basin and, probably, surround subsiding areas of the Central Polish Depression. The thickness of the Werra anhydrite within the banks in places reaches 300–400 m. In zones of anhydrite banks a separate sulfate-bearing Werra sequence is apparently distinguished; it stretches along the marginal parts of the Central European Basin. The Hartlepool anhydrite sequence developed in the East England Basin can be also regarded as a separate unit.

The Stassfurt salt sequence fills in all the interior zones of the Central European Basin. It has also been recorded both in the west (in the North Sea, in the East England, Danish, and North German Depressions) and in the east (in the North-East German, Central Polish, and Polish-Lithuanian Depressions). This sequence extends into the Weser, Subhercynian, and Thüringia Depressions, as well as into the Werra Fulda Basin. The salt deposits, 400–500 m thick, are underlain and overlain by relatively thin anhydritic units and placed into the sequence. In the middle part potash salts occur as a continuous sedimentary cover in the Mid-European Basin, the central areas of the Weser, Subhercynian and Thüringia Depressions. The base of the sequence is made up of platy dolomite and gray salt clay.

The Leine salt sequence is as widespread as the Stassfurt sequence. Its greatest thickness varies from 100–150m to 300–350 m. The sequence begins with gray salt clay overlain by platy dolomite, main anhydrite, and, finally, the Leine rock salt. Potash salt horizons are traced within the salt-bearing sequence throughout the central part of the Mid-European Basin from the East England Basin in the west to the Central Polish Depression in the east. In some places there are two horizons (Ronnenberg and Riedel), but within the Weser Basin four potash salt beds are distinguished (Ronnenberg, Bergmannssegen, Riedel, and Albert). In the Manx Furness Depression the Sandwith sulfate-bearing sequence occurs at the level of the Leine Series.

The Aller salt sequence is developed in the Weser, North German, North-East German, Altmark, Central Polish, Subhercynian, and Thüringia Depressions. Structurally it is relatively simple. A red salt clay bed overlain by a thin unit of pegmatitic anhydrite is traced at the base. Most of the succession is composed of rock salt overlain by upper clay beds. In some regions within the basin (in the Subhercynian, Weser, and East England Basins, as well as in the Danish Depression) potash salts occur. In the Weser and Subhercynian Depressions they compose the Ottoshall potash salt bed. The thickness of the Aller salt varies from 30–35 m to 150–200 m. Within the Manx-Furness Depression the Fleswic anhydrite is considered equivalent to the Aller Series; these rocks may be singled out into a separate sulfate sequence.

Thus, five salt- and sulfate sequences may be distinguished in the Central European Basin. In addition to the above sulfate sequences (Werra, Hartlepool, Sandwith, and Fleswic) sulfate red beds of the upper Rotliegende, flanking the Early Permian salt-bearing sequence in the western region of the basin, may be singled out into a separate unit under the name Ten Boer; it marks a near-shore sabkha facies (Glennie 1972; Watson and Swanson 1975; Ziegler 1975). Moreover, the peculiar red beds with lenses and inclusions of anhydrite and gypsum developed in the north-eastern part of the basin, in the Polish-Lithuanian Depression, might be regarded as a separate unit as well; these beds are known as the Suduvsk Formation and correlated with the uppermost Zechstein.

The Alpine Basin. The Permian salt-bearing deposits have been known for a long time from north- and north-eastern parts of the East Alps in Austria (Mayrhofer 1955; Schauberger 1955). They compose numerous salt domes or crop out along the major faults within the belt extending from Innsbruck in the west to Vienna in the east. These salt-bearing sediments were assumed to mark a separate salt basin which is not yet outlined. Recently a considerable amount of new data has become available throwing light on the distribution pattern of the Permian salt-bearing deposits in the East Alps and West Carpathians (Del-Negro 1960; Zhukov 1963, 1965a; Oxburg 1968; Mahel and Vozár 1971; Zhukov and Yanev 1971; Clar 1972; Falke 1972, 1974; Frank 1972; Heissel 1972a,b; Klaus 1972; Mostler 1972a,b,c; Praehauser-Enzenberg 1972; Riehl-Herwirsch 1972; Riehl-Herwirsch and Wascher 1972; Sommer 1972; Trevisan 1972; Flügel 1975; Richter and Zinkernagel 1975; Vozárová and Vozár 1975; Schauberger et al. 1976; Zhukov et al. 1976). They have shown that the salt-bearing sediments occur not only in the East Alps, but extend eastward as far as the Spišsko-Gemerskoe Rudohoři in the eastern part of the West Carpathians. Thus, it appears that the Alpine salt basin was a rather extensive seaway, apparently separated by the chain of uplifts from the Central European Basin, lying farther north.

At present, many new localities of Permian evaporite rocks are known in southern Europe. As well as in the East Alps and West Carpathians, the salt-bearing deposits were also found in north-eastern Bulgaria with a separate salt basin within the Moesian Depression. Sulfate-bearing deposits were recognized also near Rakhov in the East Carpathians and in the Dobruja Depression in Moldavia. In fact, in Permian time the entire present Alpine fold area within the East Alps, West and East Carpathians and, probably, the Balkan Mountains, was a site of evaporite sedimentation. In southernmost Europe Permian sulfate-bearing deposits are found in northern Italy and in the Carnic Alps, in the Dinarids, in the Montenegro in Yugoslavia, and in the

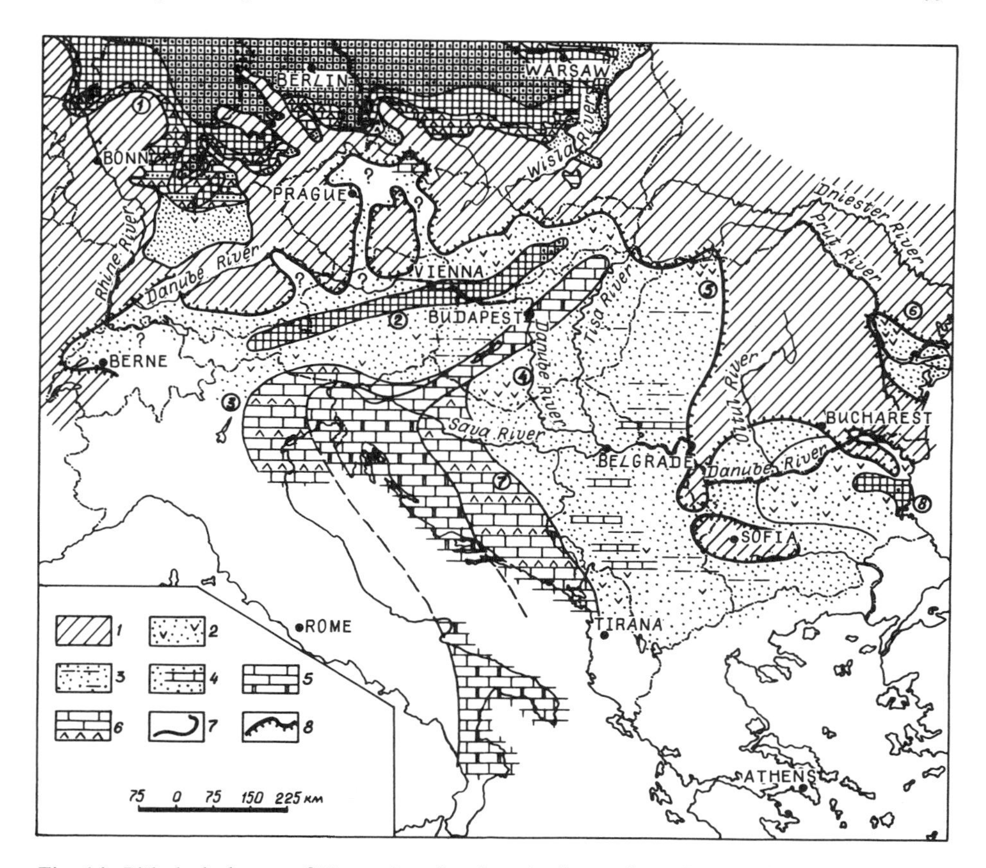

Fig. 14. Lithological map of Upper Permian deposits in southern Europe. Compiled by M.A. Zharkov and G.A. Merzlyakov from published data.

Areas of occurrence of: *1* terrigenous rocks mostly red beds, *2* terrigenous rocks, dominated by siltstone and mudstone, *3* terrigenous and carbonate rocks, *4* terrigenous rocks with lenses, inclusions, and interbeds of gypsum, *5* limestone, *6* limestone and dolomite, *7* carbonate rocks with gypsum and anhydrite, *8* carbonate rocks with approximately equal amount of anhydrite, *9* rock salt, *10* potash salt, *11* areas without deposits, *12* boundaries of the Central European Zechstein Basin, *13* boundaries of distribution of Upper Permian deposits in southern Europe.

Figures refer to basins: *I* Central European, *II* Alpine, *III* Moesian, *IV* Dobruja, *V* Rahov, *VI* Mecsek, *VII* North Italian, *VIII* Dinarids

Korab zone in north-eastern Albania, as well as in the Mecsek Mountains in southern Hungary. Figure 14 shows a peculiar pattern of evaporite distribution in southern Europe. There, in addition to the Alpine and Moesian salt basins, five sulfate basins are recognized: the North Italian, Dinarids, Mecsek, Rakhov, and Dobruja Basins.

The Alpine salt basin is discussed in more detail below.

Schauberger (1955) distinguished 15 regions in the northern part of the East Alps where Permian salt-bearing deposits are exposed (Fig. 15). They are studied in particular detail within the Hallstatt salt stock where a sequence is exposed in salt

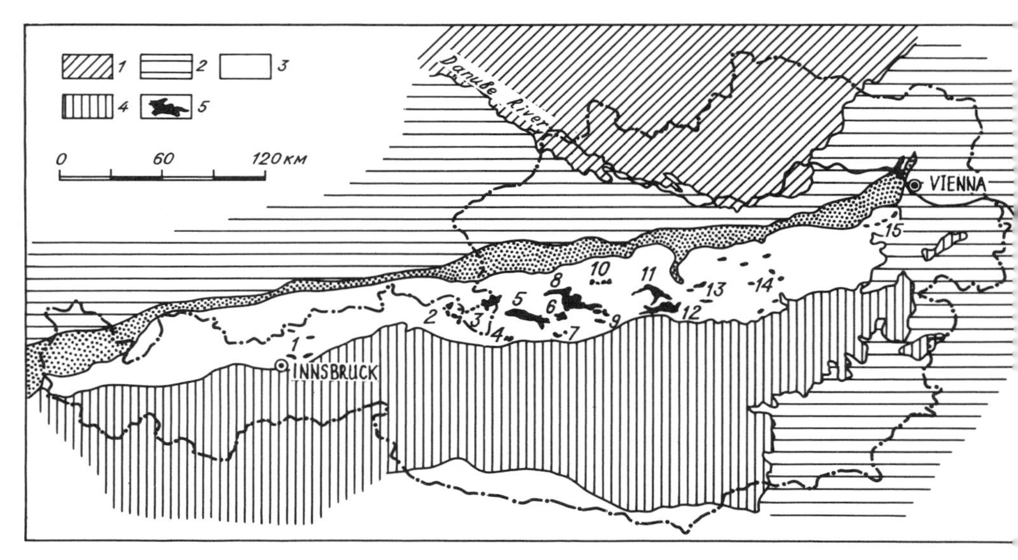

Fig. 15. Outcrops of Permian salt deposits in the northern East Alps (Schauberger 1955): *1* Bohemian Massif, *2* Tertiary and Quaternary deposits, *3* Helvetides, *4* Central Alps, *5* North and South Limestone Alps.

Outcrops of salt and sulfate deposits and their numbers: *1* Karwendel, *2* Unken-Lofer, *3* Hallein Berchtesgaden and Bad Reichenhall, *4* Werfen, *5* Golling Abtenau, *6* Halstatt, *7* Rettenstein, *8* Ischl-Altaussee-Grundsee, *9* Mitterndorf, *10* Offensee-Almsee, *11* Windischgarsten, *12* Spital-Pyrn-Bosruck, *13* Weissenbach-Gallen, *14* Rotwald-Maria-Zell, *15* Heiligenkreuz-Mödling

mines. The whole sequence is composed of various salt, sulfate-salt or sulfate-carbonate--salt rocks generally known as Haselgebirge. Several types of these rocks are recognized: gray salt, red salt, green-gray clay-salt, variegated clay-salt, bituminous-dolomite gray salt, etc. In most cases these rocks are brecciated, primary layering is rare. But the thin-bedded Haselgebirge occur in those places where rock salt alternates with anhydrite and bituminous dolomite. It should be noted that stratification of rocks within this salt sequence is obscured due to complex tectonics, and the entire sequence is strongly deformed by fluidal-tectonic processes. The mode of occurrence is also uncertain. It breaks through Triassic, Jurassic, Cretaceous, and the Neocomian sediments and is mainly assigned to the Late Permian. Schauberger (1955) places this sequence onto the level of the Bellerophon Beds of the South Alps. The base of the sequence is not exposed. The spore-pollen analysis (Klaus 1955) shows that in differently composed Haselgebirge inclusions there are found spores of both Triassic and Permian aspects. This fact makes some investigators consider the salt-bearing sequence as Permo-Triassic, whereas the others regard it as Late Permian.

The second region where a salt sequence is studied in some detail lies within the Bad Reichenhall Basin in the south of West Germany. There, nine boreholes were drilled which penetrated a considerable thickness of the salt-bearing successions. Beds of rock salt, sulfate, and carbonate are intensely folded, which points to the secondary occurrence of the sequence in the form of the lenticular stock, traced to the depth of more than 1200 m. The upper part of the stock is represented by the leached salt rocks forming a gypsum cap. Gypsification is often observed there, and all the halogenic rocks grouped under the name Haselgebirge are mostly secondary ones. Among these rocks the clayey gypsum and anhydrite-dolomitic gypsum Haselgebirge are distinguished. These rocks are mostly composed of light gray gypsum with numerous scattered inclusions and fragments of clay and dark to brown-gray anhydrite-dolomite. The thickness of the gypsum cap varies from 370 m to 500 m.

The salt-bearing sequence proper, not affected by the secondary processes, but also strongly dislocated, is mostly composed of the nonlayered salt rocks containing various inclusions: sulfate, carbonate, and clay. Proceeding from the correlation of the main components the following rocks might be distinguished: more or less pure rock salt, brown to yellow-gray; massive or thin-bedded dark-gray anhydrite; brownish-gray dolomitic anhydrite; variegated clay salt rocks, most abundant in the salt strata and composed of clay and halite gray-green to brown, often saline sandstone. A detailed description of the rocks listed was given by Schauberger et al. (1976). A characteristic feature of the salt-bearing strata of the Reichenhall Region is the extensive development of the anhydrite and anhydrite-dolomite rocks which are distinguished as the anhydrite-dolomite Haselgebirge or form units up to 15 m thick, distinct within the sequence. This allowed the assumption that the Reichenhall Region was confined to a marginal zone of the salt basin. In the central part, as in Hallstatt, the red salt rocks or the variegated clay salt rocks are most widespread; minor bituminous anhydrite-dolomite and anhydrite salt Haselgebirge are also recorded, which is typical of the more deep water parts of the basin (Schauberger et al. 1976). The marginal facies is traced from the Bad Reichenhall Trough to the salt domes in Tyrol.

Structural features of a salt-bearing sequence in the Bad Reichenhall Trough are not clear. The exposed thickness varies from 737 m to 1200 m, whereas the total normal thickness of the section studied can reach 1000 m. The sequence is overlain by the carbonate Reichenhaller Formation composed of limestone alternated with marly limestone, marly clay dolomite, and black calcareous mudstone with rare anhydrite inclusions. The carbonate rocks contain fossils of an age ranging from Early to Middle Triassic, approximately Late Scythian/Anisian. The salt sequence is nonfossiliferous. Spore-pollen analysis (Klaus 1955; Schauberger et al. 1976) suggests the Late Permian age. At the same time, the sulfur isotopic age of these sulfate rocks is Early Triassic (Scythian and Anisian). This evidence makes some investigators suggest Late Permian/Early Triassic age for this salt sequence (Schauberger et al. 1976). However, most of these salt-bearing deposits probably formed at the end of the Late Permian.

In addition to the above regions, salt and gypsum-anhydrite rocks in northern and north-eastern parts of the East Alps are known elsewhere and are worked at numerous mines in the following areas: Karwendel (north-east of Innsbruck), Unken Lofer, Hallein-Berchtesgaden, Werfen, Golling-Abtenau, Rettenstein, Ischl-Altaussee-Gründlsee, Mitterndorf, Offensee-Almsee, Windischgarsten, Spital-Pyrn-Bisruk, Weissenbach-Hallein, Rotwald-Maria-Zell, Heiligenkreuz-Mödling. The last of the above areas is situated south-west of Vienna. Thus, the salt-bearing deposits are traced as a belt extending from west to east for more than 370 km. They are represented by the same rocks as those of the salt stocks of the Hallstatt and Bad Reichenhall regions. They are also similar in age.

The easternmost salt-bearing area in the Alpine Basin, as already mentioned, lies within the Spišsko-Gemerskoe Rudohoři in the east of the West Carpathians. This region, together with the salt rock outcrops in the East Alps, is conventionally included into a single salt basin. Justification is a common composition of salt-bearing deposits and, probably, similar age. Moreover, the probable extension of the salt basin from the East Alps to the eastern boundary of the West Carpathians is, to a certain extent, confirmed by Late Permian sulfate-bearing deposits exposed in the Low Tatra and to the west in the Tribec Mountains (Zhukov et al. 1976).

In the eastern part of the West Carpathians the salt-bearing deposits are penetrated by SM-1 and SM-2 boreholes drilled near Spišska Nová Ves. Three lithological units can be distinguished in the section of SM-1 borehole (from top to base): shale-sandstone unit – 435 m; intermediate sandstone-shale unit – 32 m; salt breccia unit – 208.9 m (Mahel and Vozár 1971). The base of the salt unit is not exposed.

There are isolated anhydrite interbeds, 10–20 cm thick, mostly confined to its lower and middle parts in the upper shale-sandstone unit. Mahel and Vozár (1971) regard anhydrite as normal sediments and not as a result of diapirism. A transitional unit is subdivided into two parts; the upper part, 2 m thick, is represented by shale with anhydrite bands (1–5 cm) and the lower unit (30 m thick) composed of clayey anhydrite breccia in which the content of anhydrite varies from 50% to 70%.

In fact, the salt breccia unit is composed of rocks of the Haselgebirge type, where the halite content varies from 40% to 90%. The remaining part is represented by inclusions and clasts of shale, more rarely dolomite and anhydrite. There are isolated interbeds of shale, 1 m to 8 m thick, anhydrite, 1–2 m, clay-anhydrite breccia and dolomite in the succession. Cores taken at depths of 457.5 m and 494.0 m contain sylvine; they have a potassium chloride content of 6.8%–8.55%.

The halogenic unit is overlain by red and variegated arinaceous-argillaceous deposits containing the anhydrite and interbeds of calcareous shale. The underlying strata were penetrated by the SM-2 borehole. They are mostly represented by the same red shale often with admixture and intercalation of volcanic rocks. A salt-bearing sequence has not been recorded. The SM-2 borehole probably penetrated its equivalents composed of breccia consisting of debris of anhydrite, shale, limestone, and sandstone.

Evaporite deposits found in the Spišsko-Gemerskoe Rudohoři are confined to the interiors of the North Hemeride Syncline. Toward the margins they grade into terrigenous red beds in places with anhydrite bands. Most investigators suggest the Late Permian age for these salt and sulfate sequences (Zhukov 1965a; Mahel and Vozár 1971; Zhukov et al. 1976).

Considering the Alpine Basin in general, it can be noted that it extends for more than 650 km from west to east, 40–50 km in width. The present salt-bearing area is not less than 26,000 km^2. Though one salt sequence seemed to be developed over the entire area, the data confirming this are not available at present. At the moment two isolated regions can be outlined more or less precisely where separate salt sequences are recorded: one in the East Alps, the other in the east of the West Carpathians within the North Hemeride Syncline. The rock salt sequences amount to about 5.2×10^3 km^3.

In addition to the above salt sequences, four sulfate sequences are known within the Alpine Basin. One of them embraces the upper shale-sandstone sulfate unit of the Spišsko-Gemerskoe Rudohoři evaporite deposits and, probably, also sulfate-terrigenous deposits, framing a salt zone of the North Hemeride Syncline. The second sequence, presumably, comprises sulfate deposits overlying and replacing salt deposits in the East Alps. The third sequence forms an isolated outcrop in the Tribeč Mountains where a shale-carbonate unit with intercalations of anhydrite was recorded in a Permian succession (Zhukov et al. 1976, p. 11). The fourth sulfate sequence encloses the lower and upper red beds of the Low Tatra. These beds are represented by red conglomerate, sandstone, and siltstone; their upper parts contain bands and lenses of carbonate, gypsum, and anhydrite. Sulfate rocks dominate the upper red beds. Sometimes their thickness exceeds 30 m (Zhukov et al. 1976, pp 15–17). The stratigraphic position of the last sulfate sequence is not clear. Probably it comprises not only the Upper Permian, but Lower Permian deposits as well.

The Moesian Basin. Within the Moesian Plate in Bulgaria salt-bearing deposits were recorded near Provadia village, not far from the Sofia-Varna railway (Spasov and Yanev 1966; Lefond 1969; Yanev 1969, 1970a,b, 1971; Garetsky 1972). They make up a small salt stock about 850 m long and about 400 m wide. Up to the recent time the distribution of Permian salt-bearing deposits within the Moesian Depression was unknown. The age of salts composing this salt stock was also uncertain. They were tentatively regarded as Permian. Thus, the Moesian salt basin was only tentatively distinguished as the Late Permian (Zharkov 1974a). In the last few years new data have become available which enabled the singling out there of a separate salt basin with greater confidence and its more detailed description (Polster et al. 1976; Zhukov et al. 1976).

The evaporite series are exposed in numerous deep boreholes in Bulgaria in the vicinity of Varna, Provadiy Mirovo, Bezvoditsa, Vetrino, Zlatar, Totleben, Kaliakra, Turgovishte, Elenovo, Chereshovo, and in adjacent southern regions of Romania. From Yanev's data (1969, 1970a,b, 1971; Zhukov et al. 1976), it follows that two types of section can be distinguished in the Upper Permian: salt-bearing, and argillaceous terrigenous sulfate-bearing sections. The salt-bearing series was studied in the greatest detail using drilling data from the vicinity of Mirovo village (OP-1) and Bezvoditsa (P-75). There is a salt-bearing mudstone unit in the lower part of the section; in the upper part salts dominate. The former is composed of thin-bedded gray and ash-gray mudstone with bands of siltstone, marl, and rock salt. Thin dolomite layers are very common, as well as carbonate breccia with intercalations and lenses of anhydrite. The thickness of the lower unit varies from 150 m (Bezvoditsa) to 890 m (Mirovo). The salt-bearing unit is mostly represented by pale, yellow-to-white and pink-to-red rock salt alternated with mudstone, anhydrite, and dolomite. According to Zhukov et al. (1976) primary succession is obscured by salt tectonics. As a result, the sequence is mostly composed of brecciated rocks with mudstone, mark, and dolomite fragments enclosed in the salt matrix. Lenses of gypsum and glauberite are often recorded, as well as the carnallite veinlets and selvages. The thickness of the salt-bearing unit is over 2140 m and 380 m near Mirovo and Bezvoditsa, respectively.

Salt deposits in the Moesian Basin are traced from Mirovo and Bezvoditsa villages almost to Omurtag, i.e., for 100 km from east to west. A salt zone is surrounded by deposits of argillaceous-terrigenous sulfate type, mostly represented by siltstone, mudstone and, more rarely, by dolomite. S. Yanev distinguished three units among them. The lower one is composed of green-gray to dark gray, purple-brown to brownish-red thin-bedded mudstone, alternated with siltstone, anhydrite, and dolomite. Anhydrite often occurs as pockets and lenses. The thickness of the unit is about 640 m. It is correlated with all the salt-bearing deposits of the Mirovo and Bezvoditsa area. The overlying member is composed of massive mudstone and siltstone. The rocks are bright red, brick-red to brownish-red, more rarely dark and greenish-gray. All the rocks are calcareous to a greater or lesser extent; inclusions, lenses, and thin layers of anhydrite are often recorded there. The thickness of the unit varies from 160 m to 1093 m. The third unit crowning the Permian section is characterized by variegated rocks; it is arenaceous-argillaceous in composition, its thickness varying from 54 m to 136 m.

Within the Varna Depression the equivalents of the salt-bearing deposits are mostly represented by a sulfate-carbonate sequence which, in S. Yanev's opinion, can be distinguished as the third, independent type of the sequence occurring east of this salt accumulation area and, probably, marking a transition from a salt basin to the open sea of normal salinity, which, in his opinion, was in the place of the present Black Sea (Zhukov et al. 1976, p. 124).

An area of evaporite accumulation within the Moesian Basin was flanked on its west by an intensely subsiding depression of the East Cis-Balkan area where thick terrigenous strata were being accumulated. Red siltstone and deltaic sandstone there are about 2 km thick. The data available suggest that the Moesian salt basin was bounded by land in the north, north-west and west where source areas were located on the North Bulgarian Uplift and the South Dobruja. No data are as yet available on the presence of any source areas at the southern boundary. It is most likely that the salt basin was also connected in the south with a sea of normal salinity.

The salt-bearing deposits all over the Moesian Depression are underlain by thick Lower Permian terrigenous red beds. In places inclusions and lenses of anhydrite are sporadically observed in the upper part of red beds.

Thus, one salt-bearing and three sulfate-bearing units may be preliminarily outlined within the Moesian salt basin. A determined area under this salt sequence is 3000 km^2. If the average rock salt thickness for this area is accepted at 300 m, then at a minimum it will amount to 9.10^2 km^3. The lowermost sulfate-bearing unit may be recognized within the upper Lower Permian terrigenous red beds where anhydrite inclusions and lenses were found. The second sulfate-bearing unit is coeval with the salt-bearing sequence and comprises the argillaceous-terrigenous-sulfate-bearing type of evaporite succession. The last sulfate-bearing sequence encloses sulfate-carbonate deposits developed in the Varna Depression. It has, apparently, an age and facies equivalent to the salt sequence.

The North Italian Basin is outlined on the basis of extensive gypsum and anhydrite deposits in northern Italy. They are mostly confined to the Upper Permian Bellerophon Formation where they form two sulfate-bearing sequences: the lower one, up to 150 m thick, mostly represented by laminite gypsum, and the upper one, 200–300 m thick, composed of a cyclic alternation of gypsum and dolomite. These sequences are separated by a unit of dense fine-grained dolomite (Bosellini and Hardie 1973).

Sulfate-bearing deposits are traced there from the northern regions of Italy into the Dolomitic and Carnic Alps, and farther into the Dravt Ridge where layers of gypsum and anhydrite occur among the Late Permian Bellerophon Series (Kahler 1972; Mostler 1972a,b; Rutten 1972; Buggisch et al. 1976). They probably continue eastward into the Karawanke Range. Sulfate rocks are less abundant among the Lower Permian terrigenous red beds distinguished in the South Alps as the Groden Formation, and in North Italy as the Garden Sandstone (Oxburgh 1968, Falke 1972; Rutten 1972; Bosellini and Hardie 1973; Flügel 1975; Buggisch et al. 1976). These sulfate-bearing red beds can be regarded as a separate unit. They probably occur along the northern marginal zone of the basin. Besides the above sulfate-bearing deposits, the Lower Permian (Autunian) deposits are tentatively included into the North Italian Basin; these deposits are developed in the Bergamasche Alps in Lombardia, Italy, and are distinguished as the Collio sequence. There gypsum lenses and layers are recorded among red siltstone, marl, and siltstone (Sitter and Sitter-Koomans 1949; Falke 1972).

In the Dinarids Basin sulfate rocks are abundant among the Late Permian deposits. They occur in the Durmitor Massif where gypsum beds, 20–30 m thick, crop out (Cadet 1966, 1970a,b), as well as in the High Karst Zone around the Cvrsnica Mountains (Charvet 1970). They occur in a carbonate succession. Additionally, gypsum-bearing terrigenous red beds are recorded in the Korab Zone of the Dinarids in north-eastern Albania (Papa 1970). Miljuch (1973) outlined an extensive zone of Permian evaporite deposits in the Dinarids, embracing all the above regions. At present two sulfate-bearing sequences are distinguished within the Dinarid Basin: sulfate-carbonate and terrigenous red beds.

The Mecsek Basin is tentatively distinguished on the basis of Late Permian gypsum-bearing clay deposits developed in the Mecsek Mountains, in southern Hungary. However, the data on the thickness and areal extent of gypsum-bearing strata are very scarce. It can only be noted that gypsum layers are thin and occur among variegated terrigenous formations (Vadas 1964; Trunkó 1969; Zhukov et al. 1976).

The Dobruja Basin is confined to a depression of the same name in South Moldavia. Sulfate rocks among the Permian deposits were penetrated by several deep boreholes. They compose two sulfate-bearing sequences: the first and the second anhydrite-bearing strata separated by a bit of terrigenous red beds. They are 120 m and 110 m thick, respectively (Kaptsan and Safarov 1969b; Kaptsan et al. 1963; Bobrinsky et al. 1974).

The Rakhov Basin is conventionally distinguished as a separate basin. Zhukov (1965b; Zhukov et al. 1964, 1976) has recorded abundant sulfate rocks in the Rakhov Zone of the East Carpathians. There a gypsum-bearing sequence, about 100 m thick, is distinguished in the upper part of the Permian section; it is composed of variegated mudstone and siltstone, among which interbeds of gypsinate shale and small gypsum lenses are recorded. The age of this sequence is probably Late Permian.

The East European Basin is one of the largest Paleozoic evaporite basins. Halogenic deposits have been recorded there from the area exceeding 1,500,000 km^2. They occur in the Mezen Depression and in the Moscovian Syneclise, within the Volga-Ural Anteclise, in the central and southern Cis-Uralian Trough, in the Pre-Capsian and Dnieper-Donets Troughs. The Permian halogenic deposits are rather well known. They were discussed in many publications the list of which as of 1972 is given in the monograph by the author (1974a). Recently new works have been published providing evidence on composition, structure and environments of Permian evaporite deposits in the East European Basin (Diarov and Dzhumagaliev 1971; Ermakov 1971; Tikhvinsky and Blizeev 1972; Zamarenov et al. 1972; Fiveg 1973a,b, c,d,e; Gorbov 1973; Shafiro and Sipko 1973; Shchedrovitskaya 1973; Tikhvinsky 1973, 1974, 1976a,b; Diarov 1974, 1977; Kamashev et al. 1974; Ivanov and Voronova 1975; Shafiro 1975, 1977; Ustritsky 1975; Forsh 1976; Bogatsky et al. 1977; Ermakov and Grebennikov 1977; Ermakov et al. 1977; Golubev 1977; Ivanov et al. 1977; Kapustin et al. 1977;Khalturina et al. 1977; Kolotukhin et al. 1977; Kopnin and Zueva 1977; Kopnin et al. 1977; Krichevsky et al. 1977; Krylova et al. 1977; Kudryashov and Myagkov 1977; Levenshtein et al. 1977; Morozov 1977; Muldakulov and Marchenko 1977; Sapegin 1977; Svidzinksy et al. 1977; Tikhvinsky et al. 1977; Tretiakov 1977; Trofimova and Efremov 1977).

The salt-bearing deposits in the East European Basin are found among the Asselian, Sakmarian, and Kungurian (Lower Permian), as well as the Ufimian, Kazanian, and Tatarian (Upper Permian) strata. At present it can be stated that at least 11 separate salt sequences have been formed within the basin. The oldest Nikitovo and Slavyansk salt sequences are developed only in the Dnieper-Donets Depression. Both belong to the Asselian and are represented by the alternation of layers and units of rock salt, anhydrite, carbonate, and, more rarely, terrigenous rocks. The thickness of the first salt sequence varies from 75 m to 240–250 m, that of the second 40–600 m. Toward the margins of the Dnieper-Donets Depression salt-bearing sediments are wedging out, and the section becomes mainly sulfate-carbonate terrigenous and sulfate-terrigenous. At the level of the Sakmarian two widely spaced salt sequences are distinguished: the Kramatorsk sequence in the Dnieper-Donets Depression, and the Upper Kuloi sequence in the Mezen Depression. The Kramatorsk sequence is represented by rock and potash salts alternated with anhydrite and siltstone. The maximum thickness of the sequence is 600 m, recorded in the Bakhmut Trough. The Upper Kuloi sequence is composed of rock salt alternated with anhydrite. The thickness of these sediments exceeds 100 m.

Four individual salt sequences can be distinguished in the Kungurian: (1) the Berezniki sequence in the Solikamsk Depression, (2) a salt sequence in the Upper Pechora Depression, (3) the Iren sequence in the Chusovaya and Yuryuzan-Sylva Depressions, and (4) a salt sequence developed within the Pre-Caspian Depression, in the southern Volga-Uralian Region and in the southern Cis-Uralian Trough. The Berezniki section consists in ascending order of clay-carbonate sulfate-salt-bearing rocks, the lower rock salt unit, a zone of potash salts comprising the lower sylvinite and the upper sylvinite-carnallite horizons, and, finally, covering rock salts (Ivanov and Voronova 1975). It is reasonable to include the overlying clay-marl-salt deposits of the Ufimian (Upper Permian) into the Berezniki sequence previously distinguished by the author as a separate Solikamsk salt sequence (Zharkov 1971a, 1977). The position in the section and lithology of these deposits is similar to that of the underlying Berezniki Formation; together they form a single evaporite sequence (Ivanov and Voronova 1975). The thickness of the Berezniki salt sequence in the deepest parts of the Solikamsk Depression is 800–900 m.

The thickness of a salt sequence in the Upper Pechora Depression is 400–435 m. The underlying rock salts and a potash salt zone, 55 m thick, as well as covering rock salts, are very conspicuous within this sequence. The salt sequence in the Chusovaya and Yuryuzan-Sylva Depressions has a lower stratigraphic position in the Kungurian than those of the Solikamsk and Upper Pechora Depressions. They occur at the level of the Ledyanopeshchera and Shalashnino units of the Iren horizon. The thickness of the sequence composed of rock salt and anhydrite does not exceed 120–150 m.

A salt sequence developed in the Pre-Caspian Trough, in the southern Volga-Uralian Region, and in the southern Cis-Uralian Trough, according to Tikhvinsky (1976a,b) and Gorbov (1973) fills in a large isolated potash salt basin known as the Cis-Uralian-Pre-Caspian or the Pre-Caspian Basin. The structure of the salt sequence filling in the basin became more clear during the last years owing to the studies by Ermakov (1971), Shafiro (1972, 1975, 1977), Tikhvinsky and Blizeev (1972), Gorbov (1973), Thikhvinsky (1973, 1974, 1976a,b), Diarov (1974), Forsh (1976), and Ermakov and Grebennikov (1977). However, the age of the salt sequence remains disputable. Most of the investigators place the sequence into the Kungurian (Lower Permian), whereas Shafiro (1972, 1975) and Shafiro and Sipko (1973) suppose that its lower part might be the Artinskian, to which these authors assign the Filippovo horizon, a generally accepted lower unit of the Kungurian. This contradiction is not so great as it appears at first sight, since the discrepancy depends on how the Kungirian/Artinskian boundary is drawn in the stratotype; moreover, the correlation of the lower part of the salt sequence with the Filippovo horizon presents no difficulty. This position leads us to a conclusion that heretofore no true Artinskian salt-bearing deposits have been recorded from the Pre-Capsian Depression. The age of the upper part of the salt sequence is even more difficult to determine. This problem will be discussed in the next chapter; now, it should be only emphasized that there is some reliable evidence showing that the upper parts of the sequence might be assigned to the Ufimian (Upper Permian) (Forsh 1976). In spite of this, as yet there is much more reason to place the upper part of the salt sequence studied in the Kungurian (Shafiro 1972, 1975, 1977; Tikhvinsky and Blizeev 1972; Tikhvinsky 1973, 1974, 1976a,b).

Over much of the Pre-Capsian Depression the real thickness of the Kungurian salt sequence is obscured by salt tectonics which gave rise to numerous salt domes within which most of the salt-bearing deposits are concentrated, and interdome areas where they are either very thin or completely absent (Vasiljev 1968; Aizenshtadt and Slepakova 1977; Kapustin et al. 1977; Krichevsky et al. 1977, etc.). A normal thickness of a salt sequence in the center of the depression is estimated by most investigators at 4–5 km. It decreases to 1–2 km toward the margin. It is in the marginal, western, and northern zones that the salt sequence was penetrated by a deep well where the mode of occurrence was not affected and recently its detailed subdivision was proposed. The lower part of the sequence belongs to the Filippovo horizon composed of dolomite and anhydrite with interbeds and lenses of rock salt, whose content increases with depth, and the sequence becomes entirely salt-bearing (Tikhvinsky 1974, 1976a; Shafiro 1975). Higher, the Volgograd Formation is distinguished (Shafiro 1975, 1977; Pisarenko et al. 1977) which is represented in the Volga Region, near Volgograd, by alternation of anhydrite and rock salt, while in the northern flank zone rock salt predominates. Tikhvinsky (1974, 1976a) assigns this part of the sequence to the Lower Iren member; Shafiro (1975, 1977) places the Volgograd Formation into the Upper Artinskian.

The thickness of the overlying Kungurian salt-bearing deposits in the Volga Region near Saratov and Volgograd is about 1000 m. Shafiro subdivides this sequence into seven sedimentary cycles (rhythms), each of which begins with dolomite and anhydrite and terminates with rock salt. Potash salts are found in the upper part of some cycles: the fourth and fifth cycles contain bischofite rocks. According to Banera and Gorbov (Gorbov 1973) the Kungurian salt sequence of the Cis-Volga monocline is subdivided into ten sedimentary cycles, while Tikhvinsky (1974) assigns this part of the section to the Upper Iren member. Both in the central and northern areas of the Cis-Caspian Depression 13 sedimentary cycles grouped into the Ulagan, Elton, Chelkar, and Inderbor beds can be recognized (Tikhvinsky 1976a). All in all 11 more or less persistent potassium-bearing horizons containing one or three potash salt layers are distinguished. I.N. Tikhvinsky assumes that in the Cis-Urals-Pre-Caspian potash salt basin three potash salt zones confined to the Ulagan, Elton, and Chelkar beds, respectively, might be regionally traced.

The Ufimian salt-bearing deposits, apart from those occurring in the Solikamsk Depression, are recorded in the eastern Pre-Caspian Depression where they were penetrated by the G-2 and G-10 boreholes in the Kumsay area (Zamarenov et al. 1972). Their thickness is 20–25 m. A salt unit composed of rock salt and anhydrite occurs among variegated mudstone, siltstone, and sandstone. The areal extension of the Ufimian salt sequence in the Pre-Caspian Depression is not clear. Probably, in the west this sequence makes a single whole with the underlying Kungurian salt sequence, but at present it is more reasonable to regard them separately. Among the Kazanian deposits in the East European Basin a salt-bearing hydrochemical formation is distinguished. It occupies the lower Upper Kazanian and is widespread in the southern Volga-Ural Anteclise within the Buzuluk Depression and within the adjacent Pre-Caspian Depression. The thickness of the sequence varies from 300 m to 460 m. Anhydrite units can be traced in its lower and upper parts, whereas rock salt dominates the middle part. Recently some new evidence appeared about the Tatarian salt-bearing deposits in the eastern Pre-Caspian Depression (Zamarenov et al. 1972; Zhalybin et al. 1974; Konishchev 1976). These deposits mainly belong to the undifferentiated Late Permian formations and can be only tentatively regarded as Tatarian. Therefore, the Tatarian age is suggested for the salt sequence. In composition it is salt-bearing terrigenous red beds. Its thickness exceeds 900 m.

It is rather difficult to determine the number of the sulfate-bearing sequences within the East European Basin. This results from the fact that no studies were aimed at distinguishing sulfate, sulfate-carbonate or sulfate-terrigenous lithological units among Permian deposits of the Russian Plate and the Cis-Uralien Trough. In most cases they were included into the above salt sequences. At the same time, the sulfate-bearing deposits in the East European Basin were more widespread than salt-bearing beds. In fact, they occur through the entire Permian section and often form extensive sulfate-bearing beds enclosing small lenticular salt bodies. The thickest sulfate sequence comprising the Asselian, Sakmarian, Kungurian, Ufimian, and Kazanian deposits is developed within the Moscovian Syneclise, Mezen Depression, and Volga-Ural Anteclise. In general, it is regarded as a separate Permian sulfate-carbonate sequence. Another sulfate sequence encloses the Tatarian red beds characterized by thin intercalations and lenses of gypsum and anhydrite. This sequence is developed throughout the entire basin. Finally, the Upper Kartamysh sulfate sequence can be regarded as a separate unit; it occurs in the Dnieper-Donets Depression. This sequence comprises the uppermost Kartamysh Formation overlying marker carbonate bed Q_8 where anhydrite bands are found among terrigenous red beds.

The Chu-Sarysu Basin is situated in Central Kazakhstan. Salt-bearing deposits were found within two isolated areas: in the north-western Dzhezkasgan Depression and in the south, along the south-eastern margin of the Ulanbel-Talas Uplift, and in the south-western Chuya Depression (Ditmar 1962a,b, 1963, 1965, 1966; Bakorov 1965a,b, 1977; Shakhov 1965a,b, 1968; Kumpan 1966; Ditmar and Tikhomirov 1967; Gulyaeva et al. 1968; Gabai 1969, 1977; Orlov et al. 1969; Dobretsov and Kumpan 1973; Sinitsyn et al. 1977). At present two salt-bearing sequences are distinguished in the basin: the first is confined to the Zhidelisai and Kingir Formations, whereas the second is distinguished as the Tuzkol Formation. A characteristic feature of both sequences is a wide occurrence of terrigenous rocks and the presence of inclusions and layers of glauberite rocks.

The Karasu-Ishsay Basin lies in Uzbekistan, in the North-eastern Fergana Valley. Data on the occurrence of gypsum and anhydrite among terrigenous-carbonate deposits of the Karasu Formation were published by Biske and Kushnar (1976). The sulfate Karasu Formation is the upper Early Permian and the lower Late Permian in age.

The Darvaza Basin is outlined on the basis of gypsum-bearing deposits distribution; the deposits of probable Early/Late Permian age are distributed within the Darvaza Range in Tadzhikistan. They are either distinguished as a gypsum-bearing sequence, or as the Shakarsen Formation (Dutkevich 1937; Vinogradov 1959) composed of purple clay, fine-grained sandstone, gypsum, and limestone. The thickness of gypsum layers often attains 10–12 m, and the total thickness is 400 m.

Within the Spitsbergen Basin gypsum and anhydrite are confined to the Upper gypsum-bearing formation of the Artinskian age (Harland 1964; Sosipatrova 1967). In the East Greenland Basin a gypsum-bearing sequence occurs at the base of the Zechstein; this sequence is represented by massive gypsum alternated with limestone, dolomite, and calc-dolomite rocks. Gypsum lenses are about 50 m thick. Gypsum-bearing beds are sometimes recorded throughout the entire Zechstein section which is about 300 m thick (Maync 1964).

The Sverdrup Basin, as has already been noted, was an evaporite basin not only during the Carboniferous epoch, but also at the beginning of the Early Permian when the Mount Bayley sulfates were being formed. This sequence has been discussed earlier.

During Permian time the Arabian Basin was sulfate-bearing along its western and south-western margins where reefogenic limestone, sandstone, variegated siltstone, and clay with gypsum interbeds, lenses, and inclusions were accumulated (North East . . . 1973). In northern Iraq gypsum-bearing deposits are of Upper Permian age, whereas their lower part is known as the Chia-Sapri Formation. In south-eastern Turkey variegated gypsum-bearing rock units belong to the Upper Permian Inbirik Formation. The sulfate-bearing deposits of the Arabian Basin are as yet poorly studied. They are all tentatively regarded as a single salt sequence.

Thus, 37 salt and 48 sulfate unites are distinguished within all the Permian evaporite basins. Salt sequences were deposited in nine salt basins, most of them being confined to the Midcontinent (eleven salt sequences), Central European (five salt sequences), and East European (eleven salt sequences) basins, Two salt sequences are distinguished within each of the following basins: Supai, Peru-Bolivian, Alpine, and Chu-Sarysu; and one in the Moesian and North Mexican Basins. Thirty sulfate sequences are distinguished within the salt basins, twelve in the Midcontinent Basin, six in the Central European, four in the Alpine, three in each of the following basins: East European and Moesian Basins and one in the Supai and Peru-Bolivian Basins. During the Permian there were thirteen sulfate basins within which eighteen sulfate sequences can be distinguished: four in the North Italian Basin, two in each of the following basins: the Dinarid and Dobruja Basins, and one in each of the following basins: Rio Blanco, Parnaiba, Mecsek, Rakhov, Karasu-Ishsay, Darvaza, Spitsbergen, East Greenland, Sverdrup, and Arabian Basins.

General Conclusions

The above data allow us to recognize 97 Paleozoic evaporite basins. Thirty four of them were salt basins and 63 sulfate basins. At the same time it should be emphasized once again that the actual number of evaporite basins in the Paleozoic is not a final decision. The difficulties in solving this problem arise from several reasons.

The most important one is the absence of common methods of distinguishing the old evaporite basins. Different investigators approach this problem in different ways, which results in ambiguous understanding of boundaries and time of existence of this or that particular basin of evaporite sedimentation. The salt and sulfate basins are often distinguished within an isolated salt-bearing or sulfate-bearing sequence without analysis of their spatial distribution within a sedimentary basin in general, formed in the area discussed during a given epoch, which sometimes leads to a recognition of small salt basins, as it is often the case, e.g., with the Upper Devonian Basin of the Russian Platform, North Siberian Devonian-Early Carboniferous, East European Permian or the Permian Midcontinent. Within these major evaporite basins, in various parts of which the salt-bearing deposits were regularly accumulated, some small salt basins can be distinguished, in the Upper Devonian Basin of the Russian Platform they are: the Pripyat, Dnieper-Donets, and Timan Basins; within the North Siberian Basin – the Igarka, Khatanga, and Kempendyai; in the East European – the Solikamsk, Pechora, Mezen; Cis-Uralian – Pre-Caspian, Dnieper-Donets; in the Midcontinent Basin – the Permian, Yulesburg, Williston, and other numerous minor basins.

This approach to the establishment of the evaporite basins, including salt basins, is groundless. In our opinion, it is much more correct to consider as an evaporite basin an entire sedimentary basin of this or that epoch, in which the halogenic sedimentation was continuous or took place at regular intervals. It is in this way that we have tried to distinguish Paleozoic salt and sulfate basins. However, it was not always possible to follow this principle due to the vague history of geological evolution of some sedimentary basins. Thus, two separate salt basins were distinguished for the Devonian period of the Russian Platform: the Morsovo and the Upper Devonian Basins. However, it is most likely that in future they will have to be grouped into a single Devonian salt basin of the Russian Platform, because, apparently, during the Middle and Devonian a single major sedimentary basin was being formed in the area where halogenic sedimentation took place at regular intervals. The same problems arise in the course of investigations of the Lena-Yenisei Ordovician/Silurian and the North Siberian Devonian/Early Carboniferous basins which in future will be probably also classified together, or we would be able to establish the time of their formation more precisely.

Sometimes the dimensions of the recognized evaporite basins have been artificially enlarged, and they comprise then probably isolated areas (for instance, the Lvov Depression in the Upper Devonian Basin of the Russian Platform). In other cases they could appear artificially reduced (for instance, a series of minor evaporite basins in southern Europe which, probably, were involved in one or two of the larger evaporite basins). The uncertainty increases when sulfate evaporite basins are being distinguished. All of them are mostly determined from the presence of anhydrite and/or gypsum. At the same time, it is quite evident that in many cases these rocks could not be accumulated within the isolated basins, but within marginal near-shore areas of carbonate, terrigenous-carbonate, or terrigenous red beds basins.

The second reason lies in the fact that many of the regions of the Earth where salt and sulfate sequences might be found are as yet poorly studied. It is only during the last years that four new salt basins have been discovered: the Ordovician/Silurian Basin of the Canadian Arctic Archipelago, the Devonian Hudson Basin, and the Carboniferous Sverdrup Basin and the Mid-Tien Shan Basin. New discoveries might be expected in northern East Siberia and in central and south-western Asia, on the islands of the Arctic Ocean, in South America where rather thick sulfate-bearing deposits are known and salt-bearing deposits may occur in the deepest parts.

Thirdly, it is difficult to determine the exact number of the evaporite basins because of the vague stratigraphic position of a number of salt-and sulfate-bearing sequences. Due to this, the age of some salt and sulfate basins is accepted conventionally, as, for instance, is the case with the Ordovician/Silurian Canning or Permian North Mexican and Peru-Bolivian basins. The last two basins might be even Mesozoic, but not Paleozoic.

Finally, the fourth reason is the vague spatial relations between some coeval (or almost contemporaneous) salt or sulfate sequences which do not allow determination of whether they were being accumulated within isolated basins, or within a single sedimentary basin; the Supai Basin in North America provides such an example.

Similar remarks are also true of the number of salt and sulfate sequences established within evaporite basins. In general 90 salt and 90 sulfate units were distinguished in salt basins, and 100 sulfate units were recorded in sulfate basins. However, no general methods of their determination have been as yet worked out, and stratigraphic and areal extent of a number of evaporite strata is still uncertain. Thus, the total number of evaporite basins under discussion and that of salt sequences found within these basins is different from the earlier estimate (Zharkov 1971a,b, 1974a,b, 1977). All our estimates should be regarded as the first attempts in this respect. As to the final solution of this problem, it will be possible only through subsequent detailed peleogeographical studies aimed at the establishment and outlining of different types of old sedimentary basins and determination of the time of their existence, and only then could we say which of them were evaporitic in the Paleozoic and how many halogenic sequences were formed in these basins.

In spite of all this, the history of the Paleozoic evaporite sedimentation can probably be discussed in more detail on the basis of the number of Paleozoic salt and sulfate sequences determined and that of the Paleozoic evaporite basins (Table 1 and 2).

Let us now discuss the most general regularities of the distribution pattern of the Paleozoic evaporite basins on the continents. It should be noted that most of the evaporite sedimentary basins at present are situated within the northern hemisphere. Eighty five of the distinguished 97 basins lie in the northern hemisphere, and only 12 in the southern hemisphere. Similar relations are observed when salt and sulfate basins are discussed separately: there are 29 salt basins and 56 sulfate basins, and 5 salt and 7 sulfate basins in the northern and southern hemispheres, respectively. The distribution of the Paleozoic evaporite basins on the continents is as follows: in Eurasia (including Spitsbergen, Novaya Zamlya, and Severnaya Zemlya) there are 47 basins (13 salt basins and 34 sulfate basins), in North America (Canadian Arctic Archipelago and Greenland inclusive) 30 (16 salt basins and 14 sulfate basins), in South America 7 (2 salt basins and 5 sulfate basins), in Africa 7 (only sulfate basins), in Australia 6 basins (3 salt basins and 3 sulfate basins).

A general distribution pattern of the Paleozoic salt and sulfate basins on the continents (Fig. 16) suggests that the Eurasian continent can be subdivided into five parts on the basis of a peculiar distribution pattern of evaporite deposits: East Siberian, European, Asia Minor, Central Asia, as well as the East and South-East Asia. The East

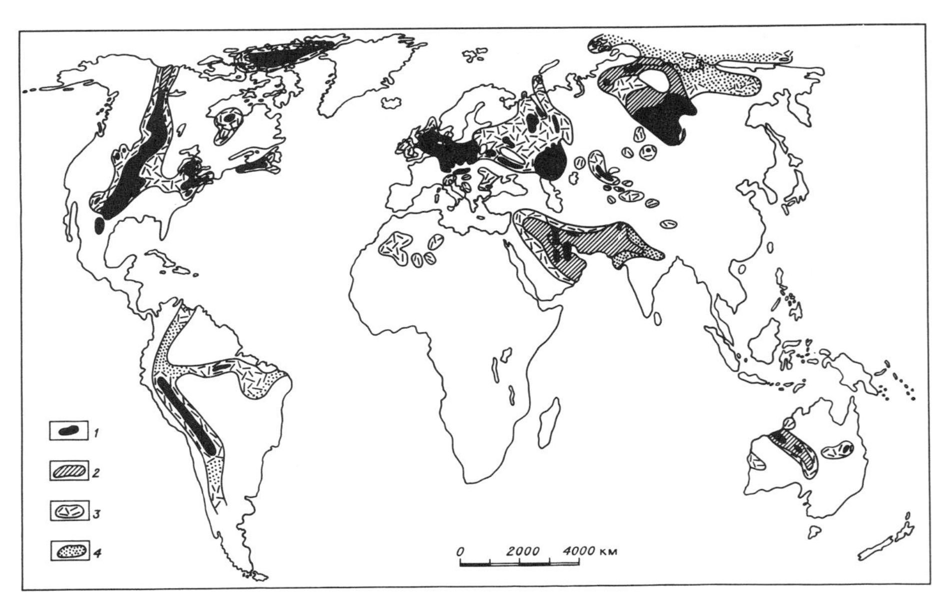

Fig. 16. Generalized map of Paleozoic evaporite deposits distribution.
Areas of development of deposits: *1, 2* salt (*1* determined, *2* inferred); *3, 4,* sulfate (*3* determined, *4* inferred)

Siberian part of the continent was the site of evaporite sedimentation only during the Early and Middle Paleozoic. During this time zones of evaporite sedimentation were displaced from the southern Siberian Platform into its northern and north-eastern regions, and then farther east into the Verkhoyansk-Kolyma Region. Since the close of the Early Carboniferous evaporites have not been deposited in north-eastern Asia. In the European part of the continent the sulfate accumulation began in the Ordovician and Silurian, but major evaporite basins, including salt basins, began to form only since the Middle Devonian. It should be noted that in the Devonian the major area of evaporite sedimentation was in eastern Europe. In the Carboniferous, and especially in the Permian, halogenic deposits began to accumulate not only in East Europe, but also over much of West Europe. So, it is established that in Europe the area of evaporite accumulation was gradually becoming more extensive, starting in the Ordovician to the close of the Permian. The regions of Asia Minor, lying south of the Alpine-Himalayan fold belt, are of interest because halogenic sedimentation took place there at the beginning and at the end of the Paleozoic. In Central Asia evaporite basins were episodically formed in the Paleozoic within a limited area. The salt accumulation was recorded in isolated, relatively small basins: Chu Sarysu, Tuva, Mid-Tien Shan. In East and South-East Asia, the data available suggest that no evaporites were accumulated there in the Paleozoic era.

In North America several areas of evaporite sedimentation existed in Paleozoic time. One of these regions was located within the Canadian Arctic Archipelago, where, beginning with the Precambrian and up to the Early Permian either salt or sulfate basins appeared periodically. The second region bounded the Canadian Shield on the west, south-west and south. Within this region evaporite basins were also periodically formed in the Cambrian, Ordovician, Silurian, Devonian, Carboniferous, and Permian; it should be emphasized that during the entire span of the time their southward displacement is observed. The third region is the central part of the Canadian Shield within the Hudson Bay. There, halogenic sediments were accumulated in the Ordovician, Silurian, and Devonian. The fourth region lies within the eastern part of the continent in the New Brunswick and Nova Scotia provinces of Canada, and in the adjacent shelf zone off the Atlantic. There an evaporite basin existed in the Carboniferous.

In South America the formation of the evaporite basins began in the Cambrian. Then, during a long time interval, comprising Ordovician, Silurian, Devonian, and Early and Middle Carboniferous, on the basis of the data available halogenic sedimentation did not take place on the continent. Since that time in many regions of South America, especially in the Piedmont Andes and in the Amazon Basin, conditions became favorable not only for sulfate, but for salt accumulation as well. In fact, the Late Paleozoic was the time of extensive evaporite sedimentation in South America.

In the Paleozoic only the north-western regions of Africa were covered by evaporites. It should be emphasized that there was exclusively sulfate accumulation. In the remaining part of the continent halogenic deposits were not accumulated in the Paleozoic. In Australia evaporites formed in the central regions mostly within the zone running from the Fitzroy and Canning Basins in the west, through the Amadeus Basin in the center of the continent to the Adavale Basin in the east. The halogenic sedimentation there was recorded from the Cambrian to the Carboniferous. The salt basins existed not only in Cambrian and Devonian, but, apparently, also in Ordovician and Silurian time.

Table 1. Distinguished Paleozoic salt basins and salt and sulfate sequences developed within them

No.	Basin	Age of Basin	Name of salt sequence and its number	Age of salt sequence	Name of sulfate sequence and its number	Age of sulfate sequence
1	2	3	4	5	6	7
1	East Siberian	Cambrian	1. Irkutsk 2. Usolye 3. Belsk 4. Angara 5. Litvintsevo 6. Mayan [a] 7. Upper Cambrian [a]	Early Cambrian Early Cambrian Early Cambrian Early Cambrian Middle Cambrian Middle Cambrian Late Cambrian	b	–
2	Iran-Pakistan	Cambrian	8. Hormoz 9. Punjab 10. Mila	Early Cambrian Early Cambrian Middle Cambrian	b	–
3	Mackenzie	Cambrian	11. Salina River	Late Cambrian	b	–
4	Amadeus	Cambrian	12. Chandler	Early Cambrian	b	–
5	Canadian Arctic	Ordovician-Silurian	13. Bay Fiord	Middle Ordovician, Late Llandeilo – Early Caradoc	1. Baumann Fiord 2. Read Bay	Early Ordovician, Arenigian Late Silurian, Ludlovian
6	Williston [a]	Ordovician-Silurian	14. Stonewall	Late Ordovician, Ashgillian (?)	3. Red River 4. Stony Mountain 5. Interlake	Late Ordovician, Late Caradoc Late Ordovician, Ashgillian Early Silurian
7	Canning	Ordovician-	15. Lower Carribady 16. Upper Carribady	Late Ordovician (?) Late Ordovician (?)	6. Stottid	Silurian (?)
8	Michigan-Pre-Appalachian	Silurian	17. Salina	Late Silurian, Late Ludlovian, Pridolian	b	–

9	North Siberian	Devonian-Early Carboniferous	18. Zubovo 19. Manturovo	Early Devonian, Middle Devonian,Eifelian,Givetian	7. Zubovo 8. Gypsum-bearing of Taimyr 9. Dezhnev	Early Devonian Early Devonian Early Devonian
			20. Fokin	Late Devonian Fammenian	10. Rusanov	Middle Devonian, Eifelian
			21. Kygyltuus	Late Devonian, Frasnian – Early Fammenian	11. Sidin	Middle Devonian, Eifelian – Givetian
			22. Nordvik	Middle Devonian, Eifelian – Givetian	12. Yukta	Middle Devonian, Givetian
					13. Atyrkan	Middle Devonian, Late Eifelian – Givetian
					14. Burkhala	Middle Devonian, Givetian
					15. Sebechan	Middle Devonian, Givetian
					16. Yayakh	Middle Devonian, Givetian
					17. Nakokhoz	Late Devonian, Frasnian
					18. Kalargon	Late Devonian, Frasnian
					19. Upper Fokin	Late Devonian, Fammenian
					20. Namdyr	Late Devonian, Fammenian
					21. Matusevich	Late Devonian
					22. Gypsum-bearing sequence of Vrangel Island	Late Devonian
					23. Gypsum-bearing sequence of Kyutingda Depression	Early Carboniferous, Tournaisian
					24. Tundrin	Early Carboniferous, Visean
					25. Kurunguryakh	Early Carboniferous, Tournaisian
					26. Sulfate-carbonate sequence of Khatanga Depression	Early Carboniferous, Visean
10	Tuva	Devonian	23. Ikhedushiingol	Middle Devonian, Eifelian – Givetian	b	–
11	Chu-Sarysu	Devonian-Carboniferous	24. Lower salt sequence	Late Devonian	27. Sulfate sequence	Early Carboniferous, Visean
			25. Upper salt sequence	Early Carboniferous, Tournaisian	28. Sulfate-bearing carbonate-terrigene sequence	Early Carboniferous, Visean-Namurian

Table 1 (continued)

No.	Basin	Age of Basin	Name of salt sequence and its number	Age of salt sequence	Name of sulfate sequence and its number	Age of sulfate sequence
1	2	3	4	5	6	7
					29. Sulfate sequence of Ulkunburul, Kashiburul, and Tekturmas Mountains	Early Carboniferous, Visean-Namurian
					30. Kysylkanat	Middle-Late Carboniferous
12	Morsovo	Devonian	26. Morsovo	Middle Devonian, Eifelian	31. Morsovo	Middle Devonian, Eifelian
			27. Narva sequence of the Pripyat Depression	Middle Devonian,	32. Middle Lopushan	Middle Devonian, Eifelian
13	Upper Devonian Basin of the	Devonian	28. Evlanovo-Liven sequence of the Pripyat Depression	Late Devonian, Frasnian	33. Shchigrov sequence of the Pripyat Depression	Late Devonian, Frasnian
			29. Dankov-Lebedyan sequence of the Pripyat Depression	Late Devonian, Fammenian	34. Voronezh-Evlanovo sequence of the Pripyat Depression	Late Devonian, Frasnian
			30. Shchigrov sequence of the Dnieper-Donets Depression	Late Devonian, Frasnian	35. Mezhsole vaya of the Pripyat Depression	Late Devonian, Fammenian
			31. Lower salt sequence the Dnieper-Donets Depression	Late Devonian, Frasnian	36. Upper Devonian sequence of the Baltic and Moscovian syneclise, and Volga-Ural Region	Late Devonian
			32. Upper salt sequence of the Dnieper-Donets Depression	Late Devonian, Fammenian	37. Ustukhta	Late Devonian, Frasnian
			33. Salt sequence of the Pre-Timan Trough	Late Devonian, Fammenian	38. Idzhid-Kamensk	Late Devonian, Fammenian

14	Canadian	Devonian	34. Lotsberg	Middle Devonian, Eifelian	39. Ernestina Lake	Middle Devonian, Eifelian
			35. Gold Lake	Middle Devonian, Eifelian	40. Chinchaga	Middle Devonian, Eifelian
			36. Prairie	Middle Devonian, Givetian	41. Muskeg	Middle Devonian, Givetian
			37. Black Creek	Middle Devonian, Givetian	42. Beaverhill Lake	Middle Devonian, Frasnian
			38. Hubbard	Middle Devonian, Givetian	43. Woodbend	Middle Devonian, Frasnian
			39. Davidson	Late Devonian, Frasnian	44. Duperow	Middle Devonian, Frasnian
			40. Dinsmore	Late Devonian, Frasnian	45. Birdbear	Middle Devonian, Frasnian
			41. Stettler	Late Devonian, Fammenian	46. Nisku	Middle Devonian, Frasnian
					47. Stettler	Late Devonian, Fammenian
15	Hudson	Devonian	42. Moose River	Middle Devonian, Eifelian	48. Upper Kenogami River and Stooping River	Early Devonian
					49. Williams Island	Middle-Late Devonian, Givetian-Frasnian
					50. Long Rapids	Late Frasnian-Early Fammenian
16	Michigan	Devonian	43. Lucas	Middle Devonian, Eifelian	b	–
17	Adevale	Devonian	44. Boree	Middle Devonian, Eifelian	b	–
18	Mid-Tien Shan [a]	Carboniferous	45. Salt sequence [a]	Early Carboniferous, Visean	51. Sulfate-bearing terrigene-carbonate sequence	Early Carboniferous, Tournaisian-Visean
					52. Chemanda	Early Carboniferous, Visean-Namurian
					53. Aktailyak	Middle Carboniferous, Bashkirian
					54. Kodzhagul	Late Carboniferous, Orenburgian
19	Sverdrup	Carboniferous-Permian	46. Otto Fiord	Middle Carboniferous, Bashkirian	55. Borup Fiord	Early Carboniferous, Namurian
					56. Antoinetta	Late Carboniferous
					57. Mount Bayley	Early Permian, Asselian

Table 1 (continued)

No.	Basin	Age of Basin	Name of salt sequence and its number	Age of salt sequence	Name of sulfate sequence and its number	Age of sulfate sequence
1	2	3	4	5	6	7
20	Williston	Carboniferous	47. Charles	Early Carboniferous, Tournaisian-Visean	58. Mission Canyon	Early Carboniferous, Tournaisian
					59. Kibbey	Early Carboniferous, Visean
21	Maritime	Carboniferous	48. Gautreau	Early Carboniferous, Tournaisian	b	–
			49. Windsor	Early Carboniferous, Visean		
22	Saltville	Carboniferous	50. MacCrady	Early Carboniferous, Visean	b	–
23	Paradox	Carboniferous	51. Paradox	Middle Carboniferous, Moscovian	b	–
24	Eagle	Carboniferous	52. Eagle Valley	Middle Carboniferous, Moscovian	b	–
25	Amazon	Carboniferous	53. Nova Olinda	Late Carboniferous, Orenburgian	60. Itaituba	Late Carboniferous, Gzhelian
26	Midcontinent	Permian	54. Ingleside	Early Permian, Sakmarian	61. Upper Minnelusa	Early Permian, Asselian, Sakmarian
			55. Broom Creek	Early Permian, Sakmarian	62. Lower Wellington	Early Permian, Sakmarian
			56. Wellington	Early Permian, Sakmarian	63. Upper Wellington	Early Permian, Sakmarian
			57. Owl Canyon	Early Permian, Sakmarian – Artinskinian	64. Wichita	Early Permian, Sakmarian
			58. Cimarron	Early Permian, Artinskinian	65. Stone Corral	Early Permian, Artinskinian

			59. Opeche sequence of Denver and Yulesburg Basins	Early Permian, Kungurian	67. Blaine	Early Permian, Kungurian
			60. Opeche sequence of Williston Basin	Early Permian, Kungurian	68. Minnekahta	Early Permian, Kungurian
			61. Salt sequence of Flowerport, Blaine, and Yelton Formations	Early Permian, Kungurian	69. Artesia	Late Permian, Ufimian, Kazanian
			62. San Andres-Artesia	Early-Late Permian	70. Glendo	Late Permian, Ufimian, Kazanian
			63. Pine	Late Permian	71. Day Creek	Late Permian, Kazanian
			64. Castile	Late Permian, Tatarian	72. Dewey Lake	Late Permian, Tatarian
27	Supai	Permian	65. Middle Supai	Early Permian, Sakmarian	73. Sulfate-bearing terrigene-carbonate sequence	Early Permian, Sakmarian, Artinskinian
			66. Upper Supai	Early Permian, Sakmarian-Artinskinian		
28	North Mexican [a]	Permian (?)	67. Salt sequence	Early Permian, Kungurian (?)	b	–
29	Peru-Bolivian [a]	Permian (?)	68. Mitu	Early Permian, Kungurian (?)	74. Capacabana	Early Permian
			69. Chuquichambi	Early Permian, Kungurian (?)		
30	Central European	Permian	70. Upper Rotliegendes	Early Permian, Artinskinian (?), Kungurian	75. Ten Boer	Early Permian, Artinskinian, Kungurian
			71. Werra	Late Permian, Kazanian	76. Werra	Late Permian, Kazanian
					77. Hartlepool	Late Permian, Kazanian
			72. Stassfurt	Late Permian, Kazanian	78. Sandwich	Late Permian, Tatarian
			73. Leine	Late Permian, Tatarian	79. Fleswik	Late Permian, Tatarian
			74. Aller	Late Permian, Tatarian	80. Suduvsk	Late Permian, Tatarian

Table 1 (continued)

No.	Basin	Age of Basin	Name of salt sequence and its number	Age of salt sequence	Name of sulfate sequence and its number	Age of sulfate sequence
1	2	3	4	5	6	7
31	Alpine	Permian	75. Salt sequence of the East Alps	Late Permian	81. Sulfate sequence of the East Alps	Late Permian
			76. Salt sequence of the Spišsko-Gemerskoe rudohoři	Late Permian	82. Sulfate sequence of the Spišsko-Gemerskoe rudohoři	Late Permian
					83. Sulfate sequence of the Tribeč Mountains	Late Permian
					84. Sulfate sequence of the Low Tatra Mountains	Late Permian
32	Moesian	Permian	77. Salt sequence	Late Permian	85. Sulfate-bearing red beds	Late Permian
					86. Sulfate-bearing terrigene sequence	Late Permian
					87. Sulfate-carbonate sequence of the Varna Basin	Late Permian
33	East European	Permian	78. Nikitovo	Early Permian, Asselian	88. Upper Kartamysh	Early Permian, Asselian
			79. Slavyansk	Early Permian, Asselian	89. Sulfate-carbonate sequence	Early-Late Permian
			80. Kramatorsk	Early Permian, Sakmarian		
			81. Upper Kuloi	Early Permian, Sakmarian	90. Sulfate-bearing red beds	Late Permian, Tatarian
			82. Berezniki	Early-Late Permian, Kungurian-Early Ufimian		
			83. Salt sequence of the Verkhnepechora Depression	Early Permian, Kungurian		
			84. Iren sequence of the Chusovaya and the Yuryuzan-Sylva Depressions	Early Permian, Kungurian		

			85. Salt sequence of the Pre-Caspian Depression, Volga-Ural Region and Cis-Uralian Trough	Early Permian, Kungurian
			86. Salt sequence of East Pre-Caspian Depression [a]	Early Permian, Ufimian (?)
			87. Hydrochemical sequence	Late Permian, Kazanian
			88. Salt-bearing red beds	Late Permian, Tatarian
34	Chu-Sarysu	Permian	89. Zhidelisai-Kingir	Early-Late Permian
			90. Tuzkol	Late Permian

a Basins and salt sequences isolated conventionally

b Sulfate sequences not isolated; sulfate rocks included into salt sequences

Table 2. Distinguished Paleozoic sulfate basins and sulfate sequences developed within them

No.	Basin	Age of Basin	Name of sulfate sequence and its number	Age of sulfate sequence
1	2	3	4	5
1	Tarim	Cambrian	91. Sulfate sequence	Early Cambrian
2	Anti-Atlas	Cambrian	92. Sulfate sequence	Early Cambrian
3	Michigan	Cambrian	93. Munising	Late Cambrian
4	Cis-Andean	Cambrian	94. Limbo	Early Cambrian
5	Anadarko	Ordovician	95. West Spring Creek	Early Ordovician, Late Arenigian
6	South Illinois	Ordovician	96. Joachim	Middle Ordovician, Llanvirn-Llandeilo
7	Georgina	Ordovician	97. Toko	Early-Middle Ordovician
8	Lena-Yenisei	Ordovician-Silurian	98. Sulfate-carbonate sequence of Norilsk Region	Ordovician
			99. Sulfate-carbonate sequence of Tunguska syneclise	Early Ordovician
			100. Irbukla-Kochak	Early-Middle Ordovician
			101. Stan	Middle Ordovician
			102. Kharyalakh	Late Ordovician
			103. Tochilnin	Early Ordovician
			104. Bratsk	Middle-Late Orcovician
			105. Utokan	Early Silurian, Late Llandoverian
			106. Kongda	Late Silurian, Ludlovian
			107. Kholyukhan	Late Silurian
9	Baltic	Ordovician-Silurian	108. Tallin	Middle Ordovician, Llandeilo
			109. Itfer	Middle Ordovician, Early Caradoc
			110. Ievsk	Middle Ordovician, Middle Caradoc
			111. Paprenyai-Pagegyai	Early-Late Silurian, Wenlockian-Ludlovian
10	Severnaya Zemlya	Ordovician-Silurian	112. Komsomolsk	Ordovician
			113. Gypsum-bearing sequence	Late Silurian
11	Moose River	Ordovician-Silurian-Devonian	114. Dolomite sequence	Late Ordovician, Late Caradoc-Ashgillian
			115. Kenogami River	Late Silurian-Early Devonian
			116. Moose River	Middle Devonian, Eifelian

Table 2 (continued)

No.	Basin	Age of Basin	Name of sulfate sequence and its number	Age of sulfate sequence
1	2	3	4	5
			117. Williams Island	Middle-Late Devonian, Givetian-Frasnian
			118. Long Rapids	Late Devonian, Late Frasnian-Fammenian
12	Pechora	Silurian	119. Kosju-Adak	Early Silurian, Llandoverian
			120. Filippyelsk	Late Silurian, Late Ludlovian
13	Dniester-Prut	Silurian	121. Pugai	Early-Late Silurian, Wenlockian-Ludlovian
14	Carnarvon	Silurian	122. Dirk Hartox	Early Silurian, Wenlockian
15	Minusinsk	Devonian	123. Abakan-Askyz	Middle Devonian, Eifelian-Givetian
			124. Beysk	Middle Devonian, Givetian
16	Kusnetsk	Devonian	125. Podonin	Late Devonian, Fammenian
17	Teniz	Devonian	126. Sulfate sequence	Late Devonian, Fammenian
18	Turgai	Devonian	127. Terrigene-carbonate sequence	Late Devonian, Fammenian
19	Moesian-Wallachian	Devonian	128. Carbonate-sulfate sequence	Middle Devonian
20	Tindouf	Devonian	129. Sulfate-bearing red beds	Middle-Late Devonian
21	Canadian Arctic	Devonian	130. Bird Fiord	Middle Devonian, Eifelian-Givetian
22	Illinois	Devonian	131. Jeffersonville	Middle Devonian, Eifelian
23	Central Iowa	Devonian	132. Kenwood	Middle Devonian, Eifelian
			133. Cedar Valley	Middle Devonian, Givetian
24	Pechora-Novaya Zemlya	Carboniferous	134. Sulfate-carbonate sequence of Pechora Depression	Early Carboniferous, Visean-Namurian
			135. Gypsum-bearing sequence of Novaya Zemlya	Early Carboniferous, Visean
25	Spitsbergen	Carboniferous	136. Lower Gypsum-bearing sequence	Middle Carboniferous, Bashkirian
26	East European	Carboniferous	137. Ozersk	Early Carboniferous, Tournaisian

Table 2 (continued)

No.	Basin	Age of Basin	Name of sulfate sequence and its number	Age of sulfate sequence
1	2	3	4	5
			138. Oka-Serpukhov	Early Carboniferous, Visean
			139. Kashira-Myachkov	Middle Carboniferous, Moscovian
			140. Kasimov	Late Carboniferous, Gzhelian
			141. Klyazma	Late Carboniferous, Gzhelian
			142. Noginsk	Late Carboniferous, Orenburgian
27	East Uralian	Carboniferous	143. Gypsum-bearing sequence of Magnitogorsk Region	Middle Carboniferous
			144. Sulfate-bearing terrigene-carbonate sequence of Bagaryak Region	Middle Carboniferous
			145. Sulfate-bearing terrigene sequence	Middle Carboniferous
28	Teniz	Carboniferous	146. Kirey	Middle Carboniferous, Bashkirian
29	Tyup	Carboniferous	147. Tyup	Middle Carboniferous
			148. Chaarkuduk	Middle Carboniferous
30	Chimkent	Carboniferous	149. Gypsum-bearing sequence	Early Carboniferous, Tournaisian
31	Aksu	Carboniferous	150. Kurukusum	Middle Carboniferous, Bashkirian-Early Moscovian
32	Achikkul	Carboniferous	151. Arktag	Middle Carboniferous, Bashkirian-Early Moscovian
33	Lhasa	Carboniferous	152. Pando	Early-Middle Carboniferous, Namurian, Bashkirian
34	Fitzroy	Carboniferous	153. Anderson	Early Carboniferous, Visean-Namurian
35	Radames	Carboniferous	154. Sulfate sequence	Early Carboniferous, Namurian
36	Illisie	Carboniferous	155. El-Adeb-Larach	Early Carboniferous, Bashkirian
			156. Tigentourin	Late Carboniferous, Gzhelian
37	Ahnet	Carboniferous	157. Sulfate sequence	Early Carboniferous, Visean
38	Reggane	Carboniferous	158. Gypsum-bearing sequence	Early Carboniferous, Visean

Table 2 (continued)

No.	Basin	Age of Basin	Name of sulfate sequence and its number	Age of sulfate sequence
1	2	3	4	5
			159. Sulfate-bearing red beds	Early Carboniferous, Namurian
39	Tindouf	Carboniferous	160. Ouarkziz	Early Carboniferous, Visean
40	Illinous	Carboniferous	161. St. Louis	Early Carboniferous, Visean
41	South Iowa	Carboniferous	162. St. Louis 163. Pella	Early Carboniferous, Visean Early Carboniferous, Visean
42	Michigan	Carboniferous	164. Michigan 165. Saginaw	Early Carboniferous, Visean Middle Carboniferous, Moscovian
43	Orogrande	Carboniferous	166. Magdalena	Late Carboniferous
44	San Juan	Carboniferous	167. Paradox	Middle Carboniferous, Moscovian
45	East Wyoming	Carboniferous	168. Minnelusa	Middle-Late Carboniferous
46	Venezuela	Carboniferous	169. Palmarito	Late Carboniferous
47	South Peruan	Carboniferous	170. Tarma	Late Carboniferous
48	Northumberland	Carboniferous	171. Gypsum-bearing calcareous sandstones	Early Carboniferous, Visean
49	North Ireland	Carboniferous	172. Roscunish	Early Carboniferous, Visean
50	Central England	Carboniferous	173. Anhydrite sequence	Early Carboniferous, Visean
51	Dobruja	Carboniferous	174. Sulfate-carbonate sequence	Early Carboniferous, Tournaisian-Visean
52	Rio Blanco	Permian	175. Patquia	Early Permian
53	Parnaiba	Permian	176. Pedra-do-Fogo	Early Permian
54	North Italy	Permian	177. Collio 178. Groden 179. Lower Bellerophon 180. Upper Bellerophon	Early Permian, Asselian, Sakmarian Early Permian, Artinskian (?), Kungurian Late Permian Late Permian
55	Dinarids	Permian	181. Sulfate-carbonate sequence 182. Sulfate-bearing red beds	Late Permian Late Permian

Table 2 (continued)

No.	Basin	Age of Basin	Name of sulfate sequence and its number	Age of sulfate sequence
1	2	3	4	5
56	Mecsek	Permian	183. Gypsum-bearing terrigene sequence	Late Permian
57	Dobruja	Permian	184. The first anhydrite sequence	Late Permian
			185. The second anhydrite sequence	Late Permian
58	Rakhov	Permian	186. Gypsum-bearing sequence	Late Permian
59	Karasu-Ishsai	Permian	187. Karasu	Early-Late Permian
60	Darvaza	Permian	188. Shakarsen	Early-Late Permian
61	Spitsbergen	Permian	189. Upper gypsum-bearing sequence	Early Permian, Artinskinian
62	East Greenland	Permian	190. Gypsum-bearing sequence	Late Permian
63	Arabian	Permian	191. Sulfate sequence	Late Permian

In general, the history of evaporite accumulation in the Paleozoic was rather unique on each of the continents. Even some regions on this or that continent occupying now similar climatic zones favorable for halogenic sedimentation differ greatly in their history of evaporite sedimentation. There is no general regularity in migration of zones of halogenic sedimentation for the continents of the southern and northern hemisphere. All this implies that evaporite sedimentation on the continents was caused not only by a climatic zonation during a certain epoch of the Paleozoic, but was also determined by tectonic history of the continents or their separate regions, and their position relative to the ancient equator.

CHAPTER II

Stratigraphic Position of Evaporites and Stages of Evaporite Accumulation

Statement of the Problem

Stages of evaporite accumulation in the geological history of the Earth have not yet been adequately studied. It is only the general regularities of the age position of evaporites that have already been discussed. Usually, investigators who studied this problem assigned evaporites or salt rocks of different composition to a system, more rarely to a series, and even more rarely to a stage, and thus they proposed certain stages of halogenic accumulation. As a result, it was ascertained that the evaporite sedimentation was discontinuous on continents, the halogenesis intensity " . . . never remaining constant, but increasing at one moment and decreasing at another" (Strakhov 1963, p. 490). The major halogenic epochs were distinguished (Early Cambrian, Middle/Late Devonian, Early Permian etc.), as well as the epochs when evaporite sedimentation was not so extensive (Ordovician, Carboniferous). These data were repeatedly quoted (Lothe 1957a, 1968; Ivanov and Levitsky 1960; Fiveg 1962; Strakhov 1962, 1963; Borchert and Muir 1964; Kozary et al. 1968; Zharkov 1971a, 1974a, 1977; Ivanov and Voronova 1972; Kalinko 1973a,b; etc.).

At the same time these regularities were established only in a general way without any detailed analysis of the data available, and thus many problems remained unsolved. The uncertainty still existed as to the differentiation in sulfate and halite accumulation, as well as those of the potash salts; the problem of the coincidence or noncoincidence of the stages of sulfate sedimentation with the epochs or rock salt accumulation, and of the latter with those of potash accumulation, is not clear. When the problem of the stages was studied, evaporites were discussed in general, the conclusions obtained were applied to the halogenic process as a whole. Moreover, as a rule, a distant correlation of the sections was not carried out, and thus the age position of evaporites for the whole Earth remains mainly undeciphered.

The evidence presented in the previous chapter suggests that a stratigraphic position of most evaporitic series in some Paleozoic basins seems quite valid. At least, in sections of certain regions the occurrence of sulfate or salt sequences was mainly determined to the rank of stage or series. As to a stratigraphic position of halogenic series, it still remains uncertain. The correlation of evaporites of different regions has not yet been carried out, though only such a correlation will enable us to establish regularities in distribution of sulfate and salt strata as to their stratigraphic position for each continent and for the Earth in general, and so to specify stages of evaporite sedimentation.

We shall make an attempt to confirm the conclusions concerning stages of sulfate, halite, and potash accumulation in the Paleozoic proceeding from correlation of sections in separate evaporite basins. But it should be noted that, firstly, the analysis will be carried out without regard for the extent and distribution area of sulfate and salt rocks which will be further used for recognition of epochs of intense evaporite sedimentation within the stages of evaporite sedimentation for the Paleozoic; secondly, interregional correlation of these deposits is still ambiguous for lack of detailed global stratigraphic charts for each system. It is because of this that we shall have to dwell upon different versions of correlation and to choose a correlation of sections of evaporite basins. This will make it possible to propose stages of evaporite accumulation in the Paleozoic. Below, we discuss an interregional correlation for each system in the Paleozoic.

Stratigraphic Position of Cambrian Evaporites

It has already been mentioned that Cambrian evaporite deposits occur within eight basins. Figure 17 shows the stratigraphic position of salt and sulfate deposits in composite sections within these basins, as well as their interregional correlation.

Let us first discuss the stratigraphic position of salt deposits. The age of some of them is not clear. This applies, for example, to the Punjab salt sequence in the Salt Range, Pakistan, and its stratigraphic equivalent, the Hormoz Formation further west in the Iran-Pakistan Basin. Most investigators assign these deposits to the Precambrian. Such a conclusion is based on the following evidence. The first occurrences of fossils in the Cambrian section within the Iran-Pakistan Basin, namely Redlichia (trilobite) was recorded from the Neobolus Shales of the Salt Range in the carbonate rocks just above the Lalun Sandstone, in North Kerman (Iran), west of the Loot Uplift. It has been widely accepted that these trilobites are indicative of lower Lower Cambrian, and thus the Neobolus Shale and its equivalents, the carbonate sequence in North Kerman, have been assigned to the Lower Cambrian. Stöcklin (1968a,b) has placed the lower boundary of the Cambrian in Iran at the base of the Lalun red beds. Krishnan (1966) correlated the Purple Sandstone of the Salt Range with the Vindhya red beds, thus considering them Precambrian. In doing so, the Hormoz and Punjab salt sequences are regarded as Precambrian.

However, in our opinion these data are not suggestive of the Precambrian age. Firstly, Redlichia is not the oldest of the Cambrian fossils recorded. On the Siberian Platform these trilobites have been found in the upper Lower Cambrian; they occur in the Olekma horizon deposits (Repina 1968), i.e., they were recorded in beds overlying thickest salt sequences (Irkutsk, Usolje, and Belsk sequences) in the Irkutsk Amphitheater. Apparently, in the Iran-Pakistan Basin beds with Redlichia are also confined to the upper Lower Cambrian, and the Neobolus Shale is correlative with Olekma horizon. Thus, it can be assumed that the Punjab salt sequence is not of Precambrian age, and occurs at the same level as the above-mentioned lower salt sequences of East Siberia. Salt deposits of the Hormoz Formation are, apparently, also Cambrian. Their stratigraphic position can be, to a certain extent, judged from

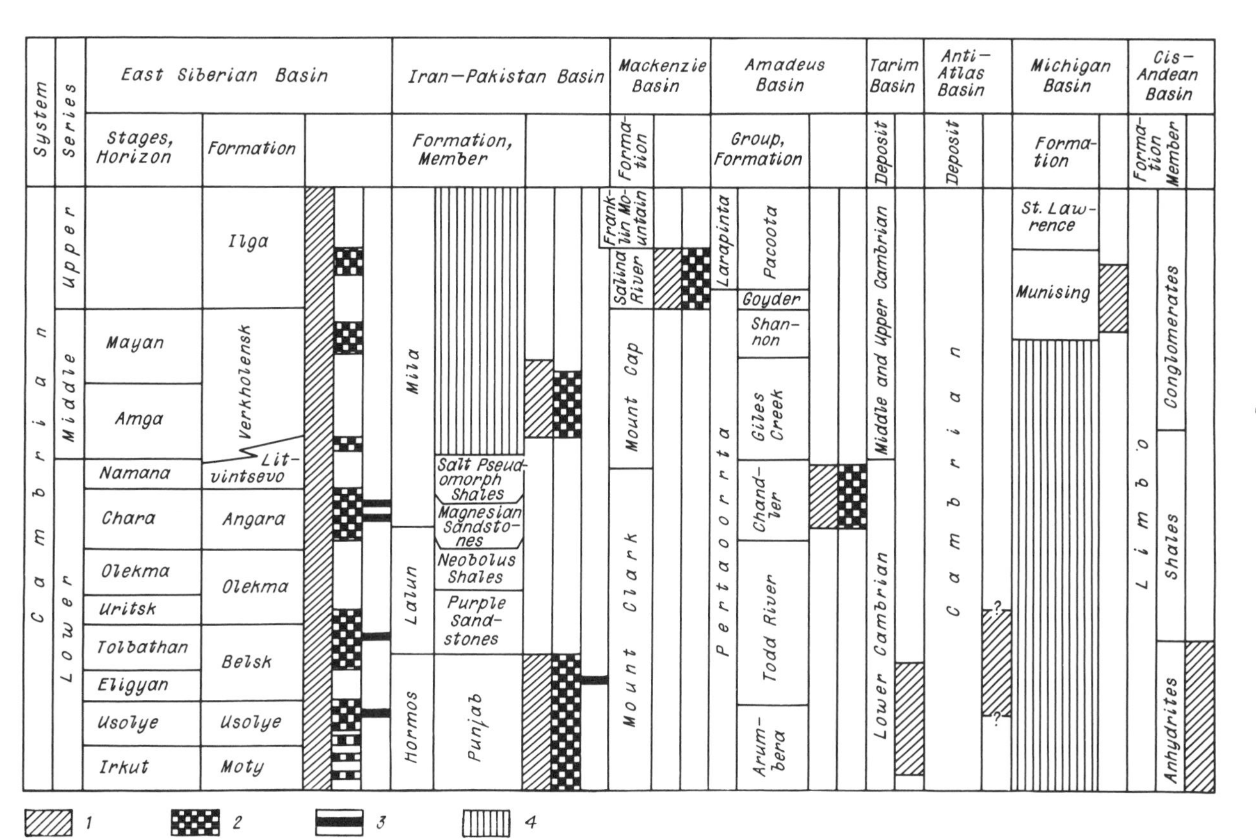

Fig. 17. Interregional correlation of sections in Cambrian evaporite basins.

Stratigraphic intervals and corresponding rocks: *1* sulfate, *2* rock salt, *3* potash salts, *4* no deposits of this age fossils and algae found in fragments from some salt domes (Kent 1970). Both the Middle Cambrian Anomocare trilobites and Riphean Conophyton stromatolites have been found there. The latter were obviously supplied by the displaced rock salt from the sub-salt beds which were overthrust on salt strata in some strongly deformed areas of the south-east Zagros. Anomocare suggests the Lower or Middle Cambrian age for Hormoz salts since limestone containing these fossils could either overlie the rock salt or alternate with it. These data make Precambrian age for the Hormoz salt, render the Punjab Series questionable, and suggest their Cambrian age.

Similar problems arise when the stratigraphic position of the lower (Irkutsk) salt sequence within the East Siberian Basin is discussed. This sequence is assumed to be Precambrian (Vendian, Yudomian), i.e., it is correlative with the Hormoz Formation and the Punjab Series assigned to the Infracambrian. In our opinion, a problem concerning the lower boundary of the Cambrian within the East Siberian and Iran-Pakistan Basins still remains disputable. It is much more correct to draw the lower boundary of the Cambrian not at the top, but at the base of the Irkutsk horizon, and thus we consider the Irkutsk salt sequence as Cambrian. In advocating this viewpoint it should be emphasized that we suggest Cambrian age for the Hormoz Salt and the Punjab Series in the Iran-Pakistan Basin.

A stratigraphic position of the salt-bearing Mila Formation in the Iran-Pakistan Basin has been established more precisely (Ruttner et al. 1968; Morgunov and Rudakov 1972). Near Shirhesth in Central Iran it occurs above the Lalun red beds. Taking this into account, as well as occurrences of trilobites, brachiopods and chiolitids, the age of the Mila Formation is determined as Middle/Late Cambrian. However, there are good grounds to believe that the lower part of this formation could be Early Cambrian.

It should be noted that the occurrence of this true Cambrian salt sequence in the northern Iran-Pakistan Basin casts stronger doubts on the Precambrian age of the Hormoz Formation. It might be that the Hormoz salts correspond, in fact, to the Mila salt unit. The Cambrian section in central and southern Iran may consist of several red sandstone units analogous to the Lalun Formation, which either occurs at the base of the Cambrian, i.e., below salt-bearing deposits, in the case of Shirhesth, or above salts, in the Cambrian. These upper red beds can be erroneously taken for the Lalun Formation, in which case the age of salt beds will be determined incorrectly. Probably, the red beds which in Shirhesth occur at the base of the section studied and are regarded as the Lalun Formation are in fact Precambrian (Vendian). All this points again to the debatable character of the Precambrian age of evaporites of the Punjab Series and the Hormoz Formation.

Within the East Siberian Basin a stratigraphic position of most salt-bearing deposits, except for the Irkutsk sequence, is rather precisely determined. Deposits overlying the Irkutsk salt sequence have the following stratigraphic position: the Usolje sequence corresponds to the Usolje horizon of the Lower Cambrian, the Belsk to the Tolbachan horizon of the Lower Cambrian, the Angara to the Chara horizon of the Lower Cambrian, the Litvintsevo to the Zeledeevo Horizon of the Middle Cambrian (Zharkov and Khomentovsky 1965; Chechel 1969). Two more sequences probably occur in the central Siberian Platform: in the Mayan stage of the Middle Cambrian and in the Upper Cambrian (Zharkov 1974a).

In the Mackenzie Basin the age of the Salina River salt sequence is not determined precisely. It is nonfossiliferous and its stratigraphic position is determined only by the mode of occurrence. The Salina River Formation rests unconformably on the Mount Cap Formation with a break at the base where the Lower and Middle Cambrian fossils have been found; thus, its age is determined as the Early/Middle Cambrian. The Salina River Formation rests unconformably on the Franklin Mountain Formation, the lower part of which is assigned to the Late Cambrian, and the upper part to the Early Ordovician. For these reasons most workers have lately placed the Salina River Formation in the lower Upper Cambrian (Aitken et al. 1973; Aitken and Cook 1974; Meijer-Drees 1975; Norford and Macqueen 1975).

The Chandler salt sequence developed in the Amadeus Basin occurs among the upper Lower Cambrian depósits. Its position is established owing to occurrence between the underlying Todd River Formation yielding Lower Cambrian archeocyatids, trilobites, brachipods, and chiolitids, and the overlying Giles Creek Formation where Middle Cambrian fossils have been found (Wells et al. 1967).

The above evidence on the stratigraphic position of the salt sequences in the Cambrian evaporite basins suggests that the age of most of them is now determined only to the rank of a series. The only exception is the East Siberian Basin; but there is a local stratigraphic chart which cannot be applied to other regions. Generally speaking, at present there is no unified subdivision for stages of the Cambrian, and thus the intercontinental correlation of the Cambrian evaporites can be more or less substantiated either within a series or within parts of a series.

The oldest Cambrian salt-bearing units are: the Irkutsk and Usolje in the East Siberian Basin, the Punjab and Hormoz in the Iran-Pakistan Basin. It can be assumed that all of them have a similar stratigraphic position, only the lower horizons of the salt-bearing sequence in the Iran-Pakistan Basin being, perhaps, older. Apparently, the formation of the salt-bearing sequences in the East Siberian and Iran-Pakistan Basins began almost contemporaneously and continued through the first half of the Early Cambrian.

The second half of the Early Cambrian was also characterized by halite accumulation within two basins: the East Siberian and Amadeus Basins. In the Iran-Pakistan Basin the salt-bearing deposits, apparently, did not accumulate at that time. The Chandler salt sequence of the Amadeus Basin occurs, probably, at the same stratigraphic level with the Angara salt sequence of East Siberia. Halite accumulation in these two widely spaced regions was contemporaneous during a considerable time interval in the second half of the Early Cambrian.

In the first half of the Middle Cambrian halite accumulation was recorded in the East Siberian and Iran-Pakistan Basins. Perhaps the Mila salt sequence should be taken as an equivalent of the Litvintsevo salt sequence. In the Late Cambrian a new evaporite basin took shape, the Mackenzie Basin, within which salt-bearing deposits of the Salina River Formation were formed. At the same time the salt-bearing beds must have also been accumulated in the central Siberian Platform.

If we consider the halite accumulation in the Cambrian in general, it might be concluded that during the Cambrian the rock salt strata were almost continuously being formed in this or that region. The data available do not allow us to distinguish shorter stages in the Cambrian during which salt accumulation could take place in all the salt basins or in most of them. No stages without halite accumulation on the Earth have been distinguished within the Cambrian. Thus, it follows that the Cambrian period was a single epoch of halite accumulation when during 115–120 m.y. salt basins always existed in one or another region of the Earth.

A similar conclusion can be drawn concerning the stages of sulfate accumulation in the Cambrian. In all the above evaporite basins sulfate rocks (anhydrite and/or gypsum) are developed not only at the same stratigraphic levels as rock salt, but even occur within a wider stratigraphic range. Thus, in the East Siberian Basin anhydrite is, in fact, recorded through the Cambrian section, beginning with the lowermost Lower Cambrian, Irkutsk horizon, and up to the Upper Cambrian. They are more widespread among the Lower and Middle Cambrian salt strata. As a rule, the lower and the upper part of the sequence is sulfate-carbonate, and the middle part salt in composition. This mode of occurrence of sulfate rocks is, for instance, typical of the Irkutsk, Belsk, Angara, and Litvintsevo sequences. Apparently, within the interior northern zones of the basin where the presence of the salt Mayan beds (Middle and Upper Cambrian) is assumed, coeval sulfate rocks will also occur. Even in the marginal south and south-western areas of the East Siberian Basin the interbeds and lenses of anhydrite and gypsum are observed among red and variegated, mostly terrigenous-carbonate deposits of the Middle and Upper Cambrian (the Verkholensk and Ilga Formation). Thus, it can be stated quite definitely that the East Siberian Basin was evaporitic throughout the entire Cambrian, while sulfate accumulation was periodical.

Sulfate rocks within the Iran-Pakistan Basin were recorded in the Hormoz, Punjab, and Mila salt sequences. They are spread throughout the entire Cambrian section, except for the Lalun red beds and the upper Mila Formation, assigned to the Upper Cambrian. In the Mackenzie Basin anhydrite (and gypsum) are confined to the Salina River Formation, i.e., occur at the level of the lower Upper Cambrian. In the Amadeus Basin sulfate rocks are found within the Chandler salt sequence, assigned to the upper Lower Cambrian.

Additionally, sulfate deposits occur in the other four Cambrian evaporite basins: Tarim, Anti-Atlas, Cis-Andean, and Michigan Basins. A stratigraphic position of these sulfate sequences is not quite clear. It can only be pointed out that in the Tarim Basin gypsum and anhydrite are apparently confined to the Lower Cambrian deposits, and in the Anti-Atlas Basin these rocks occur among the strata belonging to undifferentiated Cambrian formations. In the Cis-Andean Basin anhydrite is confined to the lower Limbo Formation where all series of the Cambrian are represented. The sulfate unit of the Limbo Formation is, probably, of Early/Middle Cambrian age. In the Michigan Basin an anhydrite sequence is placed into the Munising Group (Upper Cambrian). Summarizing data on stratigraphic position of sulfate sequences, it can be pointed out that sulfate accumulation on the Earth during the Cambrian was almost continuous. Thus, the Cambrian period is distinguished as a single epoch of halite and sulfate accumulation.

It is much more difficult to establish stages of the potash salt formation in Cambrian time. Neither economical potash salt deposits, nor even large amounts, have as yet been found in the salt strata of any of the Cambrian evaporite basins. The potash salt rocks of chloride type (halite-sylvine, sylvine, sylvine-carnallite, and carnallite) were recorded in the East Siberian Basin in the course of prospecting for potash salt rocks carried out there from 1962 to 1972 (Britan 1971; Geology and Potassium Bearing of the Cambrian Deposits . . . 1974). They occur at three stratigraphic levels: (1) the upper Usolje Formation, (2) in the salt sequence of the Belsk Formation, and (3) in the Angara Formation. The thickness of potash salt beds is not great (less 0.5 m). In the southern Siberian Platform potash rocks are confined to four potash salt horizons: Usolje, Belsk, Tynys, and Troitskiy. The last two horizons occur within the Angara Formation. In the other salt sequences of the East Siberian Basin neither potash salt rocks, nor potassium mineralization have as yet been found.

In the Iran-Pakistan Basin "dissipated" potassium mineralization has also been recorded. Potassic minerals are found among the "khalar" interbeds (rock salt with inclusions of red marl) in the Punjab Series in the Khevra, Warcha, and Kalabagh fields in the Salt Range of Pakistan (Asrarullah 1963) where sylvine, langbeinite, kainite, polyhalite, and kieserite occur (Ivanov and Voronova 1972).

Potash salts have not been recorded from other Cambrian salt basins (Mackenzie and Amadeus basins).

This evidence and, first of all, the results of potash salt exploration in the East Siberian Basin suggest that in Cambrian time conditions were favorable for potash accumulation observed in a number of salt basins. However, the periodicity of potash accumulation can be roughly outlined only for the Early Cambrian in the East Siberian Basin. Three times intervals of potash sedimentation are distinguished there: Usolje, Belsk, and Angara. It is most likely that they mark an Early Cambrian stage of potash accumulation, especially as it is also outlined within the Iran-Pakistan Basin, i.e., it is traced within both salt basins which existed in Eurasia.

The data available show that during the Middle and Late Cambrian accumulation of potash salts need not have taken place in all the salt basins. Thus, we distinguish the Middle/Late Cambrian potassium-free stage. However, this is just a tentative conclusion because of poor knowledge of the Middle and Late Cambrian salt-bearing deposits.

Stratigraphic Position of Ordovician Evaporites

In spite of the lack of a universally accepted subdivision of the Ordovician into series and stages and the debates about its lower and upper boundaries, the information available allows us to make more or less general correlation of the sections within most of the basins of evaporite sedimentation (Fig. 18). However, a number of problems dealing both with the stratigraphic position of halogenic strata within discrete basins and the correlation of sections in different regions remain unsolved and should be discussed separately.

The correlation of the Ordovician salt sequences is quite vague. This is due to the fact that it is only the stratigraphic position of the Bay Fiord sequence in the Basin of the Canadian Arctic Archipelago that has been determined rather precisely out of the four Ordovician sequences distinguished. The age of the remaining three salt units (Stonewall in the Williston Basin, the lower and the upper sequences of the Carribady

Formation in the Canning Basin) remains as yet so indefinite that even their Ordovician age seems questionable.

Most doubtful is the Ordovician age of salt sequences of the Canning Basin. In this basin the Carribady evaporites are overlain directly by thick Tandalgoo red beds tentatively assigned by Johnstone et al. (1967) to the Lower Devonian. The Middle Devonian Melinjerie Formation yielding abundant fossils occurs above these beds. The boreholes do not penetrate to the base of evaporites; thus, their relation with the underlying sequence remains uncertain. Johnstone et al. (1967), following Veevers (1967), assume that evaporites are underlain by the Middle and Lower Ordovician strata deposited along the margins of the Canning Basin. Proceeding from this assumption, they conclude that a stratigraphic position of the Carribady evaporites can fall within a range from the Upper Ordovician to the Lower Devonian. More precise determinations of their age are not available at present. Such an approach does not allow us to establish to what system (Ordovician, Silurian, or Devonian) not only the salt sequences, but the entire evaporite Carribady Series should be assigned, to say nothing of a series or a stage. However, these are not the only difficulties arising when the age of the above evaporites is being determined. Some investigators assumed that they might even be Precambrian and it would be possible to correlate them with the Bitter Springs salt sequence of the Amadeus Basin (Koop 1966). In our opinion, there might be another viewpoint according to which the Carribady evaporites are of the Cambrian age and should not be correlated with the Bitter Springs salt series, but with the Cambrian Chandler salt sequence of the Amadeus Basin.

The above evidence implies that at present the age of the salt sequence of the Carribady Formation can be determined only tentatively. The Late Ordovician age of these beds seems more likely at present. It has already been mentioned that this assumption is not well grounded. But if we assume that the Carribady Formation is 1700 m thick an falls within a range from the Late Ordovician to the Lower Devonian, then its lower part, to which both salt units are confined, might quite well be the Late Ordovician. The upper part of the formation, Stottid Member, composed mostly of variegated mudstone with interbeds, lenses, and inclusions of anhydrite, is, apparently, Silurian in age. It is this very point of view that is illustrated in the interregional correlation chart (see Fig. 18).

A stratigraphic position of the Stonewall salt sequence tentatively distinguished within the Williston Basin is also disputable. Most workers regard it as assigned to the upper unit of the Bighorn Group of the Late Ordovician, but others suggest Ordovician/Silurian for the Stonewall Formation (Anderson and Eastwood 1968). In this case it is excluded from the Bighorn Group and is often united with overlying Silurian beds of the Interlake Formation, which results in the lithological sequence of a uniform composition. There the Stonewall Formation is assigned to the Ordovician.

The Bay Fiord salt sequence in the Basin of the Canadian Arctic Archipelago is rather definitely dated as the upper Middle Ordovician. Such a stratigraphic position is suggested by its occurrence at the base of the Cornwallis Group, just above the Eleanor River Formation ranging in age from the Upper Arenigian to the Lower Llandeilo, and below the Thumb Mountain and Irene Bay Formation which are, apparently, Middle/Upper Caradocian. The age of the Irene Bay Formation was determined with the greatest degree of certainty. It yields Upper Caradocian faunal assemblage

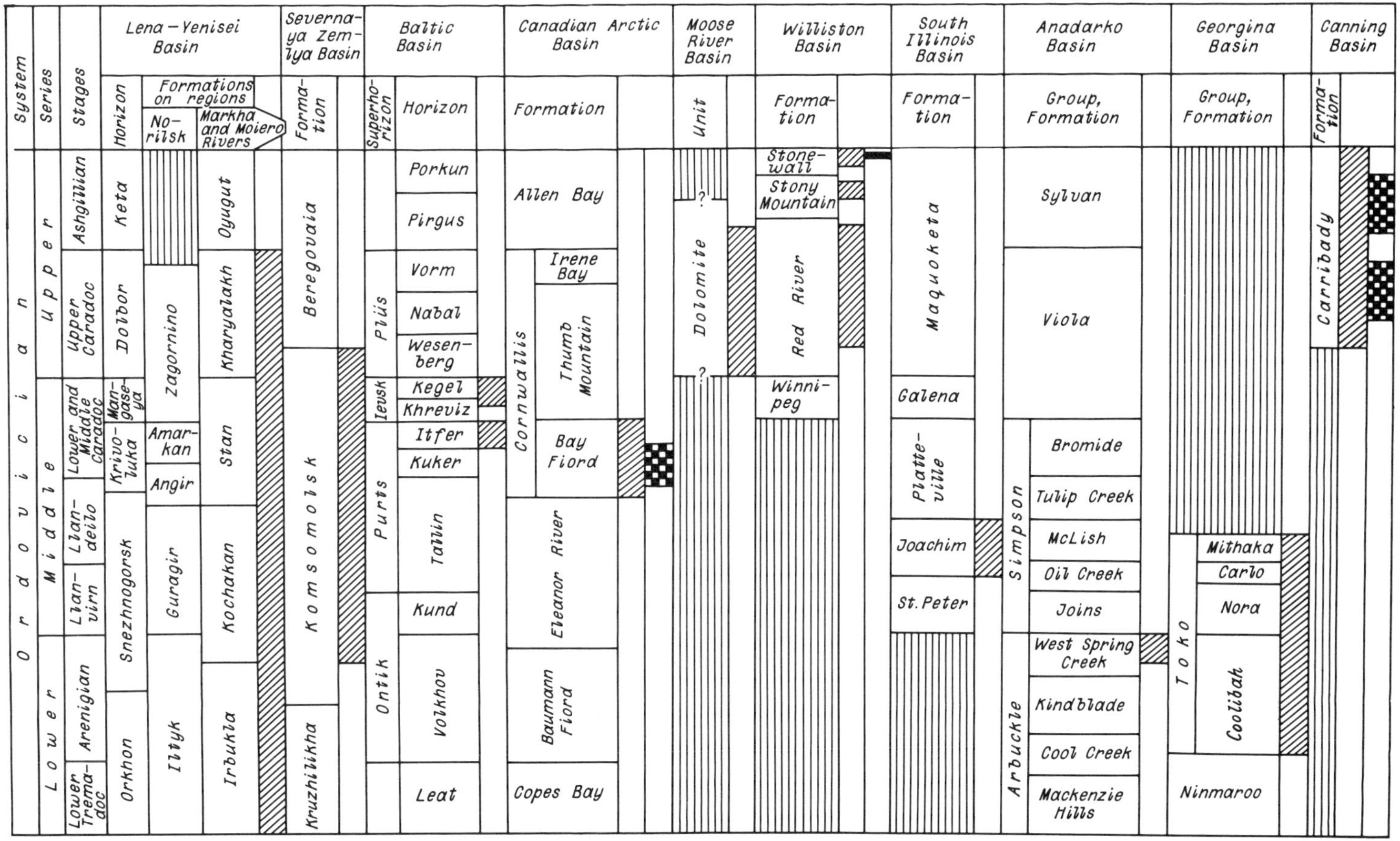

System: Ordovician

Series: Lower; Middle; Upper

Stages: Lower Tremadoc; Arenigian; Llanvirn; Llandeilo; Lower and Middle Caradoc; Upper Caradoc; Ashgillian

Lena–Yenisei Basin
- Horizon: Orkhon; Snezhnogorsk; Krivoluka; Mangaseya; Dolbor; Keta
- Formations on regions – Norilsk: Iltyk; Guragir; Angir; Amarkan; Zagornino
- Formations on regions – Markha and Moiero Rivers: Irbukla; Kochakan; Stan; Kharyalakh; Oyugut

Severnaya Zemlya Basin
- Formation: Kruzhilikha; Komsomolsk; Beregovaia

Baltic Basin
- Superhorizon: Ontik; Purts; Ievsk; Plüs
- Horizon: Leat; Volkhov; Kund; Tallin; Kuker; Itfer; Khreviz; Kegel; Wesenberg; Nabal; Vorm; Pirgus; Porkun

Canadian Arctic Basin
- Formation: Copes Bay; Baumann Fiord; Eleanor River; Cornwallis (Bay Fiord; Thumb Mountain; Irene Bay); Allen Bay

Moose River Basin
- Unit: ?; Dolomite; ?

Williston Basin
- Formation: Winnipeg; Red River; Stony Mountain; Stonewall

South Illinois Basin
- Formation: St. Peter; Joachim; Platteville; Galena; Maquoketa

Anadarko Basin
- Group, Formation: Arbuckle (Mackenzie Hills; Cool Creek; Kindblade; West Spring Creek); Simpson (Joins; Oil Creek; McLish; Tulip Creek; Bromide); Viola; Sylvan

Georgina Basin
- Group, Formation: Ninmaroo; Toko (Coolibah; Nora; Carlo; Mithaka)

Canning Basin
- Formation: Carribady

Fig. 18. Interregional correlation of sections in Ordovician evaporite basins. Symbols are the same as in Fig. 17

which allows us to draw the Caradocian/Ashgillian boundary at its top (Kerr 1967a, 1968, 1974; Thorsteinsson and Kerr 1968). This evidence, in spite of lack of fossils in the Bay Fiord Formation itself, allows us to place it into the Upper Llandeilian/ Lower Caradocian. At the same time, it is most likely that the Bay Fiord evaporites in the deepest interior parts of the basin occupy a considerable stratigraphic interval, the lower Middle Ordovician.

Thus, the data on the stratigraphic position of the Ordovician salt sequence allows us to conclude that the halite accumulation in the Ordovician began only at the end of the Llandeilo and was probably discontinuous up to the Late Ordovician. During all this time the rock salt strata were being formed episodically within separate basins, but not contemporaneously on different continents. Originally, the Canadian Arctic Basin was a salt basin; it disappeared by the end of the Middle Ordovician, then salt strata could have accumulated in the Canning Basin, and at the end of the Late Ordovician in the Williston Basin. Proceeding from this, it is quite possible to distinguish a stage which lasted from the end of the Llandeilo to the end of the Late Ordovician, as an episodical salt period. The Early Ordovician and the lower Middle Ordovician were the time when salt accumulation did not take place in any of the evaporite basins of the Earth. At present we call it a salt-free stage.

The above data suggest that potash salts have not yet been recorded in any of Ordovician salt sequences. Maybe they are poorly studied. At the same time it is highly probable that in the Ordovician salt basins, which existed episodically, conditions were not favorable for potassium accumulation. The information available makes us regard the Ordovician in general as the time when potash salts were not deposited. This stage lasted probably from the Middle Cambrian to the Late Cambrian, and apparently to the Ordovician.

Unlike halite and potassium accumulation, sulfate sedimentation was rather extensive in Ordovician time. Sulfates were being deposited in all the Ordovician evaporite basins where 18 sulfate units were formed.

Most sulfate units are confined to the Lena-Yenisei Basin. The basin being not well known, it is impossible to determine the number and stratigraphic position of the sequences precisely. Relatively thick sulfate strata were observed in the northwest of the basin, in the Norilsk Region. There sulfate rocks are developed both among the Lower (Iltyk Formation) and Middle/Upper Ordovician (Guragir, Angir, Amarkan, and Zagornino Formations) deposits; thus it is possible to unite all these rock units into a single sulfate sequence. A wide stratigraphic range of sulfate rocks is also recorded in the Vilyui, Markha, Morkoka, and Moiero Rivers Basins where sulfate sequences are confined to the Lower and Middle (Irbuklin, Kochakan, and Stan Formations) and Upper Ordovician as well (Kharyalakh Formation). In other regions of the Lena-Yenisei Basin sulfate deposits fill within a narrower stratigraphic interval. Thus, in the central Tunguska Syneclise they have been established only among Lower Ordovician deposits (sulfate-dolomite strata), whereas in the Berezovaya Depression they are confined to the Tochilnaya Formation, the upper part of which is of Lower Ordovician age. Within the Irkutsk Amphitheater interbeds and lenses of gypsum and anhydrite have been recorded in the Bratsk Formation (Middle/Late Ordovician). This evidence allows us to conclude that sulfate accumulation within the Lena-Yenisei Basin was continuous during most of the Ordovician.

In the Severnaya Zemlya Basin a stratigraphic position of sulfate sequences is not yet clear. They comprise dolomite-marly and gypsum-limestone members of the Komsomol Formation which, because of their occurrence above a horizon with Angarella, belong mostly to the Middle and partly to the Upper Ordovician (Egiazarov 1970). Sulfate rocks in the Baltic Basin occur in the Middle Ordovician. Their lowermost (Tallin) sulfate sequence is probably Llanvirnian, and the two upper units (Itfer and Ievsk) Caradocian (Selivanova 1971). Gypsification of rocks is also observed in the Porkun horizon.

Within the Basin of the Canadian Arctic Archipelago sulfate deposits are developed at two stratigraphic levels. The lower Baumann Fiord sulfate sequence is assigned to the Early Ordovician (Arenigian) (Mossop 1973) on the basis of brachiopods found in limestone of unit "B" occurring between two anhydrite units "A" and "C". It has already been noted that sulfate rocks confined to the Bay Fiord salt sequence are placed in the upper part of the Middle Ordovician (Kerr 1974, 1975).

In the Moose River Basin sulfate rocks (gypsum and anhydrite) are confined to an undifferentiated dolomite sequence. Fossils found in this sequence allow us to correlate them with those of Bad Cache Rapids Group occurring to the north within the Hudson Basin (Sanford et al. 1968; Norris and Sanford 1969; Cumming 1975). Fossils found in the lower Bad Cache Rapids Group appear to be identical to those of the Red River Formation in the Williston Basin, characteristic of the Upper Caradocian (Nelson 1964). Thus, the dolomite sequence of the Moose River Basin is mostly Late Ordovician, its greater lower part belonging to the Caradocian and the upper part to the Ashgillian.

Cumming (1975) demonstrated that Ordovician basins – Moose River, Hudson, and Williston – were formed within a single climatic zone, the northern arid zone, and probably were connected with each other. A similarity in lithology of sediments filling the basins accounts for a comparatively easy correlation. The section of the Williston Basin is more complete than that of the Moose River Basin. There are, besides the Late Caradocian sulfate Red River sequence, two overlying sequences (Stonewall and Stony Mountain) assigned to the Ashgillian.

In the South Illinois Basin sulfate rocks are confined to the Joachim Formation which is of Llanvirnian/Llandeilo age. The sulfate West Spring Creek Formation in the Anadarko Basin is regarded as the Upper Arenigian. Within the Georgina Basin gypsum in the form of inclusions and lenses occurs through the entire Toko Group section ranging from the Early to Middle Ordovician.

The above data show that sulfate accumulation was continuous in the Ordovician period, sometimes being contemporaneous in several evaporite basins. Thus, in the Early Ordovician sulfate was deposited in North Eurasia, in the Lena-Yenisei Basin, in central North America in the Anadarko Basin, and in Australia in the Georgina Basin. In many evaporite basins (Lena-Yenisei, Severnaya Zemlya, Baltic, Canadian Arctic, South Illinois, and Georgina) sulfate rocks accumulated during the Middle Ordovician. In the Late Ordovician sulfate accumulation was recorded in the Lena-Yenisei, Moose River, Williston, and, probably, Canning Basins. All this points rather definitely to a continuous sulfate accumulation in evaporite basins of the Earth in the Ordovician. The Ordovician period in this respect differs but slightly from the Cambrian, and, thus, it might be concluded that the Cambrian and the Ordovician constituted a single epoch in the history of the Paleozoic sulfate accumulation.

Stratigraphic Position of Silurian Evaporites

The Silurian period is characterized by a rather limited salt accumulation. Only one Silurian salt basin is definitely distinguished: the Michigan-Pre-Appalachian Basin. The remaining evaporite basins are regarded as sulfate basins, though in two of them (Dniester-Prut and Lena-Yenisei) salt deposits might have been accumulated.

Figure 19 shows the correlation chart of sections within the Silurian evaporite basins. Let us discuss this chart in detail and try to distinguish stages of halite, potassium, and sulfate accumulation.

The salt strata of the Michigan-Pre-Appalachian Basin are Silurian and assigned to the Salina Group, its stratigraphic position is not quite clear, yet may be due to the different interpretations of the age of Devonian/Silurian boundary beds. According to the latest assumption (Berry and Boucot 1970), the Salina Series occur in the Přidolian (Silurian) in Europe which is distinguished as a new super-Ludlovian unit of the Silurian (Sokolov and Polenova 1968), as well as the upper part of the Ludlovian. The Ludlovian/Přidolian boundary is drawn approximately either at the top or in the middle of the Vernon Shale. Thus, it may be assumed that the salt-bearing sediments in the Michigan-Pre-Appalachian Basin are of the Late Ludlovian/Přidolian age.

Late Ludlovian/Přidolian was the only time interval in the Silurian when halite accumulation took place on the Earth. However, this stage can now be regarded as only an episodical salt stage, since halite accumulation is known only from one region – the Michigan-Pre-Appalachian Basin in North America, whereas on other continents it was not recorded. One should bear in mind that if the assumption concerning the presence of Late Silurian salt strata in the Lena-Yenisei Basin is confirmed, the Late Ludlovian/Přidolian will be regarded as a stage of salt accumulation on the globe. The data available show that no salt was deposited in any evaporite basins during the Early Silurian and most of the Ludlovian age (Late Silurian). Proceeding from these data a salt-free Llandoverian/Early Ludlovian stage can be almost definitely determined in the history of the Paleozoic halite accumulation.

The data concerning stages of potassium accumulation in the Silurian are as follows. Within a single salt basin of the Silurian – the Michigan-Pre-Appalachian Basin – potash salts (sylvine rocks) are found in the lowermost part (subunit A_1) of the Salina Formation (Matthews 1970). The thickness of the potash salt distribution zone is 28 m, its area amounting to about 36,000 km^2. Judging by its stratigraphic position, the potash zone probably occurs in the upper part of the Ludlovian. Since potash salts have not been recorded from other stratigraphic levels within salt sequences of the Michigan-Pre-Appalachian Basin, we distinguish there only one, Late Ludlovian stage of potassium accumulation. It is quite natural that the Late Ludlovian potassium stage cannot be traced in other evaporite basins of the then not salt basins. Thus, in the history of potassium accumulation the Late Ludlovian stage was only a time of episodic potassium accumulation. The Silurian period might now be subdivided into three parts as to potassium sedimentation: Llandoverian/Early Ludlovian potassium-free time, Late Ludlovian episodical potassium time, and Přidolian time which is potassium-free. Llandoverian/Early Ludlovian time crowned the preceding potassium-free stage which began in the Middle Cambrian, continued through the Ordovician into the Early Silurian, and terminated in the second half of the Ludlovian

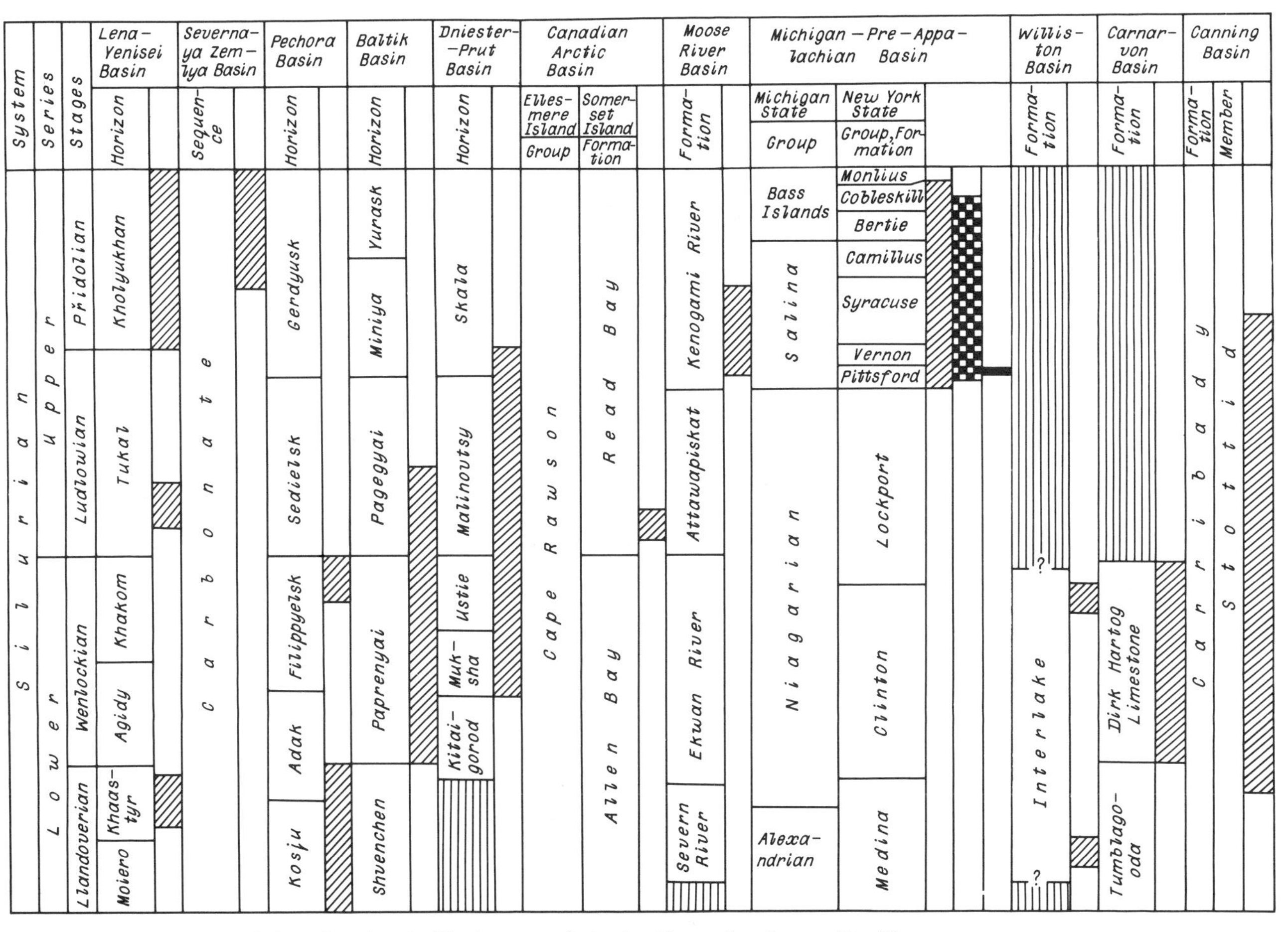

Fig. 19. Interregional correlation of sections in Silurian evaporite basins. For explanation see Fig. 17

age (Late Silurian). In its turn, a new potassium-free stage began with the Přidolian which, as will be shown below, continued into the Early Devonian.

Many more data are available to allow us to determine stages of sulfate accumulation in the Silurian, in spite of the fact that a stratigraphic position of many sulfate sequences in the evaporite basins remains not clear.

Recently the age position of sulfate formations in the Lena-Yenisei Basin has been determined. The results obtained by Yu.I. Tesakov, N.N. Predtechensky, and many others were used as a basis for a unified stratigraphic chart for the Silurian of the Siberian Platform which is compared to the unified stratigraphic chart of the Silurian. The following horizons have been distinguished as the regional age time units: Moiero, Khaastyr, Agidy, Khakom, Tukal, and Kholyukhan. The Moiero and Khaastyr, the Agidy and Khakom, the Tukal, Kholyukhan horizons are assigned to the Llandoverian, Wenlockian, Ludlovian, and Přidolian, respectively, Additionally, within some regions formations were distinguished which were given names of their own. It appeared that three Silurian sulfate units can be distinguished within the Lena-Yenisei Basin: the Utokan, Kongda (or Yangada) and Kholyukhan. The Utokan sequence occurs in the Khaastyr horizon and is of Late Llandoverian age. Sulfate rocks of the Kongda Formation are confined to its lower part within which a sulfate sequence of the same name is recognized; these beds occur in the lower Tukal horizon and are assigned to the Ludlovian. The Kholyukhan sequence is of Přidolian age. Proceeding from the data available we can outline in the first approximation the periodicity of sulfate accumulation in the Lena-Yenisei Basin. The accumulation of sulfate rocks took place during the Late Llandoverian, Lower Ludlovian and the Přidolian. The main regions of sulfate sedimentation in the Llandoverian were the Nyuya and Berezovaya Depressions, and in the Late Silurian the northern and central Siberian Platform.

Within the Severnaya Zemlya Basin the sulfate strata occur among the uppermost Silurian carbonate deposits. Such a stratigraphic position is determined on the basis of fossils found in underlying rocks below a Ludlovian gypsum unit (Egiazarov 1970). It can be assumed that, according to the recent subdivision into stages for the Silurian, these strata will occur in the Pridolian stage (see Fig. 19).

Both sulfate units in the Pechora Basin are apparently confined to the Lower Silurian. Their age can be determined only approximately for lack of detailed stratigraphic study of the sulfate dolomite series in the Upper Pechora Basin, and their correlation with the more north-eastern and eastern sequences where the regional stratigraphic horizons are distinguished (Pershina 1962, 1966) is carried out in different ways. The evidence available suggests that the lowermost sulfate sequence found in the Pechora Basin occurs at the base of the Silurian, probably in the Llandoverian (Filippova 1973). In the Pechora Basin and in the North Urals the Kosju horizon and the lower Adak horizon are also Llandoverian. Since the lower sulfate sequence is nameless, we shall name it Kosju Adak for the convenience of the descriptions. The second sulfate sequence is recorded in the upper Wenlockian (Filippova 1973); it corresponds to the Filippjelsk horizon of the Pechora Basin. The sequence is named after the horizon.

A stratigraphic position of sulfate deposits in the Baltic Basin has been rather reliably determined (Pashkevichius 1961; Gailite et al. 1967; Silurian of Estonia 1970;

Aaloe et al. 1976). They are confined to the Paprenyai horizon and to the lower Pagegyai horizon, i.e., they occur in the Upper Wenlockian and Lower Ludlovian. Thus, sulfate strata in the Baltic Basin are of Early/Late Silurian age.

Presently there is no adequate stratigraphic chart for the Silurian of the Dniester-Prut Basin and for the Podolia. There sulfate beds occur in the Muksha, Ustje, Malinovtsy horizons and the lower Skala horizons. The age of these horizons is rather disputable. Most investigators (Stratigraphy of Sedimentary . . . 1964; Edelstein 1969; Tesakov 1971; Reference Section . . . 1972) assign the Muksha and the Ustje horizons to the Wenlockian. However, Tsegelnyuk (1976) places them into the lower Ludlovian. According to Nikiforova et al. (Reference Section . . . 1972) and Yu.I. Tesakov the Malinovtsy horizon is the lower Ludlovian, whereas Tsegelnyuk (1976) regards it as the Upper Ludlovian, and Edelstein (1969) as the upper Wenlockian stage. In a few years the Skala horizon was either wholly placed above the Ludlovian, into the Přidolian (Upper Silurian) (Tsegelnyuk 1976), or was assigned to the Upper Ludlovian, or to the Přidolian (Tesakov 1971; Reference Section . . . 1972). But, in spite of all these contradictions, evaporite deposits of the Dniester-Prut Basin, as a whole, are definitely regarded as the Wenlockian/Ludlovian. They were formed during the second half of the Early Silurian and the first half of the Late Silurian.

Within the Canadian Arctic Archipelago sulfate rocks confined to the lower part of the Read Bay Formation probably occur on Somerset Island at the same level as those of the Cape Rawson Group on Ellesmere Island. The Read Bay Formation is mostly of the Late Silurian age, only its uppermost strata being Devonian (Fortier et al. 1963; Christie 1964; Berry and Boucot 1970). Thus, it can be assumed that the gypsum sequences of the Read Bay Formation and of the Cape Rawson Group are of Ludlovian age.

The Kenogami River Formation within the Moose River Basin is subdivided into three units: the lower one, composed of dolomite with interbeds of white gypsum or anhydrite; the middle one, represented by red and green gypsinate marl, siltstone, sandstone, dolomite, and limestone; and the upper, Early Devonian, composed of dolomite and dolomite breccia with interbeds of gypsum and anhydrite (Sanford et al. 1968; Norris and Sanford 1969). Thus, sulfate rocks are distributed through the Kenogami River Formation. Its lower and middle parts comprise Late Silurian deposits known as Cayugan; it is mainly correlative with the Přidolian of Europe (Berry and Boucot 1970). These parts of the Kenogami River Formation are correlated with the Salina and Bass Island groups of the Michigan-Pre-Appalachian Basin, which confirms its mainly Přidolian age. It is the lower Kenogami River Formation that occurs in the Upper Ludlovian. This stratigraphic evidence suggests that the Kenogami River sulfate sequence formed at the close of the Ludlovian and Přidolian ages (Late Silurian and lower Lower Devonian).

Most investigators place the Interlake sulfate sequence in the Williston Basin in the Niagaran stage using the stratigraphic standard of the U.S.A. This stage covers approximately the Upper Llandoverian, the Wenlockian, and the lower Ludlovian. Since the section of the Interlake Formation in the Williston Depression is incomplete, we can suggest Llandoverian/Wenlockian age. The sulfate sequence in the Carnarvon Basin may be only tentatively assigned to the Wenlockian. Within the Canning Basin the sulfate-bearing deposits of the Stottid sequence may be only conventionally regarded as Silurian.

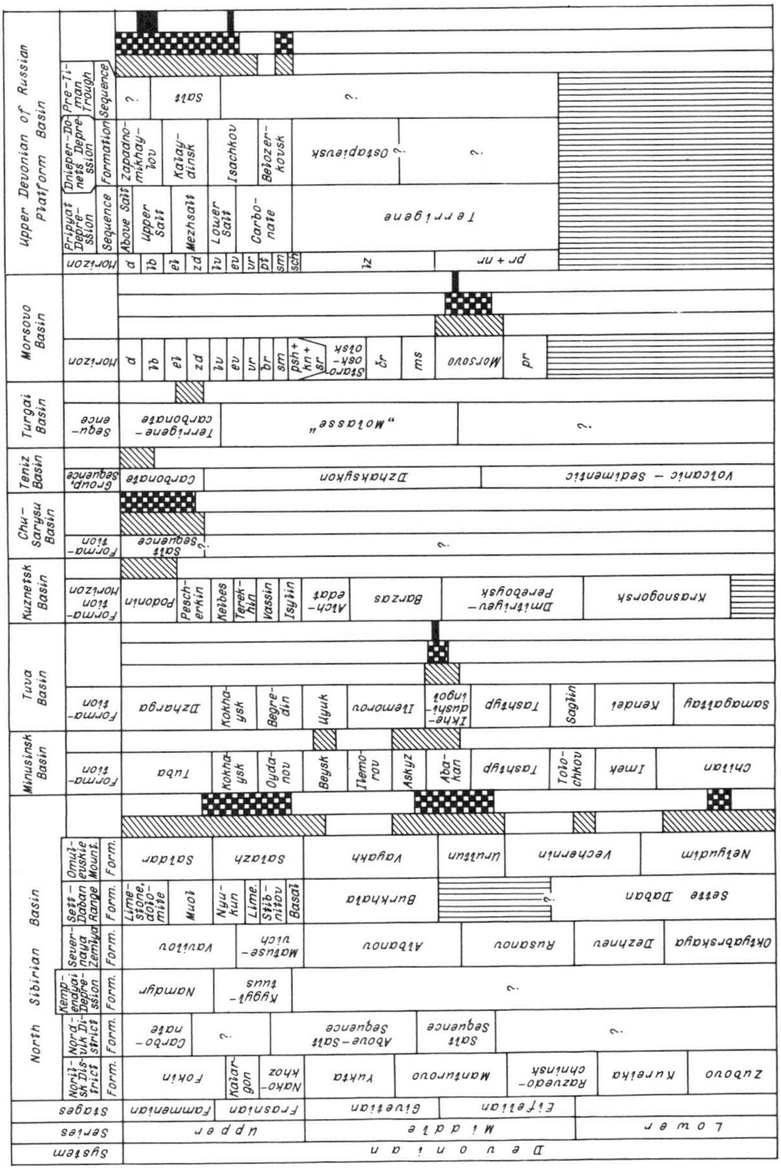

System: Devonian
Series: Lower; Middle; Upper
Stages: Eifelian; Givetian; Frasnian; Fammenian

North Sibirian Basin
Norilsk District, Form.: Zubovo; Kureika; Razvedochninsk; Manturovo; Yukta; Nakokhoz; Kalargon; Fokin
Nordvik District, Form.: ?; Salt Sequence; Above-Salt Sequence; ?; Carbonate
Kempendyai Depression, Form.: ?; Kygyltuus; Namdyr
Severnaya Zemlya, Form.: Oktyabrskaya; Dezhnev; Rusanov; Albanov; Matusevich; Vavilov
Sett-Daban Range, Form.: Sette Daban; ?; Burkhala; Basal; Stibnitov; Lime.; Nyukun; Muol; Limestone, dolomite
Omulevskie Mount., Form.: Nelyudim; Vechernin; Urultun; Vayakh; Salazh; Saldar

Minusinsk Basin, Formation: Chilan; Imek; Tolochkov; Tashtyp; Abakan; Askyz; Ilemorov; Beysk; Oydanov; Kokhaysk; Tuba

Tuva Basin, Formation: Samagaltay; Kendei; Saglin; Tashtyp; Ikhedushiingol; Ilemorov; Uyuk; Begredin; Kokhaysk; Dzharga

Kuznetsk Basin, Formation, Horizon: Krasnogorsk; Dmitriyev-Pereboysk; Barzas; Alchedat; Isylin; Vassin; Terekhin; Kelbes; Peschеrkin; Podonin

Chu-Sarysu Basin, Formation: ?; Salt Sequence ?

Teniz Basin, Group, Sequence: Volcanic – Sedimentic; Dzhaksykon; Carbonate

Turgai Basin, Sequence: ?; "Molasse"; Terrigene-carbonate

Morsovo Basin, Horizon: pr; Morsovo; ms; čr; Staroosk-olsk; psh + kn + sr; sm; br; vr; ev; lv; zd; el; lb; d

Upper Devonian of Russian Platform Basin
Horizon: pr + nr; lz; sch; sm; bt; vr; ev; lv; zd; el; lb; d
Pripyat Depression, Sequence: Terrigene; Carbonate; Lower Salt; Mezhsalt; Upper Salt; Above Salt
Dnieper-Donets Depression, Formation: ?; ?; Ostapievsk; Belozerkovsk; Isachkov; Kalaydinsk; Zapadnomikhaylov
Pre-Timan Trough, Sequence: ?; Salt; ?

Fig. 20. Interregional correlation of sections in Devonian evaporite basins. Horizons: *pr* Pyarnu; *nr* Narova; *ms* Mosolov; *cr* Chernoyarsk; *lz* Luzhsk; *psh* Pashiysk; *kn* Kynov; *sr* Sargaev; *sm* Semiluki; *br* Bureg; *vr* Voronezh; *ev* Evlanov; *lv* Liven; *zd* Zadonsk; *el* Elets; *lb* Lebedyan; *d* Dankov. The same symbols as in Fig. 17

A proposed stratigraphic position for the sulfate sequences allows us to conclude that during the Silurian the accumulation of sulfate deposits was continuous in one of the Silurian evaporite basins. Thus, during the Llandoverian age sulfate was deposited in three basins: Lena-Yenisei, Pechora, and Williston; during the Wenlockian age in five basins: Pechora, Baltic, Dniester-Prut, Williston, and Carnarvon; during the Ludlovian age also in five basins: Michigan-Pre-Appalachian, Lena-Yenisei, Baltic,

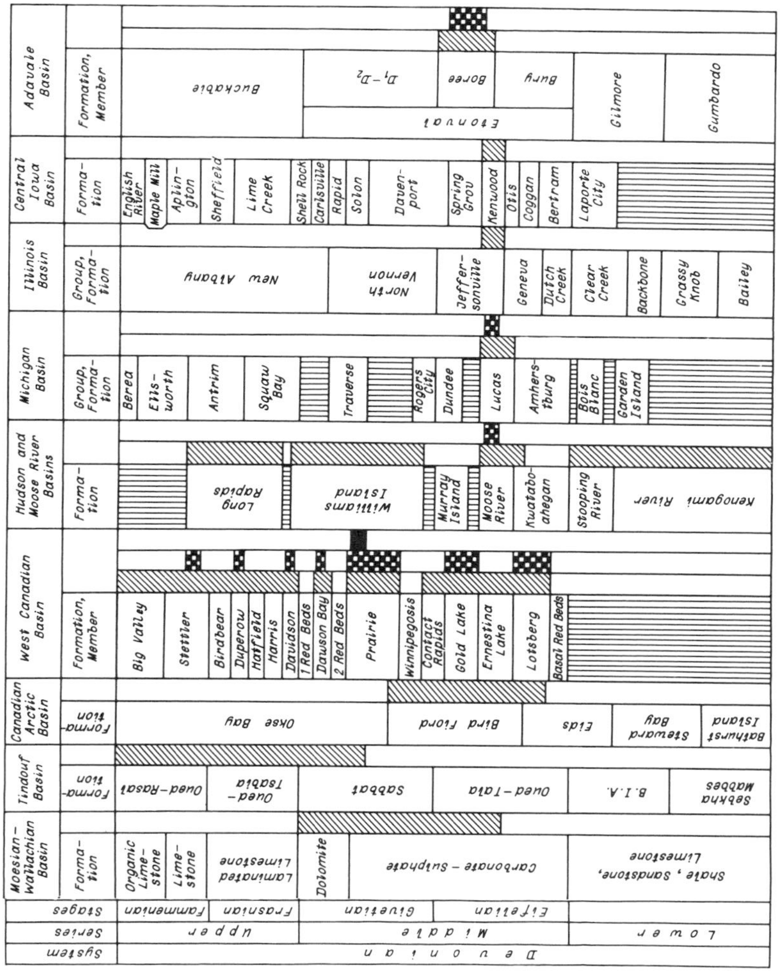

Fig. 20 (continued)

Dniester-Prut, and the Canadian Arctic Archipelago. During the Přidolian age sulfate rocks accumulated in four basins: Michigan-Pre-Appalachian, Lena-Yenisei, Severnaya Zemlya, and Moose River. It should be borne in mind that during most of the Silurian sulfate accumulation took place, probably, within the Canning Basin. Thus, it can be stated that the Silurian period was the time of almost continuous sulfate deposition; together with the Cambrian and the Ordovician it is one single epoch of sulfate sedimentation.

Stratigraphic Position of Devonian Evaporites

It has already been established that 75 halogenic units were formed in 19 Devonian evaporite basins, namely 26 salt and 49 sulfate units. Figure 20 shows their stratigraphic position and interregional correlation of composite sections of the evaporite basins.

Let us first discuss the peculiar age position of rock salt strata and stages of halite accumulation for the Devonian. Only one salt unit has been found as yet; it is the Zubovo Sequence in the North Siberian Basin, presumably of Early Devonian age. This discovery makes us change our previous assumptions that the Early Devonian was a salt-free stage (Zharkov 1974a,b, 1976, 1977). This was, apparently, an episodically salt-bearing time, since salt accumulation took place only within one evaporite basin. But this conclusion is not decisive. Some corrections may be introduced in the course of further studies, since the age of the Zubovo salt sequence is still uncertain; most likely it should be assigned to the Late Silurian.

The extensive halite accumulation in the Devonian actually began in the Eifelian age (Middle Devonian) when salt deposition commenced on some continents.

True Eifelian salt sequences are known as yet only from North America, having been found in the West Canadian, Michigan, and Hudson Basins. Within the West Canadian Basin two lower salt units are assigned to the Eifelian, Lotsberg, and Gold Lake. In the Michigan Basin the Lucas salt sequence occurs in the middle Eifelian. Within the Hudson Basin the Moose River salt sequence is of the same age.

The Eifelian age of the salt strata in other evaporite basins is only tentatively determined as yet. Thus, within the Adavale Basin the assignment of the Boree salt member to the Eifelian is more or less likely, proceeding from its occurrence in the lower Middle Devonian sequence D_2 ranging through the whole of the Givetian (Tanner 1967).

As to the age of salts in the Morsovo Basin, this problem is still rather disputable. Some workers (Tikhy 1964, 1967; Rzhonsnitskaya 1967; The Devonian System, Book 1 . . . 1973) assign the Morosovo and Narova horizons to the Givetian, others (Lyashenko 1962; Tikhomirov 1967; The Devonian . . . 1967; Kislik et al. 1976a) to the Eifelian, and still others to the Late Eifelian/Early Givetian (Kislik et al. 1977). These contradictions are mostly caused by differences in correlation of halogenic deposits of the Morsovo horizon, containing quite a peculiar fossil assemblage, with horizons of the Volga-Ural area; they differ in lithology and a faunal assemblage as well. In our opinion (Zharkov 1971b; Blagovidov et al. 1972), Eifelian carbonate rocks of the Biya horizon of the Volga-Ural area are equivalents of the Morsovo salt sequence. Thus, we agree with those investigators who place the Morsovo horizon of the Moscovian Syneclise and correlatable deposits of the adjacent areas on the Russian Plate into the Eifelian. It is this stratigraphic position of the Morsovo and Narova sequences in the Morsovo Basin that is shown in Fig. 20.

The next highest level of halite accumulation is established in the North Siberian and Tuva Basins. Within the North Siberian Basin salt deposits of the Manturovo Formation of the Norilsk Region are assigned to the upper Eifelian/Lower Givetian (Menner 1962, 1965, 1967a,b; Menner and Fradkin 1973; Matukhin and Menner 1974; Fradkin et al. 1977). It is assumed that the Nordvik salt sequence in the Khatanga Trough occupies the same stratigraphic level. The Late Eifelian/Early Givetian age has also been postulated for the Ikhedushiingol salt sequence of the Tuva Basin (Meleshchenko et al. 1973). However, it is sometimes assumed that the Ikhedushiingol sequence in Tuva and the Manturovo sequence in the north-eastern Siberian Platform might be dated as the Eifelian, and therefore the Eifelian/Givetian boundary might be drawn at the top. Some investigators regard the above units as the Givetian.

Thus, it might appear that salts in the North Siberian Basin will either be exclusively Eifelian or Givetian. The salt accumulation in the Tuva Basin could also have taken place in Eifelian or Givetian time. Taking into account the observed contradictions, we place the above salt sequences (Manturovo, Nordvik, and Ikhedushiingol), as is usually accepted, into the Late Eifelian/Early Givetian.

Thus, the Eifelian salt sequences can occur not only in North America, but in Australia, Asia, and Europe as well. It might be tentatively noted, taking into account all the above reservations, that the Eifelian age was the beginning of a new stage of halite accumulation. During the Eifelian salt strata were formed in some evaporite basins. During the Early/Middle Eifelian salt was deposited in the West Canadian, Michigan, Hudson, and Adavale Basins, whereas in the Late Eifelian in the West Canadian and, probably, North Siberian and Tuva Basins.

The Givetian salt deposits were reliably determined in the West Canadian Basin where they are represented by the Prairie, Black Creek, and Hubbard salt sequences. It should be noted that if the Prairie and Black Creek salt sequences formed during the first half of the Givetian, the Hubbard evaporites of the upper Dawson Bay Formation accumulated during the second half of the Givetian and, probably, even at the end of this stage. During the Early Givetian, salt accumulation might have taken place in the North Siberian and Tuva Basins where, as has already been pointed out, some parts of the Manturovo, Nordvik, and Ikhedushiingol salt sequences are placed into the lower Givetian. Thus, it can be concluded that the stage of halite accumulation, which began in the Eifelian, continued through the Givetian, since salt strata also deposited at that time in a number of evaporite basins: the West Canadian, North Siberian, and Tuva Basins.

The Frasnian salt deposits are found in the West Canadian Basin and in the Upper Devonian Basin of the Russian Platform; they probably occur as well in the North Siberian Basin. Over the Russian Plate the Frasnian is represented by the Evlanovo-Liven salt sequence of the Pripyat Depression and the lower salt sequence of the Dnieper-Donets Depression. Some investigators suggest that within the Dnieper-Donets Depression there might be other salt sequences, Shchigrov, confined to a horizon of the same name. It will occur in the Lower Frasnian. Salt deposits, composing the Seregov Dome in the Timan Trough, are also probably of the Frasnian age, though they might as well belong to the higher stratigraphic horizons of the Late Devonian, for instance to the Lebedyan and Dankov. In the West Canadian Basin the Davidson and Dansmore sequences are of Frasnian age, their areal distribution being comparatively limited. In the Siberian Platform the Frasnian is, probably, represented by the lower Kygyltuus salt sequence, its upper part, on the basis of drilling data, was assigned to the Famennian (Fradkin et al. 1977). On the whole, the Frasnian age was the time when rock salts accumulated in this or that evaporite basin on the Earth. They were mostly formed during the Late Frasnian because we can only suggest the presence of the Early Frasnian salt deposits in the Dnieper-Donets Depression, in the West Canadian Basin their development being rather limited.

The Famennian rock salt units are known in Eurasia and North America. They have been recorded from the Kependyai Trough and Norilsk Region in the North Siberian Basin, where the upper Kygyltuus and the Fokin salt sequences are Famennian; the Upper Devonian Basin of the Russian Platform within which the Dankov-

Lebedyan salt sequence of the Pripyat Depression and the upper salt sequence of the Dnieper-Donets Depression is of the Famennian age; the West Canadian Basin where the Stattler salt sequence is regarded as the Famennian. Moreover, a lower salt sequence in the Chu Sarysu Basin might be Late Devonian (Famennian) in age (Kislik et al. 1977; Sinitsyn et al. 1977). Such a stratigraphic position of the salt sequences among the Famennian deposits shows that the halite accumulation at that time was almost continuous and sometimes contemporaneous in several regions on different continents of the Earth.

The above data allow us to conclude that both in the Early Devonian and in the Late Silurian halite accumulation of the Earth was episodical and recorded only within certain evaporite basins. Thus, the time from the Late Ludlovian to the end of the Early Devonian can now be united into the Late Ludlovian-Early Devonian episodical salt stage. From the Eifelian to the end of the Devonian a new salt stage is distinguished, characterized by an almost continuous halite sedimentation in a number of salt basins on different continents.

The stages of potassium accumulation in the Devonian are determined much more precisely than those in the Early Paleozoic epochs. This is due to the fact that in some Devonian salt basins thick potash salt horizons have been found, whereas they have not yet been reported from the older basins. Thus, it is assumed that the Middle and Late Devonian was the first stage of vigorous potassium accumulation on the Earth (Kislik et al. 1971; Zharkov 1974b, 1976). Potash salt beds of chloride composition (sylvine and carnallite) are now established within four salt basins: West Canadian, Upper Devonian of the Russian Platform, Morsoco, and Tuva. The oldest potash salts in the Devonian have been recorded in the Morsovo Basin within the Moscovian Syneclise (On the Potassium . . . 1970). There, a potassium horizon composed of rock salt with inclusions and crystals of sylvine and carnallite is traced near Yartsevo (Smolensk Region). The initial content of potash minerals could attain 25%–30%. This oldest potassium horizon on the Russian Plate occurring within the Morsovo salt sequence is of the Eifelian age.

Potash salts in the Tuva Basin have a higher stratigraphic position (Lepeshkov et al. 1958; Ivanov and Levitsky 1960; Pastukhova 1960, 1965; Zaikov et al. 1967; Minko 1972; Kolosov et al. 1977). They are confined to the Ikhedushiingol salt sequence of the Late Eifelian/Late Givetian age where interbeds of sylvine-halite rocks are recorded, their thickness is of the first tens of centimeters (Zaikov et al. 1967). Also, a complex iron haloid-rinneite was found there (Kolosov and Pustylnikov 1967). Potash salts in the West Canadian Basin are always Givetian (Holter 1969). They compose three horizons within the Prairie Formation: Esterhazy, Belle Plaine, and Patience Lake found in the Saskatchewan part of the basin. These horizons consist of rock salt, sylvine, and carnallite rocks. The lower (Esterhazy), middle (Belle Plaine), and upper (Patience Lake) horizons are 18–21 m, 15–18 m, 10–15 m thick, respectively. Within the Upper Devonian Basin of the Russian Platform potash salts are found among the Frasnian and Famennian deposits.

Frasnian potash salts have been recorded in the Dnieper-Donets and Pripyat Depressions. In the north-western Dnieper-Donets Depression they were penetrated by deep boreholes along the crest of the Romny Salt Dome, presumably composed of the Late Frasnian lower salt sequence. Two sylvine-halite beds, 6 to 80 m thick, were penetrated by a borehole. Gamma-ray logging points to a possible presence of the potash salt in the Borkov area (Britchenko et al. 1977). In the western Pripyat Depression several potash salt beds are distinguished within the Evlanovo-Liven salt sequence with a limited distribution (Eroshina et al. 1976; Kislik et al. 1976b; Kislik et al. 1977). Famennian potash salts are extremely widespread in the Pripyat Depression.

The above evidence suggests that potash salts of the chloride type (sylvine, carnallite, and mixed ones) occur in the Middle and Upper Devonian salt sequences. In the Early Devonian no potassium accumulation was recorded. Thus, it might be concluded that potassium accumulated on the Earth periodically from the Eifelian to the end of the Famennian. This allows us to distinguish rather precisely the Middle/Late Devonian potassium stage in the history of the Paleozoic halogenesis. At the same time, during different ages of the Devonian period potassium accumulation took place only within individual basins. Thus, in the Eifelian it was confined to the Morsovo Basin, in Late Eifelian/Early Givetian time to the Tuva Basin, in Givetian to the West Canadian Basin, and in Frasnian and Famennian ages to the Upper Devonian Basin of the Russian Platform. Thus, the outlined Midde/Late Devonian stage might be regarded as episodical potassium time. It was separated from the preceding Late Ludlovian stage of episodical potasium by a potassium-free stage, comprising, evidently, not only the Přidolian age (Late Silurian), but Early Devonian as well.

Sulfate accumulation during the Devonian was registered in all 19 evaporite basins. In many of them sulfates periodically deposited either during the Devonian, or during most of this period. The sulfate sedimentation was often contemporaneous within some, and sometimes many, basins.

In the Early Devonian sulfate strata formed within three evaporite basins: North Siberian, Hudson, and Moose River. In the North Siberian Basin three sulfate units of this age are distinguished: the Zubovo sequence developed over much of the northwestern Siberian Platform, and in its central regions in the Kotui, Moiero, Olenek and Vilyui Basins, and in the Tunguska Syneclise; a gypsum unit found on Taimyr; the Dezhnev on the Severnaya Zemlya. The Zubovo sequence and gypsum unit, formed probably during the first half of the Early Devonian, whereas the Dezhnev sequence in the second half. The Lower Devonian in the Hudson Basin is represented by a sulfate sequence comprising the upper Kenogami River Formation and the Stooping River Formation. In the Moose River Basin the upper Kenogami River sulfate sequence is dated as Lower Devonian. In both basins sulfate rocks accumulated practically throughout the Early Devonian.

In the Eifelian age sulfate sedimentation was much more extensive. It took place in 13 basins in Asia, Europe, North America, and Australia. The Eifelian sulfate strata (either reliable or tentative) are known from the following evaporite basins: North Siberian (Rusanov), Morsovo (sequence of the same name and Lopushany), West Canadian (Ernestina Lake and Chinchaga), Moose River (sequence of the same name), Illinois (Jeffersonville), and Central Iowa (Kenwood). In a number of basins sulfates in the Eifelian time deposited during the formation of salt strata. They are the following basins: Morsovo, West Canadian, Hudson, Michigan, and Adavale where sulfate rocks compose the Morsovo, Lotsberg, Gold Lake, Moose River, Lucas salt sequences and the Boree salt member. Further, the lower parts of many sulfate units of the Late Eifelian/Early Givetian or Efelian/Givetian age accumulated in the evaporite basins during the Late Eifelian or at the end of it. The lower beds of the following sequences are regarded as being of this age: the Sidin and Atyrkan, the Abakan-Askyz carbonate-sulfate, the Bird Fiord in the North Siberian, Minusinsk, Moesian-Wallachian, and Canadian Arctic Basins, respectively.

The Late Eifelian sulfate accumulation in a number of salt basins was contemporaneous with deposition of salt sequences, their age is at present determined as Eifelian/Givetian, exemplified by the North Siberian and Tuva Basins where sulfate rocks are developed in the Eifelian part of the Manturovo, Nordvik, and Ikhedushiin-gol sequences.

Sulfate rocks are also rather widespread among Givetian deposits. They were found in all ten evaporite basins: the North Siberian, Minusinsk, Tuva, Moesian-Wallachian, Tindouf, Canadian Arctic, West Canadian, Hudson, Moose River and Central Iowa Basins. In the North Siberian Basin the upper Sidin and Atyrkan sequences, the Yukta, Burkhala, Sebechan, and Vayakh sulfate sequences occur at this stratigraphic level, as well as sulfate rocks, confined to the upper Manturovo salt sequence. In the other basins the Givetian is represented by the upper Abakan-Askyz and Belsk sequences (Minusinsk Basin); sulfate rocks of the upper Ikhedushiingol sequence (Tuva Basin); the upper carbonate-sulfate sequence (Moesian-Wallachian Basin); lower sulfate variegated sequence (Tindouf Basin); the uppermost Bird Fiord sequence (Canadian Arctic Basin); sulfate Muskeg sequence and sulfate rocks of the Prairie, Black Creek, and Hubbard salt sequences (West Canadian Basin); the lower Williams Island sequence (Hudson Basin); the lowermost Williams Island sequence (Moose River Basin); and the Cedar Valley sulfate sequence (Central Iowa Basin).

At present ten Frasnian sulfate units are known: the Nakokhoz and Kalargon, Shchigrov, Voronezh-Evlanovo, Ust-Ukhta, Beaverhill Lake, Woodbend, Duperow, Birdbear, and Nisku sequences in the North Siberian, Upper Devonian of the Russian Platform, and West Canadian Basins, respectively. Additionally, the Frasnian stage is represented by the lower Matusevich sequence and a gypsum sequence of the Vrangel Island in the North Siberian Basin, the lower Upper Devonian sulfate sequence of the Baltic area, Moscovian Syneclise, and Volga-Ural area in the Upper Devonian Basin of the Russian Platform, the upper members of the Williams Island sulfate sequences in the Hudson and Moose River Basins, the lower Long Rapids sequence in the same basins, the middle part of a sulfate variegated sequence in the Tindouf Basin, as well as sulfate rocks of the lower Kygyltuus salt sequence of the North Siberian Basin, Evlanovo-Liven salt sequence of the Pripyat Depression, probably of the Shchigrov and the lower salt sequences of the Dnieper-Donets Depression, a salt sequence of the Timan Trough in the Upper Devonian Basin of the Russian Platform, the Davidson and Dinsmore salt sequences in the West Canadian Basin. Thus, the sulfate accumulation during the Frasnian was recorded within six evaporite basins.

During the Famennian age sulfate rocks accumulated in ten basins: the North Siberian, Kuznetsk, Chu-Sarysu, Teniz, Turgai, Upper Devonian of the Russian Platform, Tindouf, West Canadian, Hudson, and Moose River Basins. They are in the North Siberian Basin the Upper Fokin, Namdyr, upper Matusevich, and, probably, the upper gypsum sequence on Vrangel Island, as well as sulfate rocks of the Fokin and Kygyltuus salt sequences; in the Kuznetsk Basin the Podonin sequence; in the

Chu-Sarysu sulfate deposits of a lower salt sequence; in the Teniz a sulfate sequence; in the Upper Devonian Basin of the Russian Platform the Mezhsolevaya sequence of the Pripyat Depression, the Idzhid-Kamensk sequence and the upper Upper Devonian beds of the Baltic area, the Moscovian Syneclise, and the Volga-Ural area; sulfate deposits of the Dankov-Lebedyan salt sequence of the Pripyat Depression and an upper salt sequence of the Dnieper-Donets Depression; in the Tindouf Basin the upper sulfate variegated unit; in the West Canadian Stattler and sulfate rocks of the Stattler salt sequence; in the Hudson and Moose River Basins the upper Long Rapids.

The age position of Devonian sulfate sequences and sulfate rocks confined to salt strata suggests rather clearly that sulfate accumulation during the Devonian was continuous, halogenic sediments often being deposited contemporaneously in a number of evaporite basins on different continents. Thus, it can be quite definitely concluded that the epoch of sulfate sedimentation, begun in the Cambrian, continued to the end of the Devonian.

Stratigraphic Position of Carboniferous Evaporites

It is rather difficult to make interregional correlation of Carboniferous deposits in general and evaporites in particular, due to the lack of universally accepted subdivision of the Carboniferous. Three equally important charts are used in the Soviet Union, Western Europe, and North America. Here we use the subdivision accepted in the USSR. Figure 21 shows comparison of this chart with the American and West European charts and the correlation of sections in most evaporite basins. This shows rather distinctly a stratigraphic position of salt and sulfate units distinguished, as well as that of potash horizons, and it allows the precise study of the stages of halite, potassium, and sulfate accumulation in the Carboniferous.

The stages of halite sedimentation in the Carboniferous were more periodical than during the preceding Paleozoic epochs which is reflected in the age of salt sequences. As a rule, salt deposits occur at four stratigraphic levels: (1) Tournaisian and Visean (Lower Carboniferous), (2) Bashkirian (Middle Carboniferous), (3) Upper Moscovian (Middle Carboniferous), and (4) Orenburgian (Upper Carboniferous).

The Tournaisian and Visean salt sequences are distinguished within five evaporite basins in North America and Asia. In three basins (Williston, Maritime, and Saltville) they occur approximately at the same stratigraphic level: the Upper Tournaisian and Visean (Osagian and Meramecian of the Mississippian system). This position is characteristic of the Charles salt sequence in the Williston Basin, MacCrady in the Saltville Basin, as well as Gautreau and Windsor in the Maritime Basin. An upper salt sequence in the Chu-Sarysu Depression is probably older and can be dated as Early Tournaisian (Sinitsyn et al. 1977). Much less definite is a stratigraphic position of the salt sequence tentatively outlined within the Mid-Tien Shan evaporite basin. Not only their Paleozoic age is doubted, but even their Carboniferous age is debatable. As was mentioned above, some investigators (O.I. Karas et al.) place salt deposits into the Bashkirian (Middle Carboniferous), emphasizing their occurrence among beds of the Chemanda Formation. The others (Poyarkov 1972) assign the above formation to the Visean

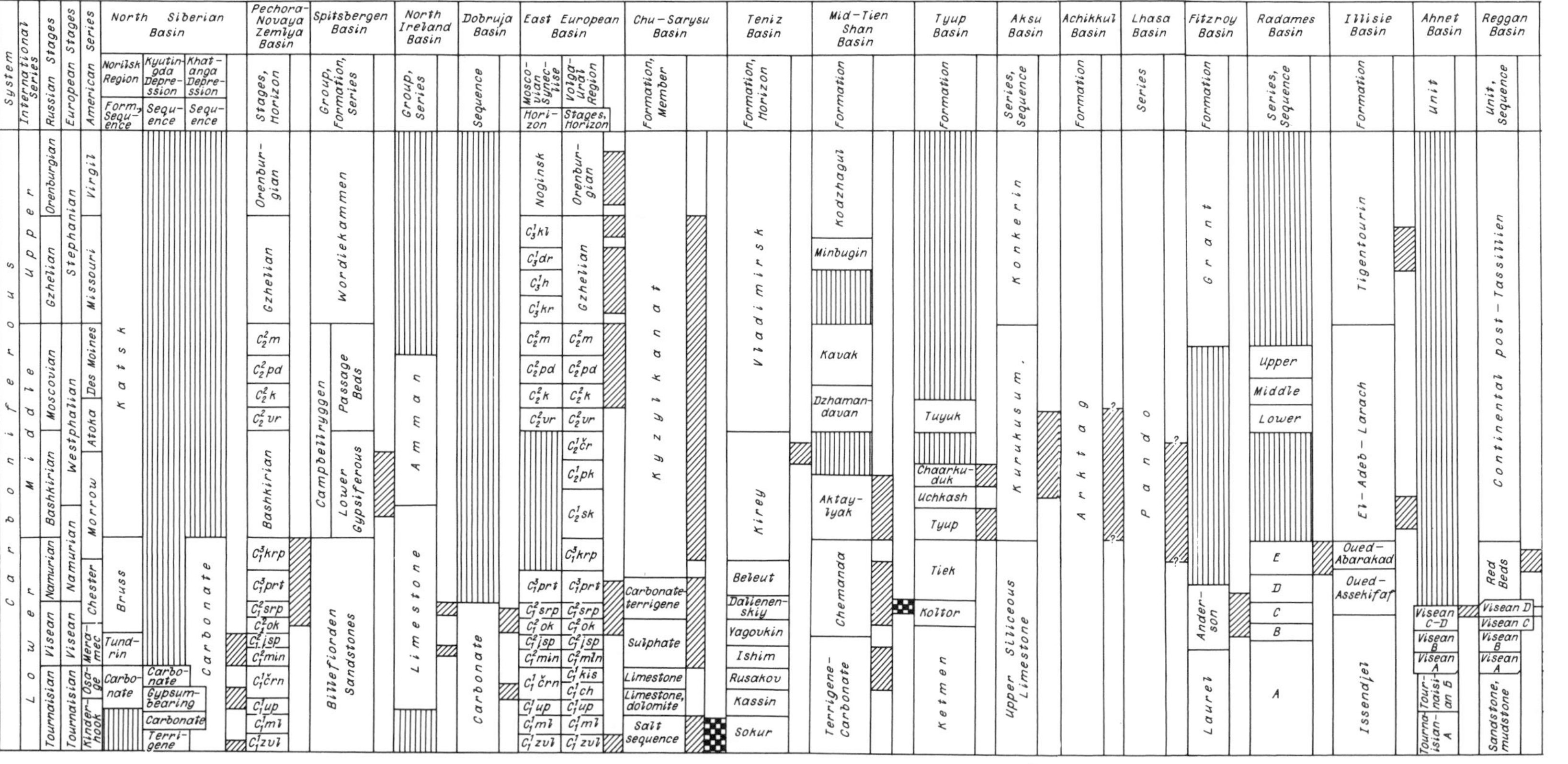

Fig. 21. Interregional correlation of sections in Carboniferous evaporite basins.

Superhorizons and horizons: C_1^1 *zvl* Zavolzhsk, C_1^1 *ml* Malevsk, C_1^1 *up* Upa, C_1^1 *črn* Chernyshin, C_1^1 *ch* Cherepets, C_1^1 *kis* Kizelov, C_1^2 *mln* Malinov, C_1^2 *jsp* Yasnaya Polyana, C_1^2 *ok* Oka, C_1^2 *srp* Serpukhov, C_1^3 *prt* Protvin, C_1^3 *krp* Krasnaya Polyana, C_2^1 *sk* North Keltma, C_2^1 *pk* Kama, C_2^1 *čr* Cheremsha, C_2^2 *vr* Verey, C_2^2 *k* Kashira, C_2^2 *pd* Podolian, C_2^2 *m* Maychkov, C_3^1 *kr* Krevyakino, C_3^1 *h* Khamovnich, C_3^1 *dr* Dorogomilovsk, C_3^1 *kl* Klyazma.

Formations: *Ham* Hampton, *Bur* Burlington, *Keo* Keokuk, *War* Warsaw, *Spe* Spergen, *S-L* Saint Louis, *Pel* Pella, *N-A* New Albany, *Har* Harrodsburg, *Sal* Salem, *S-G* Saint Genevieve, *Cyp* Cypress, *T-S* Tar Springs, *Men* Menard, *Kin* Kinkaid, *Car* Carbondale. Symbols are the same as in Fig. 17

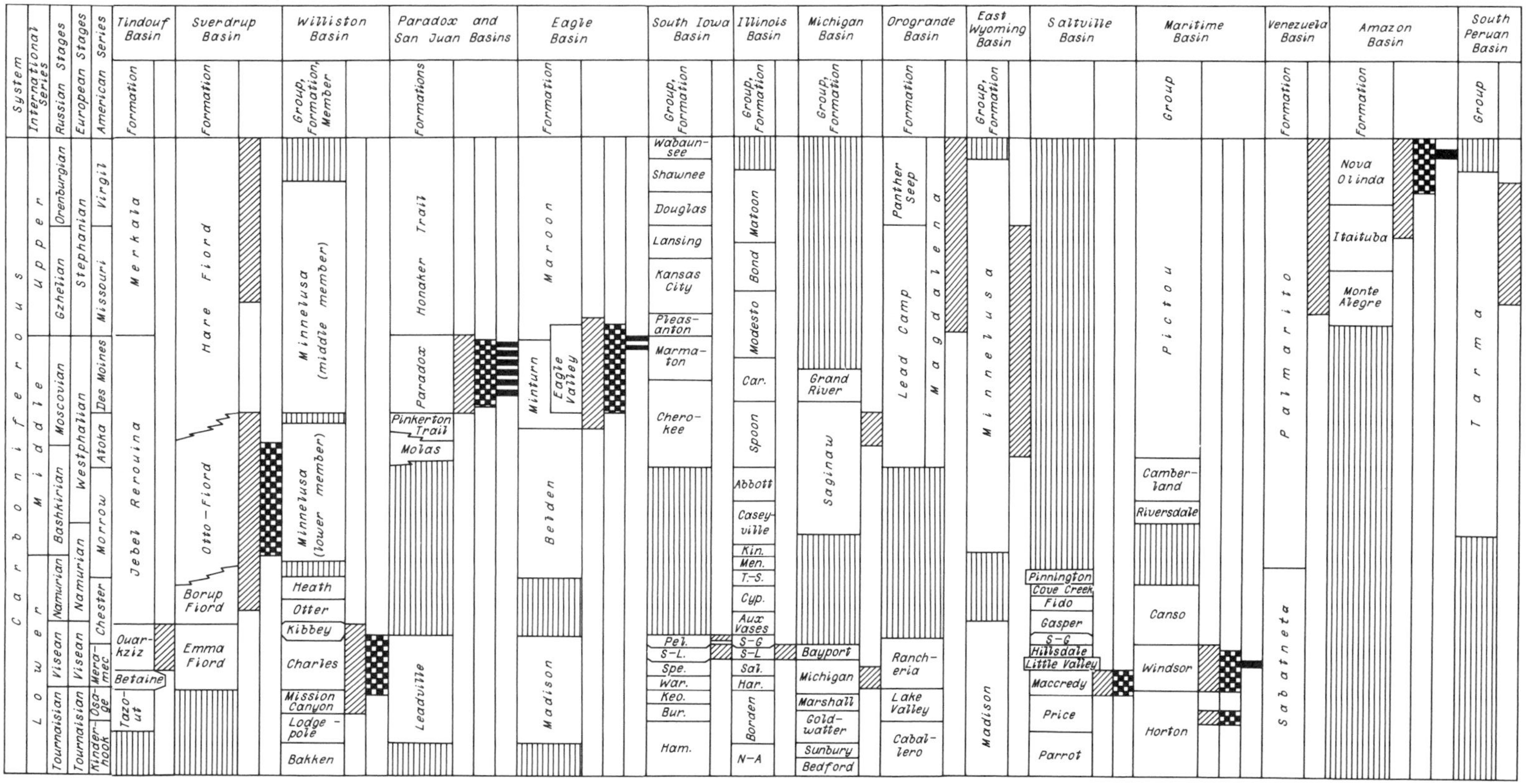

System	International Series	Russian Stages	European Stages	American Series
Carboniferous	Upper	Orenburgian	Stephanian	Virgil
		Gzhelian		Missouri
	Middle	Moscovian	Westphalian	Des Moines
				Atoka
		Bashkirian		Morrow
	Lower	Namurian	Namurian	Chester
		Visean	Visean	Meramec
		Tournaisian	Tournaisian	Osage
				Kinderhook

Basin	Heading	Units (top to bottom)
Tindouf Basin	Formation	Merkala; Jebel Rerouina; Ouarkziz; Betaine; Tazout
Sverdrup Basin	Formation	Hare Fiord; Otto-Fiord; Borup Fiord; Emma Fiord
Williston Basin	Group, Formation, Member	Minnelusa (middle member); Minnelusa (lower member); Heath; Otter; Kibbey; Charles; Mission Canyon; Lodge-pole; Bakken
Paradox and San Juan Basins	Formations	Honaker Trail; Paradox; Pinkerton Trail; Molas; Leadville
Eagle Basin	Formation	Maroon; Eagle Valley; Minturn; Belden; Madison
South Iowa Basin	Group, Formation	Wabaunsee; Shawnee; Douglas; Lansing; Kansas City; Pleasanton; Marmaton; Cherokee; Pel.; S-L.; Spe.; War.; Keo.; Bur.; Ham.
Illinois Basin	Group, Formation	Matoon; Bond; Modesto; Car.; Spoon; Abbott; Caseyville; Kin.; Men.; T.-S.; Cyp.; Aux Vases; S-G; S-L; Sal.; Har.; Borden; N-A
Michigan Basin	Group, Formation	Grand River; Saginaw; Bayport; Michigan; Marshall; Goldwater; Sunbury; Bedford
Orogrande Basin	Formation	Panther Seep; Lead Camp; Magdalena; Rancheria; Lake Valley; Caballero
East Wyoming Basin	Group, Formation	Minnelusa; Madison
Saltville Basin		Pinnington; Cove Creek; Fido; Gasper; S-G; Hillsdale; Little Valley; Maccredy; Price; Parrot
Maritime Basin	Group	Pictou; Cumberland; Riversdale; Canso; Windsor; Horton
Venezuela Basin	Formation	Palmarito; Sabaneta
Amazon Basin	Formation	Nova Olinda; Itaituba; Monte Alegre
South Peruan Basin	Group	Tarma

Fig. 21 (continued)

and Namurian (Lower Carboniferous), which is also shown in the unified stratigraphic chart. In this paper we adhere to the latter point of view, so salt sequences of the Mid-Tien Shan Basin are assigned to the Visean.

Salt deposits of Bashkirian age are known at present only from one evaporite basin, the Sverdrup Basin, located in the Canadian Arctic Archipelago. The halogenic Otto Fiord Formation developed there is of Morrowan age (Bashkirian) in type locality, only its upper beds being probably Early Moscovian in age. However, the lower and upper boundaries of the formation are not isochronous throughout the basin. Thus, in south-western Melville Island fossils were found in limestones of the upper part of the formation, pointing to the Namurian age. Generally, the information available led W. Nassichuk (1975) to conclude that the deposits united into the Otto Fiord Formation vary in age from the Namurian to the end of the Bashkirian, and their upper beds might be Early Moscovian. At the same time, the salt sequence proper is probably Bashkirian, as is shown in the chart (see Fig. 21).

The Late Moscovian age of salt sequences is established within two adjacent evaporite basins, Paradox and Eagle. Salt deposits developed there belong to the upper Desmoinesian, and correspond to the uppermost Middle Carboniferous. It is only in the Amazon Basin that salt formations are distinguished among the Upper Carboniferous; there the Nova Olinda salt sequence is assigned to the Late Pennsylvanian, probably Virgilian (Orenburgian).

The data on the stratigraphic position of the above salt units suggest that the Carboniferous period in the history of halite accumulation may be reliably subdivided, as we previously assumed (Zharkov 1974a,b; 1977), into five stages. But some corrections should be introduced into the old chart; because during the last years Bashkirian salt deposits were found in the Sverdrup Basin, whereas before no salt rocks of this age were recorded, and thus the Bashkirian age was regarded as a salt-free stage. Now we have to distinguish this age as a time of episodical salt accumulation and regard it together with the Moscovian.

The first stage still covers most of the Early Carboniferous comprising the Tournaisian and Visean. Its main feature is that halite accumulation during this time took place not only in North America, but, probably, in Asia as well (in the Chu-Sarysu and Mid-Tien Shan Basins). In North America salt deposits formed within three widely spaced regions and not within individual basins, as during the Middle and Late Carboniferous. The Early Carboniferous stage of halite accumulation, as well as that of sulfate accumulation, as will be shown below, in some evaporite basins was closely associated with the Devonian stage, as if crowning it. All this implies that a stage of halite accumulation began in the Middle Devonian, continued to the end of Late Devonian, and also covered the Early Carboniferous to the end of the Visean.

Thus, this major stage might be called the Middle Devonian/Early Carboniferous salt stage. The next, second stage of the Carboniferous period is distinguished as a salt-free stage. It corresponds to the Namurian age (Early Carboniferous). The third stage, corresponding to the Bashkirian and Moscovian (Middle Carboniferous) is distinguished as the Middle Carboniferous stage of episodical salt accumulation, since the formation of salt deposits took place within three basins: first in the Sverdrup, and then, during Late Moscovian time, in two relatively small adjacent basins, Paradox and Eagle. The fourth stage is again a salt-free stage. It corresponds to the Gzhelian age (Late Carbonfierous). Finally, the fifth stage, comprising the Orenburgian age (Late Carboniferous), might be regarded, like the Middle Carboniferous, as an episodical salt stage, for the halite sedimentation for that time was recorded only in the Amazon evaporite basin in South America.

Potassium accumulation in the Carboniferous period was confined to two time intervals, the Visean age and the Late Moscovian time. In both cases potash salts accumulated within individual salt basins: in the Visean this was the Maritime Basin, and in the Late Moscovian mostly the Paradox Basin and, to a lesser extent, the Eagle Basin.

In different regions of the Maritime Basin (Malagash, Pugwash, Cumberland in Nova Scotia) potash salts were found in the Windsor salt sequence: carnallite, sylvine, and sylvine-carnallite interbeds of polyhalite rock are not so numerous, neither are the inclusions of rinneite (Evans 1970). Potash salts are extremely abundant in the Paradox Basin. They are present in 18 evaporite cycles, in 11 of them their content attaining economic estimates. Potash salts are represented by sylvine and carnallite rocks. Polyhalite and kieserite rocks are also found, as well as rinneite. The thickness of some potash horizons (bed No. 19) is great (up to 200 m). The potash salt reserves in the Paradox Basin are considered to be the largest in the world (Hite 1961). Within the Eagle Basin, which is as yet poorly known, two potassium beds composed of sylvine, sylvine-carnallite and carnallite are distinguished. Their thickness is 1.8 m and 2.1 m (Mallory 1971).

The rest of the Carboniferous, except the Visean age and Late Moscovian time, was not characterized by potassium accumulation even in those basins where extensive halite accumulation took place. This evidence allows a subdivision of the Carboniferous into five stages based on the history of potassium sedimentation: (1) Tournaisian, potassium-free, (2) Visean, episodically potassium, (3) Namurian/Middle Moscovian, potassium-free, (4) Late Moscovian, episodically potassium, and (5) Late Carboniferous, potassium-free. The last stage, as will be seen, continued into the Asselian age (Early Permian).

Sulfate sedimentation in the Carboniferous period was recorded on all the continents, except for Antarctica, for which no data are available. In many evaporite basins sulfate accumulation was periodically reiterated, sulfate deposits almost always formed contemporaneously in several basins.

In the Tournaisian sulfate formation was recorded in eight evaporite basins: North Siberian, East European, Dobruja, Chu-Sarysu, Chimkent, Mid-Tien Shan, Williston, and Maritime. In some of them several sulfate units were deposited during this period. Thus, within the North Siberian Basin sulfates accumualted in the Kyutingda and Kempendyai Troughs; in each of them separate sulfate sequences formed a gypsum and the Kurunguryakh sequence. Within the East European Basin sulfate precipitation during the Tournaisian covered a vast area. There the Ozersk sulfate sequence was formed. It should be noted that halogenic sedimentation was closely related to the preceding Devonian epoch. Sulfate accumulation in the Chu-Sarysu Basin during the formation of the upper salt sequence has also continued since the Devonian period. With the Tournaisian, evaporite sedimentation starts within the Chimkent and Mid-Tien Shan Basins in Asia and in the Maritime Basin in North America. Finally, after a short break in evaporite sedimentation in the Williston Basin at the end of the Devonian, when this basin was a part of a big West Canadian Basin, in the Tournaisian first sulfate sediments (Mission Canyon sequence) and then salt sediments (Charles sequence) were again deposited. The Dobruja sulfate basin took shape at the end of the Tournaisian.

During the Visean sulfate strata formed within 19 evaporite basins: North Siberian, Pechora-Novaya Zemlya, Northumberland, North Ireland, Central England, Dobruja, East European, Chu-Sarysu, Mid-Tien Shan, Fitzroy, Ahnet, Reggane, Tindouf, Williston, South Iowa, Illinois, Michigan, Saltville, and Maritime. The Visean, generally, marked the end of evaporite accumulation in the North Siberian Basin, sulfates during that time deposited in the Norilsk region and, probably, in the Khatanga Trough. During the subsequent period of geological history no halogenic sedimentation took place either in this region, or over the entire north-east of Asia.

After the Visean sulfate deposition ceased for a long time in a number of evaporite basins in North America: Maritime, Saltville, Illinois, South Iowa, and Williston which did not become halogenic during the Carboniferous. Finally, evaporite sedimentation never occurred in Australia from Visean time to the close of the Paleozoic (probably Early Namurian) when Fitzroy was a sulfate basin. In contrast to this, in many other basins either sulfate accumulation renewed in the Visean, or evaporites became more widespread in some of them. Thus, the Visean age became the time of the most extensive halogenic sedimentation in most of north-western Africa. It began concurrently in three sedimentary basins: Ahnet, Reggane, and Tindouf. In the Pechora-Novaya Zemlya Basin sulfate accumulation repeated in the Late Visean, succeeding the Late Devonian halogenic stage when the Pechora zone was a part of the Upper Devonian salt basin in the Russian Platform, and the subsequent cessation of evaporite sedimentation in Tournaisian, Early and Middle Visean time. In the Michigan Basin a break in the evaporite sedimentation was even longer. It covered the time interval from the Givetian age (Middle Devonian) to the end of the Tournaisian age (Early Carboniferous), but later, in the Visean, sulfate sedimentation there was renewed, and the sulfate Michigan sequences formed. In the East European, Chu-Sarysu and, probably, Mid-Tien Shan Basins, within which evaporites were being periodically deposited during most of the Carboniferous, the Visean age was the next period of sulfate formation, the second succeeding the Tournaisian.

Thus, the Visean was, to a certain extent, the turning point in the history of halogenic sedimentation. For the extensive areas on different continents this time is either associated with cessation of evaporite accumulation (Australia, East and North-East Asia, western and south-western North America), or with a considerable increase of sulfate sedimentation (north-western Africa). It is only within the zone extending from Novaya Zemlya through Eurasia to the Mid-Tien Shan that halogenic accumulation was regularly repeated in the course of the Carboniferous, continuing in some basins since the Devonian.

In the Namurian sulfate strata were being formed within eight evaporite basins: Pechora-Novaya Zemlya, East European, Chu-Sarysu, and, probably, the Mid-Tien Shan, Lhasa, Rhadames, Reggane and Sverdrup Basins. A spatial position of these basins suggests that in the Namurian sulfate accumulation was mostly confined either to those regions where evaporite sedimentation continued since the earlier stages, as is the case within the above zone, extending through Eurasia, or to the regions which became zones of halogenic sedimentation only in the Visean (north-western Africa). A new region of evaporite sedimentation, Sverdrup Basin, appeared in the Canadian Arctic Archipelago, but it is known that halogenic basins existed there in the Ordovician, Silurian, and Devonian; thus it may be assumed that this region was already present during the preceding periods. The Sverdrup Basin might well fall within the extension of the Eurasian zone of the evaporite basins, and they might be united into a single zone of halogenic accumulation which, at the end of the Early Carboniferous, probably, extended from some regions of Central Asia to the Canadian Arctic Archipelago.

In the Bashkirian age (Middle Carboniferous) sulfate sedimentation was also recorded in the same two major regions of the Earth: in the north-western part of Africa where sulfate deposits (El Adeb Larache sequence) accumulated in the Illisie Basin and within the "Eurasian-Canadian" halogenic sedimentation zone where ten evaporite deposits were distinguished forming a chain: Lhasa, Achikkul, Aksu, Tyup, Mid-Tien Shan, Chu-Sarysu, Teniz, East Uralian, Spitsbergen, and Sverdrup Basins. On the whole, sulfate accumulation in the Bashkirian was recorded in ten sulfate basins and one salt basin within which 12 sulfate units deposited.

In the Moscovian the spatial distribution of the sulfate formation regions was somewhat different from that of the lower Middle Carboniferous. Sulfate sediments were no longer accumulated in the basins of north-east Africa. Sulfate strata in the "Eurasian-Canadian" zone of halogenic accumulation formed within a much smaller number of basins (probably, only in six basins: Sverdrup, East European, Chu-Sarysu, and, possibly, East Uralian, Aksu, and Achikkul), whereas in western North America four new evaporite basins appeared: Paradox, Eagle, San Juan, and East Wyoming, where considerable thicknesses of sulfate sediments were deposited. Sulfate sedimentation was renewed over a small area within the Michigan Basin. Thus, during the Moscovian sulfate formations were recorded in 11 evaporite basins (three salt basins and eight sulfate basins).

During the Gzhelian age (Late Carboniferous) the formation of sulfate strata continued in the East European and Chu-Sarysu Basins in Eurasia, in the Illisie Basin in north-western Africa, in the Sverdrup Basin in the Canadian Arctic Archipelago, in the basins of East Wyoming and during a short period of time in the Eagle Basin in North America. Four new evaporite basins appeared at that time: Orogrande in North America, Venezuela, Amazon, and South Peruvian Basins in South America. All these basins were sulfate-bearing in Gzhelian time. It should be noted that in South America evaporite sedimentation was renewed after a very long break and lasted during the Ordovician, Silurian, Devonian, Early and Middle Carboniferous. During the Gzhelian sulfate sediments accumulated in ten basins.

In the Orenburgian age sulfate sedimentation was recorded in seven basins in Eurasia (East European and, probably, Mid-Tien Shan), North America (Sverdrup and Orogrande) and South America (Venezuela, Amazon, and South Peruvian).

Thus, sulfate accumulation was also extremely widespread during the Carboniferous, being almost continuous and often contemporaneous in many evaporite basins on different continents. The epoch of sulfate sedimentation which began in the Cambrian, no doubt, continued into the Carboniferous.

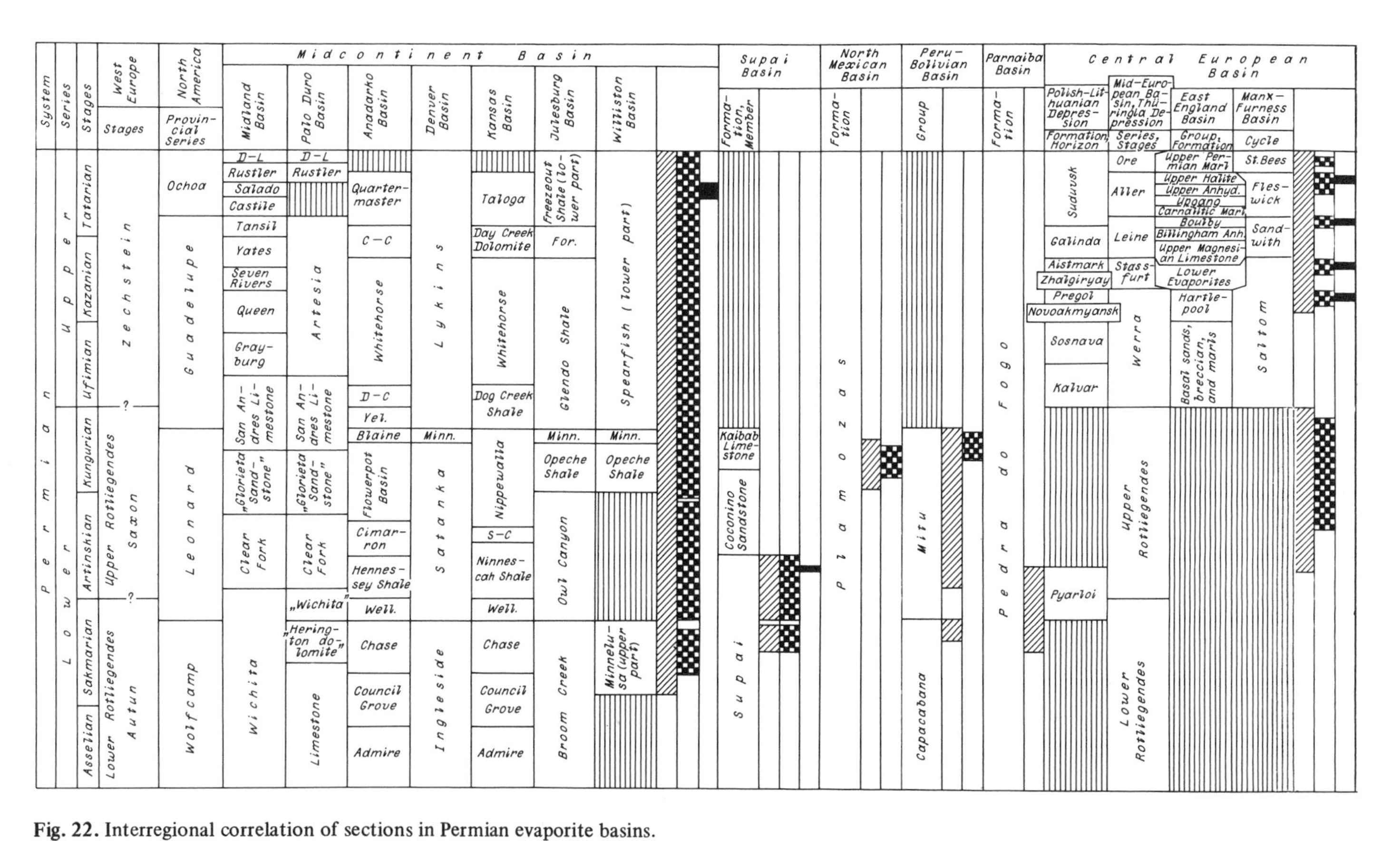

Fig. 22. Interregional correlation of sections in Permian evaporite basins. Formations: *D-L* Dewey Lake, *C-Ch* Cloud Chief, *For* Forelle, *D-C* Dog Creek, *Yel* Yelton, *Minn* Minnekahta, *S-C* Stone Corral, *Well* Wellington. Symbols are the same as in Fig. 17

System	Series	Stages	Alpine Basin (Sequence)	North Italian Basin (Formation)	Dinarids Basin (Formation, Sequence)	Mecsek Basin (Formation, Sequence)	Moesian Basin (Sequence)	Rakhov Basin (Formation)	Dobruja Basin (Member)
Permian	Upper	Tatarian	Saliferous	Bellerophon	Carbonate	Gypsum-bearing	Salt-bearing	Gypsum-bearing	The first anhydrite; Terrigene; The second anhydrite
		Kazanian					Salt-bearing terrigene		
		Ufimian	Red Beds		Gröden	Yakobkhed			
	Lower	Kungurian	Volcanic-Sedimentic	Garden Sandstone		Conglomerate, sandstone, coal	Red Beds	? Volcanic-Sedimentic	
		Artinskian		Verrucano	Kosna				
		Sakmarian	Red Beds			Sandstone, shale		? Basal rocks	
		Asselian							

Stages	East European Basin: Solikamsk Basin (Formation, Horizon)	Belaya Trough (Formation)	Buzuluk Depression (Formation, Horizon)	Pre-Caspian Depression (Sequence, beds)	Dnieper-Donets Depression (Formation)	Mezen Depression (Formation, Horizon)
Tatarian		Nugush	Kutuluk; Malokinelsk; Amakan; Bolshekinel; Soksk	Sandstone, mudstone, shale, salt	Dronov: Korenev; Shebelinsk; Peresazhsk	Vyatka; Severodvinsk; Urzhum
Kazanian	Belebeevo		Sosnovsk; Hydrochemical; Kalinovsk	Limestone, anhydrite, shale, salt		Dolomite, mudstone, shale; Limestone, mudstone
Ufimian	Sheshma; Solikamsk			Mudstone, siltstone, salt		Vikhtovsk
Kungurian	Berezniki; Filippovo	Halogenic	Iren; Filippovo	Inder; Chelkar; Elton; Ulagan; Volgograd; Filippovo		Upper Kuloi
Artinskian	Sargin; Irginsk	Limestone, mudstone	Sulphate-carbonate	Sulphate-carbonate		
Sakmarian	Sterlitamak; Tastub	Sterlitamak; Tastub	Sterlitamak; Tastub		Kramatorsk	
Asselian	Limestone	Shikhan; Limestone	Kokhansk; Belsk; Pervomaysk	Carbonate	Slavyansk; Nikitovo; Kartamysh	Dolomite, limestone, anhydrite

Stages	Chu-Sarysu Basin (Formation)	Karasu-Ishsay Basin (Formation)	Darvaza Basin (Formation)	Spitsbergen Basin (Formation, member)	East Greenland Basin (Sequence)	Sverdrup Basin (Formation)
Tatarian	Kingir; Tuzkol	Karasu	Kaftarmol		Gypsum-bearing	
Kazanian						Trold Fiord
Ufimian	Sorkol		Shakarsen			
Kungurian	Zhidelisai; Karakyr	Kelematin	Ioilikharsk	? Brachiopod Cherts		
Artinskian			Gundara	Upper Gypsiferous		Assistance; Sabine Bay
Sakmarian			Safetdaron	Wordiekammen		Tanquary
Asselian			Charyndarin			Mount Bayley

Fig. 22 (continued)

Stratigraphic Position of Permian Evaporites

The Permian was one of the periods most favorable for evaporite accumulation through the entire Paleozoic history. 37 salt and 48 sulfate units formed during this time (Fig. 22).

There is no adequate stratigraphic chart for the Permian for a number of basins, many problems dealing with correlation of sulfate sequences are highly disputable, and, thus, a stratigraphic position of individual salt and sulfate units is not clear even within individual basins. The distant correlation is extremely difficult since different charts are used for West Europe, the USA, and the USSR, their correlation remaining disputable up to now. Thus, it is necessary to discuss, at least briefly, some problems of stratigraphy in order to explain a proposed correlation chart.

Within the Midcontinent Basin Permian deposits are subdivided, according to the stratigraphic chart of the Permian accepted in the USA, into four series or stages: the Wolfcampian, Leonardian, Guadalupian, and Ochoan. There is no universally acknowledged correlation of these units with that adopted in the Soviet Union. Many investigators consider the Wolfcampian as an equivalent of the lower sub-series of the Lower Permian in the USSR, or of the Asselian and Sakmarian stages; the Leonardian is correlated with the upper sub-series of the Lower Permian comprising the Artinskian and Kungurian; the Guadalupian is correlated with Ufimian and Kazanian, whereas the Ochoan corresponds to the Tatarian (Upper Permian) (Permian System . . . 1966). However, American geologists (McKee et al. 1967; Oriel 1967a,b) assume that there is no precise correlation of the boundary beds within the stages of the Permian in the American and Soviet charts. They suggest that the Wolfcampian might be correlated with the Asselian and most of the lower Sakmarian; the Leonardian is an equivalent of the Upper Sakmarian, the entire Artinskian and Lower Kungurian; the Guadalupian is correlated with the upper Kungurian, as well as with the Ufimian, Kazanian, and Lower Tatarian; the Ochoan with the Upper Tatarian. This correlation is accepted in the present paper. The problem of the boundary between the Lower and Upper Permian in the USA has not yet been solved. It is drawn either at the base of the Guadalupian, or within its lower, or even middle part. Depending on this, the Guadalupian is either totally assigned to the Upper Permian, and then the top of the Leonardian coinciding with the top of the Kungurian, or the Lower Guadalupian is placed into the Lower Permian and correlated with the Upper Kungurian (see Fig. 22).

The subdivision of the Permian for the Midcontinent Basin is very complicated due to the mineral composition which is extremely different for various sedimentary strata and fast facies changes. Usually within a small region and, as a rule, within each structural zone of the Midcontinent, local stratigraphic units are distinguished; these are the formations marked by "nonpersistent" age boundaries, and often disappear or appear under new names. A stratigraphic position of these formations is not persistent. It is most difficult to determine a stratigraphic position of salt and sulfate units. They occur as rather small lenses and are rarely traceable for long distances. The most difficult problem of the Permian stratigraphy in the Midcontinent Basin is the age determination of salt sequences placed between the Leonardian and Guadalupian. These sequences (San-Andres, Blaine, and their equivalents) are sometimes assigned to the Lower, and sometimes to the Upper Permian. The age of Pine salt sequence in the Williston Basin also has not yet been finally determined; it can belong totally to the Guadalupian, though in the chart (see Fig. 22) it corresponds to the Ochoan. A stratigraphic position of the other rock salt units in the Midcontinent Basin is more or less reliable.

Let us dwell on some unsolved problems of the Permian stratigraphy of salt deposits in the East European Basin.

Firstly, it should be noted that rock salt in the Dvina-Sukhona region in the northern Russian Platform (the Mezen Depression) is tentatively assigned to the Sakmarian (Lower Permian). Its stratigraphic position is not yet clear. These salt units enclosed into the Upper Kuloi Formation are often assigned to undifferentiated deposits of the Sakmarian, Artinskian, and Kungurian (Pakhtusova 1963a; Zoricheva 1966, etc.) or to the undifferentiated deposits of the Upper Sakmarian and the Artinskian (Gorbatkina and Strok 1971). However, more and more investigators tend to assume that the Upper Kuloi Formation of the Mezen Depression is not younger than Sakmarian (Kashik et al. 1969). Tikhvinsky (1971) places these salt beds in the Sterlitamak horizon (Sakmarian). The last viewpoint is illustrated by the proposed correlation chart.

The second problem to be solved in the stratigraphy of the Permian deposits of the East European Basin is the boundary between the Upper and Lower Permian and the closely related problem of a stratigraphic position of salt sequences presently dated as Kungurian. It is known that the Kungurian type localities are on the Ufimian Plateau and in the Sylva-Iren area of the Permian Kama area. The Lower Kungurian is represented there by the Filippovo Formation, and the upper one by the Iren Formation subdivided into seven members, in ascending order Ledyanopeshchera, Nevolnino, Shalashnino, Elkino, Demidkovo, Tyuya, and Lunezh. There is no rock salt in the section, but against its background equivalents of the Filippovo and Iren Formations are distinguished in the rest of the East European Basin, not only within the adjacent regions of the Cis-Uralian Trough, but also farther west (in the Moscovian Syneclise) and to the south (Pre-Caspian Depression). It is generally acknowledged at present that the thickest salt series in all the above regions are mostly confined to the uppermost Lunezh Formation of the Iren horizon. This conviction issues, first of all, from the fact that in depressions nearest to the type localities (Chusovaya and Solikamsk) rock and potash salts were associated with the Lunezh Formation. However, in our opinion, such a correlation is far from being evident. For instance, possibly the equivalents of the Lunezh Formation are, in fact, placed into deposits underlying the salt. The salt sequence proper of the Solikamsk Depression, known as the Berezniki Formation, might occur above the Lunezh Formation of the Iren horizon at the level of a clay-marly unit of the lower Solikamsk horizon or, in other words, belong not to the Kungurian, but to the Ufimian, i.e., it might be not the Lower Permian, but Upper Permian if we accept the stage subdivision of the Permian now proposed and agree that the series boundary is drawn at the base of the Ufimian. Such a correlation touches upon a highly important problem of the age of a thick salt sequence, including potash, in the southern Volga-Ural area, in the Cis-Uralian Trough, and in the Pre-Caspian Depression. Most of the sequence belongs to the Iren horizon of the Kungurian and is even regarded as the Upper Lunezh, whereas in fact it might occur within the Ufimian (Upper Permian). In 1960 Forsh (1976) taking into account a somewhat different stratigraphic position of these beds, expressed an idea that most of the salt sequence of the Pre-Caspian Depression was Late Permian. Not long ago he confirmed this viewpoint once again (Forsh 1976) through detailed correlation of the Ufimian deposits from Buguruslan in the north to the Krasnokut borehole 26 in the Pre-Caspian Depression in the south; thus, much of the salt series evidently comprising the Elton, Chelkar, and Inderbor beds, according to Tikhvinsky (1974, 1976a,b), is Ufimian. Thus, a stratigraphic position of the salt sequence of the Pre-Caspian Depression assigned to the Kungurian remains uncertain and calls for further investigation. In the present paper this sequence is regarded as Kungurian. Salt deposits of the Ufimian age are distinguished in the Solikamsk Depression and presumably in the eastern Pre-Caspian Depression.

It is not yet clear whether there are any Artinskian salt deposits in the East European Basin. They were recorded in the western Pre-Caspian Depression (Latskova 1961, 1967; Fiveg and Banera 1968; Shafiro 1972, 1975, 1977; Shafiro and Sipko 1973); whereas others assign these beds to the Kungurian (Gorbov 1973; Thikhvinsky 1973, 1974, 1976a,b; Ermakov and Grebennikov 1977). It has already been noted that such contradictions result from a different understanding of the extent of the Kungurian in the type locality and are in particular due to the assignment of the Filippovo horizon either to the Kungurian or to the Artinskian. The Filippovo horizon being at present assigned to the Lower Kungurian, salt deposits developed at this level in the Pre-Caspian Depression should be assigned to the Kungurian as well. The age of some salt sequences found in the eastern Pre-Caspian Depression is taken as conventional. This concerns the Ufimian and Tatarian salt strata.

One of the most puzzling problems of the Permian stratigraphy which still remains disputable is a correlation of the Permian subdivision accepted in Western Europe and that of the USSR. Within the Central European Basin Permian formations are divided into the Rotliegende subdivided further into the Autunian (Lower Rotliegende) and Saxonian (Upper Rotliegende) and assigned to the Lower Permian, and the Zechstein, corresponding to the Upper Permian. It is usually assumed that the Autunian corresponds to the lower sub-series, and the Saxonian to the upper subseries of the Lower Permian (Permian System . . . 1966). In this connection salt strata found in the Rotliegende in the North German and East England Depressions are, most likely, Kungurian and, probably, Artinskian.

As to the Zechstein deposits, the data available (Suveizdis 1963; Permian System . . . 1975) suggest that they might be correlated with the Ufimian, Kazanian, and Tatarian (Upper Permian). The lower Werra Series is, probably, Ufimian, but its salt unit and its equivalent in the Polish-Lithuanian Depression – the Pregol Formation – should, evidently, be dated as Kazanian (Suveizdis 1963). According to this correlation-the Stassfurt, Leine, and Aller salt sequences should be of Tatarian stage (Upper Permian) (see Fig. 22). At the same time, it should be borne in mind that due to the probable Late Permian age of many salt units in the East European Basin, which are now assigned to the Kungurian, as was mentioned above, it is highly probable that they are correlatives with the Zechstein deposits, which will affect the interregional correlation.

At present it is still very difficult to correlate Permian deposits in southern Europe with coeval formations of the Central and East European Basins. Thus, the salt sequence of the Alpine Basin is more likely Upper Zechstein (Kazanian and Tatarian stages); most investigators suppose that this sequence occurs above the Zechstein salt series and probably contributes to the Upper Permian sequence (Klaus 1955, 1972; Schauberger et al. 1976), or partly belongs to the Lower Triassic (Werfenian). A stratigraphic position of salt and sulfate sequences in the North Italian, Dinarids, Mecsek, Rakhov, Dobruja, and Moesian Basins is not quite clear.

It has already been noted that in some salt basins even Permian age of the salt strata is tentatively determined. These are salt sequences developed in the North Mexican and Peru-Bolivian Basins; they might well be younger, Mesozoic. In the correlation chart these strata are conventionally shown as Early Permian, namely Kungurian, since they are confined to rocks referred to, presumably, as Leonardian (Benavides 1968; Salas 1968a,b).

A stratigraphic position of salt strata in the Chu Sarysu Basin is not clear either. On the whole, they are dated as Permian. But at present it is impossible to solve the problems not only at the rank of a stage, but even at the rank of the Permian system. It can only be assumed that the Zhidelisai and the Karakyr salt formations would be the Early Permian (Artiskian/Kungurian), and the Kingir, Sorkol, and Tuzkol formations Early/Late Permian (Late Kungurian/Ufimian and, possibly, Early Kazanian).

The age of many sulfate strata is also uncertain. They have been discussed in the section dealing with evaporite basins.

Now let us consider a peculiar age distribution of Permian salt strata on the Earth and stages of halite accumulation in Permian time. It should be remembered that stages of the Permian halite accumulation discussed below are, of course, more or less reliable; they result from a proposed interregional correlation of the Permian beds; if one correlates in a different way, which is also quite possible, the conclusion concerning the stages of halite sedimentation will differ greatly.

When discussing a stratigraphic position of Permian salt deposits it should be taken into account that many salt units on the Earth first appeared as early as the Sakmarian age. During that time rock salt strata accumulated in Europe, in the Dnieper-Donets Depression, and probably in the Mezen Depression within the East European Basin, as well as in North America within the Midcontinent and Supai Basins. It might be assumed that in the Sakmarian at least eight salt units were formed, four of them (Ingleside, Broom Creek, Wellington, and lower Owl Canyon) accumulated within the Midcontinent Basin, two (Middle Supai and lower Upper Supai) within the Supai Basin and two (Kramatorsk and Upper Kuloi) within the East European Basin. Thus, the Sakmarian age was the time of rather extensive halite accumulation over vast areas of North America and Eurasia. In contrast, the Asselien age was the time of a rather limited halite accumulation. The Asselian salt deposits (the Nikitovo and Slavyansk sequences) are reliably known only from one region, the Dnieper-Donets Depression in the East European Basin. Peculiarities of halite sedimentation suggest that the Asselian is very similar to the Orenburgian age (Carboniferous), as if they formed together a single stage of episodical salt accumulation. Beginning with the Sakmarian, a new stage of halite formation can be clearly distinguished.

During the Artinskian, salt accumulation in North America continued in a number of regions within the Midcontinent Basin, where this age is apparently shown by the Cimarron salt sequence and the upper Owl Canyon salt member, and in the Supai

Basin where most of the upper Supai salt sequence is apparently Artinskian. In Eurasia salt strata formed within the Central European Basin (lowermost Upper Rotliegende) and possibly within the Chu-Sarysu (the Lower Zhidelisai and Kingir sequences).

Most of the Permian salt units are of Kungurian age. They occur in the Midcontinent, Central and East European, Chu-Sarysu Basins and probably also in the North Mexican and Peru-Bolivian Basins. The Kungurian deposits are represented by the Opeche salt sequence of the Denver and Julesburg Depressions, the Opeche salt sequence of the Williston Depression, Flowerport, Blaine, and the Yelton, lower San Andres-Artesia salt sequence in the North Mexican Basin, Mitu and Chuquichambi, the upper Rotliegende, Berezniki, a salt sequence of the Upper Pechora and Pre-Caspian Depressions, Iren sequence of the Chusovaya and Yuryuzan-Sylva Depressions, and middle Zhidelisai-Kingir. On the whole, the Kungurian salt sequences occur in Eurasia, and North and South America, i.e., they are widely spread on the Earth.

The Ufimian salt deposits are known from the Midcontinent Basin, where they are exemplified by parts of the San Andres-Artesia and Pine salt sequences. The Ufimian salt deposits are also recorded in the East European Basin (the upper Berezniki sequence and probably a salt sequence developed in the eastern Pre-Caspian Depression). Salt-bearing formations may also be developed in the Chu-Sarysu and Moesian Basins. The Ufimian was the time in which halite accumulation was more limited than in the Kungurian. In this connection it should be once again emphasized that this general pattern is determined by the accepted mode of correlation. If it proves more reasonable that many salt units now assigned to the Kungurian are, in fact, Ufimian, then, naturally, halite accumulation in the Ufimian could be much more extensive than in the Kungurian. Salt deposits of Kazanian age were found in the basins of Midcontinent (the upper San Andres-Artesia and Pine sequences), Central European (Werra and Stassfurt sequences), and East European Basins (a Hydrochemical sequence). They might also be developed within the Alpine, Moesian, and Chu-Sarysu Basins. The rock salt in the Tatarian beds has been found in five basins: Midcontinent (Castile salt sequence), Central European (Leine and Aller salt sequences), Alpine (salt sequences of the Eastern Alps and Spišsko-Gemerskoe Rudohoři), Moesian, and presumably East European Basins.

Thus, most of the Permian, except the Asselian, appears to make up a single salt stage when salt strata deposited contemporaneously in several regions of the Earth.

Let us analyze the stages of potassium accumulation during the Permian period. At present potash salt horizons have been distinguished within five Permian salt basins: East European, Central European, Alpine, Midcontinent, and Supai. The oldest of them are confined to the Dnieper-Donets Depression of the East European Basin. There, the potash salts are confined to the Kramatorsk salt sequence and are probably of Sakmarian age. The potassium mineral inclusions and interbeds occur almost throughout the section of the Kramatorsk Formation. Thus, the first (lowermost) complex is characterized by polyhalite impregnations and layers of kieserite-halite, carnallite-kieserite, and carnallite rock, whereas the second consists of sylvinite interbeds, and the third is made up of a layer of variegated sylvinite (3.35 m); within the fourth complex two potassium beds, 7.5 m and 5.5 m thick, are distinguished, and the upper part of the fifth complex is represented entirely by the rock and potash salts; it is known as the Chasov-Yar sylvinite-carnallite horizon, about 60 m thick

(Halogenic . . . 1968). It is only within the East European salt basin that Sakmarian potash salts were found and, thus, the Sakmarian age might be regarded as the episodical potassium time in the history of potassium sedimentation. The Asselian was potassium-free, since no evidence of potassium series was reported for that time.

It is only in the Supai Basin that potash salts have been found among Artinskian (Early Permian) salt deposits. There they have been recorded in the upper Supai Formation within the third salt zone characterized by thin layers of sylvine rocks (Peirce and Gerrard 1966). Thus, the Artinskian, even to a greater degree than the Sakmarian, seems to be the time of minor episodical potassium accumulation.

The Kungurian is often regarded as the age of extensive potassium accumulation. This assumption issues, first of all, from the extremely extensive development of potash salt horizons within the East European Basin where the largest potassium deposit in the world (Solikamsk) is located, as well as thick potassic horizons traceable for the entire upper Pechora and Pre-Caspian Depressions. At the same time, the potassium accumulation in the Kungurian took place only in the East European Basin, whereas in the other salt basins of this time no potash salts have as yet been found and, probably, they did not deposit within these basins. Even this allows the conclusion that the Kungurian age was episodical potassium time, as well as the two preceding ages all over the globe. However, it differs considerably from the Artinskian and Sakmarian in a more extensive area and intensity of potassium sedimentation, which will be discussed in detail in the next chapter.

Within the East European Basin potash salts were found in three isolated areas: the Solikamsk, Upper Pechora, and Pre-Caspian Depressions. In the Solikamsk Depression potash salts cover the area exceeding 3000 km^2. The thickness of a potassium sequence attains 150 m. This unit is represented by the alternation of potash and rock salt beds, and is usually subdivided into two horizons: the lower, sylvinite, and the upper, sylvinite-carnallite horizons. The thickness of the first one varies from 7–8 m to 30–40 m and that of the second one 20 m to 115 m. Four major potassic beds are distinguished within the sylvinite horizons where, as in the sylvinite-carnallite horizon, they are nine in number. In the upper Pechora Depression the thickness of a potash salt sequence composed of sylvinite and carnallite rocks alternated with rock salt and halopelite varies from 9 m to 55 m. Three potassium zones are distinguished within the Pre-Caspian Depression: Ulagan, Elton, and Chelkar, comprising not less than 11 potassium horizons. The thickness of the potassium zone of the salt sequence attains several hundred meters. Thick sequences of bischofite rocks have also been recorded in many regions. Potash salts, usually represented by carnallite, sylvine-carnallite, and sylvine, polyhalite, kieserite, and other potassic rocks of sulfate composition, were also reported. At present it is assumed that all the above horizons in the Solikamsk, upper Pechora, and Pre-Caspian Depressions belong to the uppermost Kungurian, being enclosed in the upper Iren horizon. It has already been noted that the age of these deposits is questionable. Some workers consider them Ufimian (Upper Permian).

The development of potash salts of Ufimian age is very limited. They have by now been found only within the Solikamsk Depression, in the East European Basin. The Ufimian time in potassium sedimentation in this basin, so to speak, crowned the preceding Kungurian cycle of evaporite sedimentation, forming thus, in fact, a single stage. For the Earth in general the Ufimian age was undoubtedly episodical potassium time.

It is also only within one salt basin, the Central European, that potash salts occur among the Kazanian (Upper Permian) salt deposits. But they are extremely widespread within this basin. Potash salts of this age are confined to two stratigraphical levels: the Werra and Stassfurt series. In the Werra Series they occur in the Werra Fulda, Lower Rhine, and Polish-Lithuanian Depressions. Two potassic layers are known within the Werra Fulda Depression: Thüringian and Gessen. Their thickness varies from 2 m to 10 m and from 2 m to 15 m, respectively. They are represented by sylvine, carnallite, kieserite, and other potassic rocks. In the Lower Rhine Depression, in the middle Werra Series, there is a potassic zone, about 10 m thick, composed of potassic rocks of chloride and sulfate types, and rock salt. In the Polish-Lithuanian Lowland polyhalite rocks are developed. Potash salts in the Stassfurt Series are spread through almost all of the Central European Basin. They are distinguished as a bed of the same name, with thicknesses from several meters to about 40 m. In the interior zones the bed is composed of carnallite, and in the marginal parts it is represented by various sulfate and chloride salts: carnallite, sylvine, kieserite, langbeinite, loeweite, kainite, polyhalite, and mixed salts. With its abundance of potash salts the Kazanian age is similar to the Kungurian, but is has to be distinguished as episodical salt time, since potassium sedimentation took place in a single, though large, basin.

Among the Tatarian deposits of the Upper Permian potash salts have been reported from three basins: the Central European, Alpine, and Midcontinent Basins. Within the Central European Basin four potassic layers are known (Ronnenberg, Bergmanssegen, Riedel, and Albert) confined to the Leine Series, and one potassic zone, probably corresponding to the Aller Series in the East England Depression. Potash salts of the Leine Series are spread through the interior zone of the basin, whereas those of the Aller Series occur only in the west. The thickest potassic beds are the Ronnenberg (4–5 m), Riedel (6–10 m), and that of the Aller Series in Yorkshire (4–8.6 m). Within the Alpine Basin potash salts in the form of sylvine inclusions and interbeds in the rock salt have been recorded from the Spišsko-Gemerskoe Rudohoři. They are tentatively assigned to the Tatarian. In the Midcontinent Basin potash salts are developed in the south, in the West Texas Region. These salts are mainly confined to the Salado Formation where a potassic zone, McNutt, is distinguished, its thickness being 40 m to 180 m. It is represented by sylvine, carnallite, langbeinite, kieserite, kainite, leonite, and polyhalite alternated with rock salt; glauberite and tenardite rocks were also found. In spite of a much more extensive distribution of potash salts of Tatarian age which must have accumulated in Europe and in North America, the Tatarian age is still considered as an episodical potassium time. This conclusion has to be drawn, for one cannot be quite certain that potassic horizons in all three basins are of the same age. A contemporaneous potassium accumulation could probably have taken place only at the very end of the Tatarian age, but even in this case in each of the basins it was rather limited as compared to rock salt accumulation, and lasted for a very short time. Potassium sedimentation in the Tatarian age was also episodical all over the globe.

Thus, potassium accumulation during the Permian period began with the Sakmarian age and continued to the end of the Permian. During all this time potassium accumulation was episodical mostly in one of the salt basins, and it is only in the Late Tatarian age that it became more extensive and was probably contemporaneous in three basins. These data allow us to distinguish two unequal stages within the Permian period based on the history of potassium sedimentation: the Asselian potassium-free stage and the stage of episodical potassium accumulation comprising the rest of the Permian from Sakmarian to the end of the Tatarian age; the second stage may be named Permian. It should be remembered that the Asselian potassium-free age succeeded Late Carboniferous time when potash accumulation did not take place on the Earth either. Thus, this major potassium-free stage might be called Late Carboniferous/ Early Permian.

Sulfate sedimentation during the Permian period was recorded in Eurasia, North and South America, being, in fact, continuous. Even in the Asselian age of the Early Permian when salt strata were episodically accumulated, sulfate was deposited in four evaporite basins: the East European, Sverdrup, North Italian, and Midcontinent Basins. During this time in the East European Basin the Upper Kartamysh sulfate sequence formed in the Dnieper-Donets Depression and the lower part of a sulfate sequence on the rest of the basin territory. In Asselian time deposition of the Mount Bayley sulfate sequence terminated in the Sverdrup Basin. Probably in the North Italian Basin the formation of the Collio sulfate sequence began. Later sulfate sedimentation became even more extensive and was contemporaneous in many evaporite basins. Thus, in the Sakmarian age sulfate deposits accumulated in five salt basins and two sulfate basins, in the Artinskian in seven salt basins and two sulfate basins, in the Kungurian in seven salt basins and four sulfate basins, in the Ufimian in five salt basins and four sulfate basins, in the Kazanian and Tatarian in five salt basins and five sulfate basins (see Tables 1 and 2). Thus, the Permian period, like the preceding Paleozoic periods, was characterized by a continuous sulfate accumulation, and so they were united into a single epoch of sulfate sedimentation which began in the Cambrian.

Stages of Sulfate, Halite, and Potassium Accumulation in the Paleozoic

The above data on the age of sulfate and salt strata, as well as potash salt horizons and their distant correlation (Fig. 23) imply that the history of sulfate, halite, and potassium accumulation in the Paleozoic differed in some peculiar features and periodicity. The existing theories concerning the stages of halogenesis are to be revised, because evaporite deposition up to now was considered only in general terms. First, peculiar features of each stage of sulfate, halite, and potassium sedimentation should be discussed; and only then could one restore a general picture of the evaporite deposition periodicity.

The data available point, undoubtedly, to an almost continuous pattern of sulfate sedimentation in the Paleozoic. From the Cambrian to the Permian sulfate sediments were deposited in this or that basin, sometimes a deposition being contemporaneous within several basins. Sulfate accumulation in the Paleozoic took place at all stages and, thus, the Paleozoic era in general appears to be a single sulfate epoch.

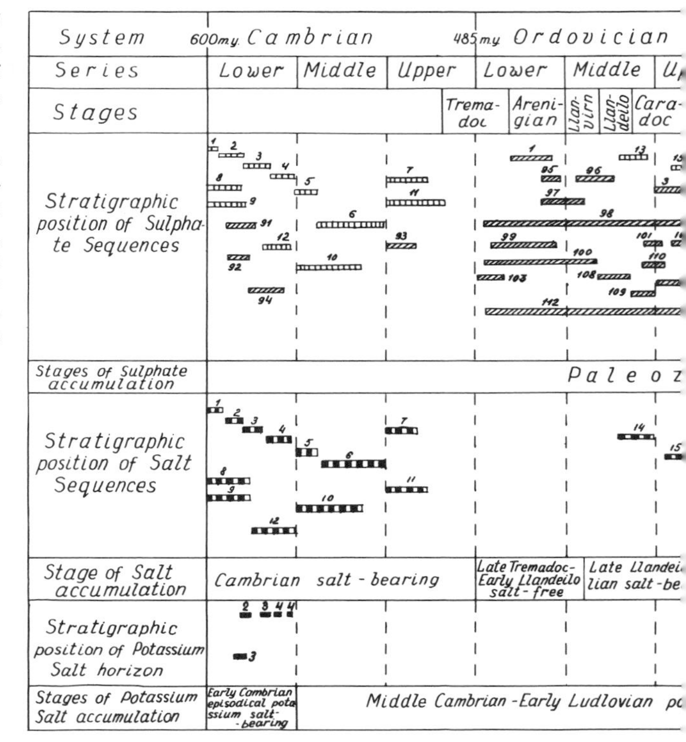

Fig. 23. Stratigraphic position of salt and sulfate beds and potassium horizons; global stages of evaporite accumulation for the Paleozoic.

1 sulfate rocks in salt sequences and their numbers, *2* sulfate sequences and their number, *3* salt sequences and their number, *4* potassium horizons confined to salt sequences and a number of the latter. Figures refer to numbers of salt and sulfate units

It was probably in the Precambrian and Vendian that the epoch of a continuous sulfate sedimentation began bracketing the time from the Mesocenozoic through the Phanerozoic. The intensity of sulfate accumulation was probably different, which will be exemplified by Paleozoic time for which the volume and area of sulfate accumulation are discussed. However, it might be noted that there is one distinct landmark in the Phanerozoic shown by the abrupt change in the sulfate formation process. It corresponds to the Lower/Middle Devonian boundary. Up to this landmark during the Cambrian, Ordovician, Silurian, and Early Devonian sulfate strata formed within a few evaporite basins, and since the Eifelian sulfate sedimentation became much more extensive and took place in many regions. Figure 23 clearly shows the above landmark. We emphasize it because, as will be seen from the subsequence discussion, the Early/Middle Devonian boundary was a turning point in the history of halite and potassium accumulation, i.e., the boundary was one of the most distinct landmarks in the history of the Peleozoic evaporite sedimentation.

It should be noted that a continuous epoch of sulfate accumulation might be distinguished for the Earth in general, using the data available for the continents.

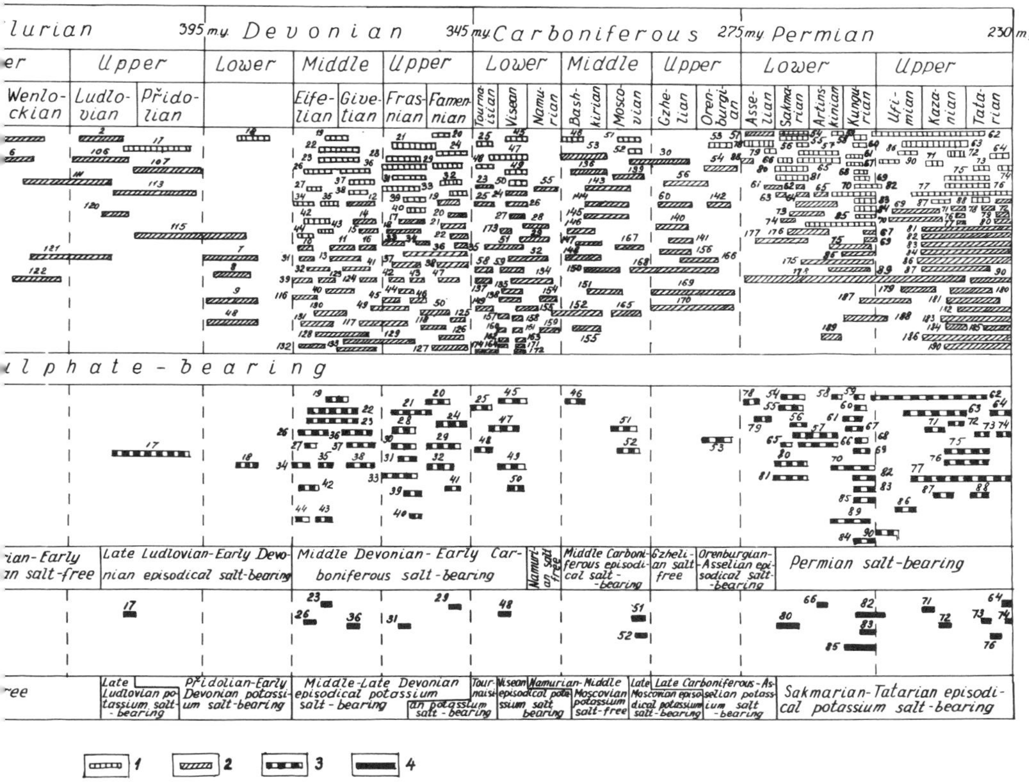

On some continents sulfate sedimentation was not so continuous, and may be divided into more or less distinct stages (Fig. 24). It is only for Eurasia that a single sulfate epoch can be established. The history of sulfate formation on other continents was quite different. Thus, in North America the Paleozoic sulfate sedimentation probably began in the Late Cambrian and was not recorded earlier. This allows us to subdivide the Paleozoic epoch of the evolution of North America into two unequal stages based on the history of sulfate accumulation: the Early/Middle Cambrian sulfate-free and Late Cambrian/Permian sulfate stages. In South America the formation of sulfate series was recorded in the Early Cambrian, Late Carboniferous, and Early Permian. Thus, at present the history of Paleozoic sulfate formation in South America is subdivided into four stages: The Early Cambrian episodical sulfate, Middle Cambrian/Middle Carboniferous sulfate-free, Late Carboniferous/Early Permian sulfate, and Late Permian sulfate-free stages. The stages of sulfate sedimentation in Africa were slightly different. There, the Early Cambrian episodical sulfate stage is also distinguished. Beginning with the Middle Cambrian to the end of the Early Devonian sulfate accumulation did not occur in Africa, and this time corresponds to a sulfate-free

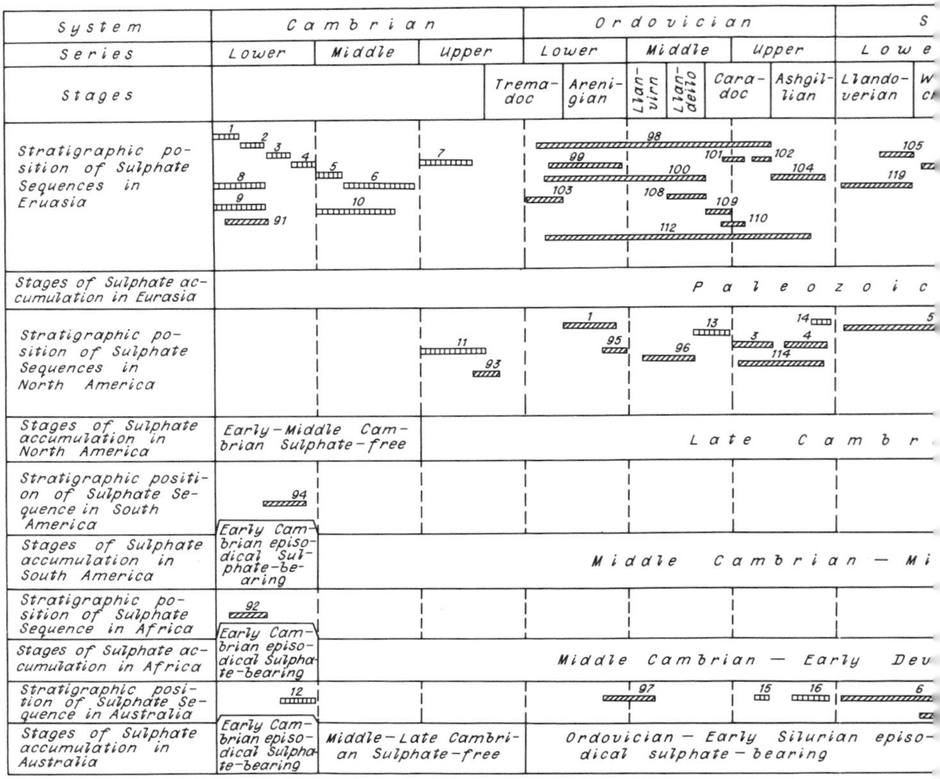

System	Cambrian			Ordovician	S
Series	Lower	Middle	Upper	Lower / Middle / Upper	Lowe
Stages				Tremadoc, Arenigian, Llanvirn, Llandeilo, Caradoc, Ashgillian	Llandoverian, W
Stratigraphic position of Sulphate Sequences in Eruasia	1, 2, 3, 4, 8, 9, 91	5, 6, 10	7	98, 99, 100, 101, 102, 103, 104, 108, 109, 110, 112	105, 119
Stages of Sulphate accumulation in Eurasia	Paleozoic				
Stratigraphic position of Sulphate Sequences in North America			11, 93	1, 95, 13, 96, 3, 4, 14, 114	5
Stages of Sulphate accumulation in North America	Early-Middle Cambrian Sulphate-free		Late Cambr		
Stratigraphic position of Sulphate Sequence in South America	94				
Stages of Sulphate accumulation in South America	Early Cambrian episodical Sulphate-bearing	Middle Cambrian — Mi			
Stratigraphic position of Sulphate Sequence in Africa	92				
Stages of Sulphate accumulation in Africa	Early Cambrian episodical Sulphate-bearing	Middle Cambrian — Early Dev			
Stratigraphic position of Sulphate Sequence in Australia	12			97, 15, 16	6
Stages of Sulphate accumulation in Australia	Early Cambrian episodical Sulphate-bearing	Middle-Late Cambrian Sulphate-free		Ordovician — Early Silurian episodical sulphate — bearing	

Fig. 24. Stratigraphic position of Paleozoic sulfate sequences and stages of sulfate accumulation on continents. For explanation see Fig. 22

stage. A new, in this case sulfate stage, is distinguished spanning the interval from Middle Devonian to the end of the Carboniferous. The Permian period in Africa was sulfate-free. Six stages in the history of the Paleozoic sulfate accumulation can be distinguished in Australia: the Early Cambrian episodical sulfate, Middle/Late Cambrian sulfate-free, Ordovician/Early Silurian episodical sulfate, Late Silurian/Early Devonian sulfate-free, Middle Devonian/Early Carboniferous episodical sulfate, and Middle Carboniferous/Permian sulfate-free stages.

The history of Paleozoic sulfate accumulation suggests that continents are of two groups. The first probably comprises Eurasia and North America where sulfate sedimentation was either continuous, or was longer in the Paleozoic. The second group arbitrarily comprises South America, Africa, and Australia. On these continents the history of sulfate sedimentation in the Paleozoic was characterized by more or less distinct, as a rule different, stages. Thus, it can be noted that a common epoch of Paleozoic sulfate accumulation for the Earth in general is established mainly on the basis of the history of sulfate accumulation on two continents, Eurasia and North America. In spite of this, some features in common are typical of sulfate sedimenta-

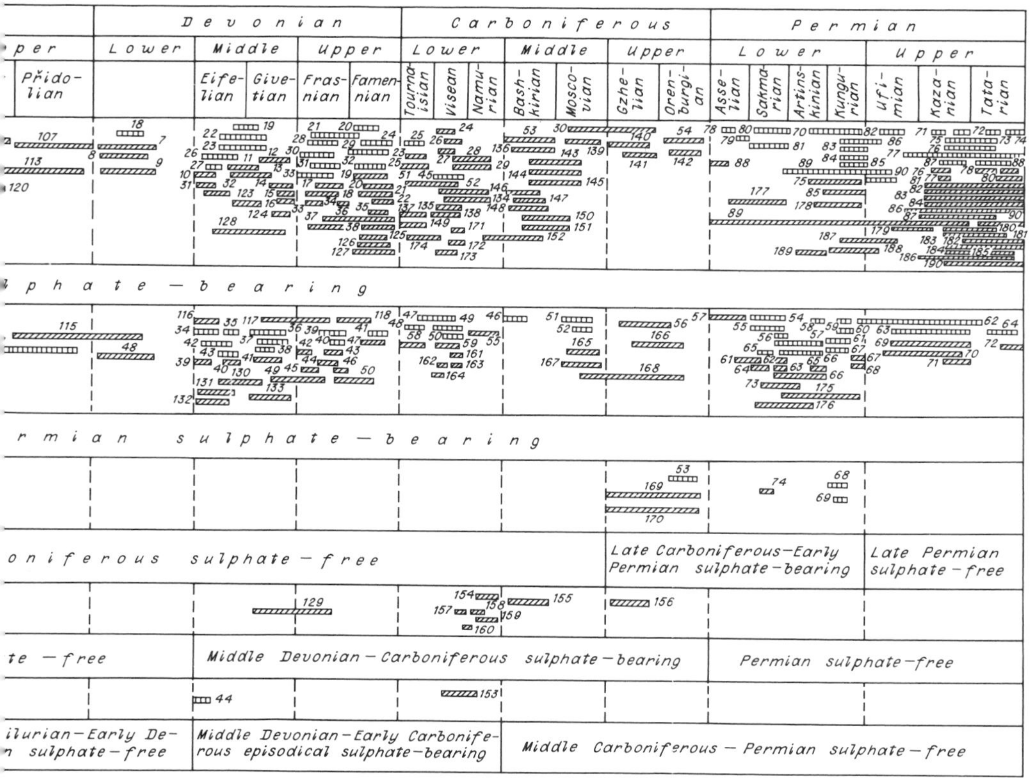

Devonian | Carboniferous | Permian

Lower | Middle | Upper | Lower | Middle | Upper | Lower | Upper

Přidolian | Eifelian | Givetian | Frasnian | Famennian | Tournaisian | Visean | Namurian | Bashkirian | Moscovian | Gzhelian | Orenburgian | Asselian | Sakmarian | Artinskinian | Kungurian | Ufimian | Kazanian | Tatarian

...phate-bearing

...rmian sulphate-bearing

...oniferous sulphate-free | Late Carboniferous–Early Permian sulphate-bearing | Late Permian sulphate-free

...te-free | Middle Devonian–Carboniferous sulphate-bearing | Permian sulphate-free

...ilurian–Early Devonian sulphate-free | Middle Devonian–Early Carboniferous episodical sulphate-bearing | Middle Carboniferous–Permian sulphate-free

tion on all the continents studied. Thus, for instance, in the Early Cambrian sulfate accumulation took place in Eurasia, South America, Africa, and Australia, i.e., it was widespread on the Earth.

The Middle Devonian time of sulfate formation is traced rather distinctly on different continents, being observed in Eurasia, North America, Africa, and Australia. The same applies to Early Carboniferous time. It should be emphasized that, beginning with the Middle Devonian, the number of the continents where sulfate accumulation was recorded increased as compared to that of Paleozoic time, except for Early Cambrian. Thus, the Paleozoic sulfate epoch could be subdivided into three major stages based on sulfate accumulation intensity: Early Cambrian, Middle Cambrian/Early Devonian, and Middle Devonian/Permian. This conclusion will be given in more detail in the next chapter.

Let us pass to the stages of halite accumulation in the Paleozoic. Figure 23 clarly shows that in the Paleozoic history of Earth evolution we can distinguish stages of almost continuous salt deposition in this or that region. There was an alternation of salt accumulation stages either with salt-free stages when rock salt did not deposit, or with episodical salt stages when salt strata formed within isolated evaporite basins.

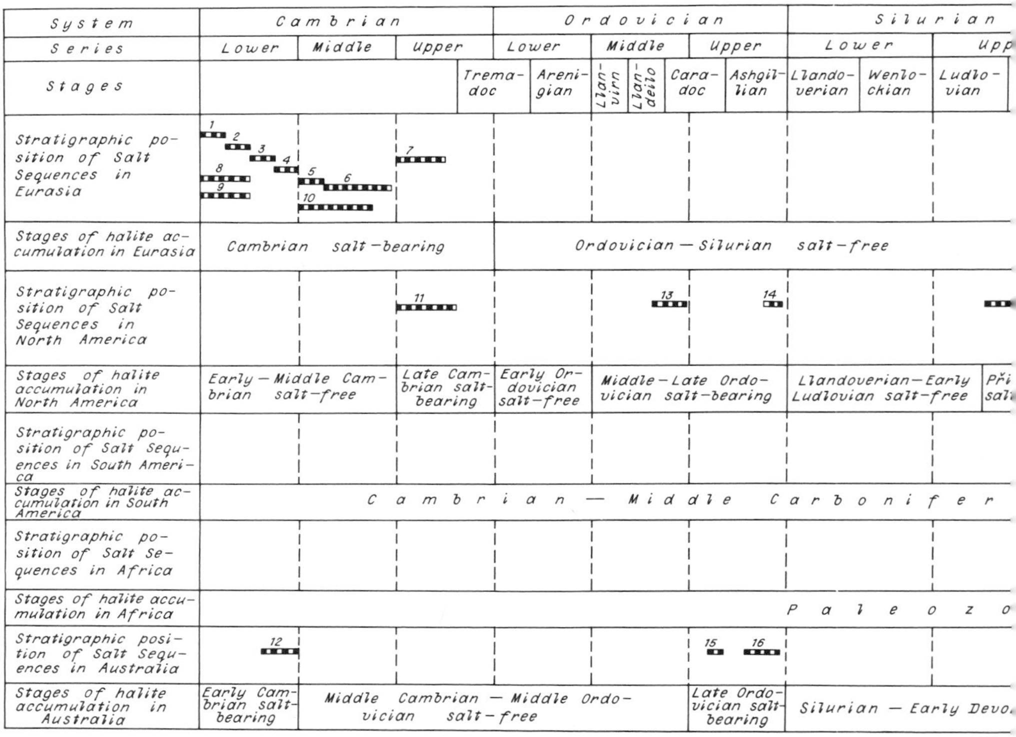

System	Cambrian			Ordovician						Silurian		
Series	Lower	Middle	Upper	Lower		Middle		Upper		Lower		Upp
Stages				Tremadoc	Arenigian	Llanvirn	Llandeilo	Caradoc	Ashgillian	Llandoverian	Wenlockian	Ludlovian
Stratigraphic position of Salt Sequences in Eurasia	1, 2, 3, 4, 8, 9	5, 6, 10	7									
Stages of halite accumulation in Eurasia	Cambrian salt-bearing			Ordovician — Silurian salt-free								
Stratigraphic position of Salt Sequences in North America			11				13		14			
Stages of halite accumulation in North America	Early – Middle Cambrian salt-free		Late Cambrian salt-bearing	Early Ordovician salt-free		Middle – Late Ordovician salt-bearing				Llandoverian – Early Ludlovian salt-free		Při sal
Stratigraphic position of Salt Sequences in South America												
Stages of halite accumulation in South America	Cambrian — Middle Carbonifer											
Stratigraphic position of Salt Sequences in Africa												
Stages of halite accumulation in Africa	Paleozo											
Stratigraphic position of Salt Sequences in Australia	12							15	16			
Stages of halite accumulation in Australia	Early Cambrian salt-bearing	Middle Cambrian — Middle Ordovician salt-free						Late Ordovician salt-bearing		Silurian — Early Devo		

Fig. 25. Stratigraphic position of Paleozoic salt sequences and stages of halite accumulation on continents. For explanation see Fig. 23

At present 11 stages[5] can be recognized in the history of the Paleozoic halite accumulation: (1) Cambrian, (2) Late Tremadocian/Early Llandeilian, (3) Late Llandeilian/Ashgillian, (4) Llandoverian/Early Ludlovian, (5) Late Ludlovian/Early Devonian, (6) Middle Devonian/Early Carboniferous, (7) Namurian, (8) Middle Carboniferous, (9) Gzhelian, (10) Orenburgian/Asselian, (11) Permian.

The Cambrian was the time when halite accumulation was almost continuous on the Earth during ca. 115–120 m.y. The Early Ordovician and the lower Middle Ordovician, from the Late Tremadocian to the Early Llandeilian, are considered as a salt-free stage which lasted about 30–35 m.y. The Late Llandeilian/Ashgillian stage is distinguished as an episodical salt-bearing time on the basis of salt strata occurrence in two evaporite basins, the Canadian Arctic and Canning Basins, the latter being arbitrarily assigned to the Ordovician. The duration of this stage was not less than 30 m.y. The next, Llandoverian/Early Ludlovian (20–25 m.y.) was again a salt-free stage. The Late Ludlovian/Early Devonian stage is at present distinguished as an episodical salt time since salt strata did not form contemporaneously in two widely

5 The number of stages and their age range are different from earlier reported (Zharkov 1974a,b, 1976, 1977) owing to the discovery of some new Ordovician and Carboniferous salt strata

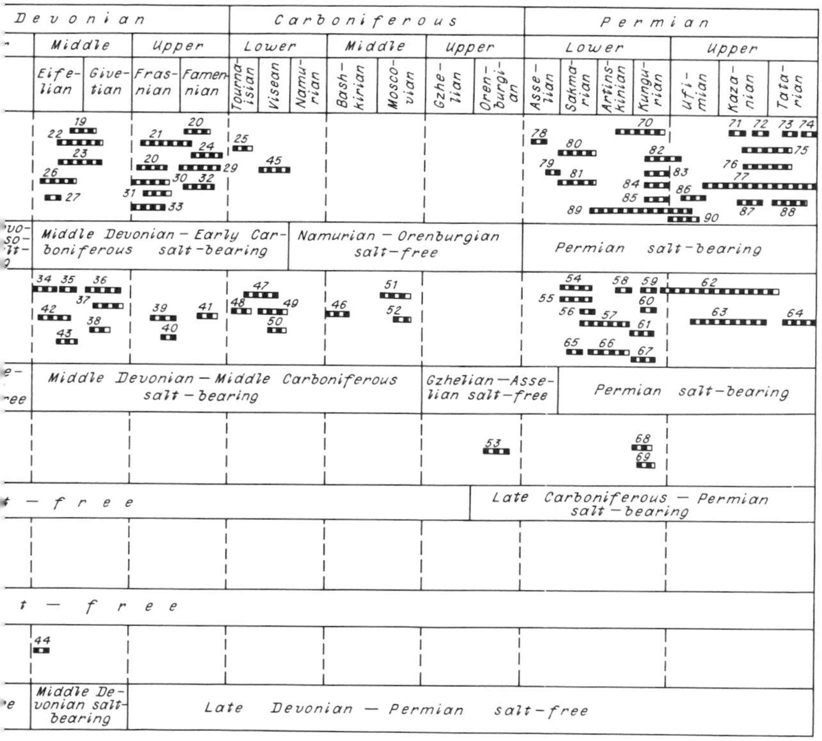

spaces basins: the Michigan-Pre-Appalachian and North Siberian Basins. This stage lasted for about 25–30 m.y. The Middle Devonian/Early Carboniferous stage spanned the time from the Eifelian to the end of the Visean age, i.e., about 55 m.y. During this time salt accumulation on the Earth was continuous. The next four stages (Namurian, Middle Carboniferous, Gzhelian, and Orenburgian/Asselian) were rather short-term. The first lasted for about 10 m.y. and was salt-free, the second (about 15–20 m.y. was episodical salt-bearing, the third (10 m.y.) was salt-free, and the fourth (15–17 m.y.) represented episodical salt-bearing stages. Finally, the last, Permian stage, which began in the Sakmarian time and probably ended at the close of the Paleozoic, lasted 35–40 m.y. and was salt-bearing.

Thus, three salt stages can be distinguished in the history of Paleozoic halite accumulation – Cambrian, Middle Devonian/Early Carboniferous, and Permian; four salt-free stages – Late Tremadocian/Early Llandeilian, Llandoverian/Early Ludlovian, Namurian and Gzhelian; four episodical salt stages – Late Llandeilian/Ashgillian, Late Ludlovian/Early Devonian, Middle Carboniferous and Orenburgian/Asselian.

Three major epochs of about the same length can be recognized within the history of Paleozoic halite accumulation; the Early Paleozoic, corresponding to the Cambrian period when during 115–120 m.y. halite accumulation was continuous in this or that region of the Earth: the Middle Paleozoic (110–115 m.y.), comprising the Ordovician, Silurian, and Early Devonian, marked by the absence of or episodical salt deposition; the Late Paleozoic covering the time from the Eifelian age (Middle Devonian) to the end of the Paleozoic (130–140 m.y.) when salt accumulation on the Earth was again continuous with several short breaks in Carboniferous time. Speaking about the history of halite accumulation it is also interesting to find out whether the stages distinguished in the Paleozoic are sustained, at least generally, on each of the continents, or whether these stages differed so greatly between continents that the established general pattern of Paleozoic halite accumulation seems to be very approximate. For this purpose in Fig. 25 stratigraphic positions of Paleozoic salt strata are shown separately for Eurasia, North and South America, Africa, and Australia. The above data allow us to draw the following conclusions.

The global stages of the Paleozoic halite accumulation are established on the basis of the age distribution pattern of salt strata, mostly on two continents; Eurasia and North America, since there the number of such units is the greatest. The stages of halite accumulation in the Paleozoic were quite different on all the continents. In Eurasia six stages are distinguished (Cambrian salt-bearing, Ordovician/Silurian salt-free, Early Devonian episodical salt-bearing, Middle Devonian/Early Carboniferous salt-bearing, Namurian/Orenburgian salt-free, and Permian salt stages), whereas in North America there are ten stages (Early/Middle Cambrian salt-free, Late Cambrian salt-bearing, Early Ordovician salt-free, Middle/Late Ordovician salt-bearing, Llandoverian/Early Ludlovian salt-free, Přidolian salt-bearing, Early Devonian salt-free, Middle Devonian/Middle Carboniferous salt-bearing, Gzhelian/Asselian salt-free, Permian salt-bearing), in Australia six (Early Cambrian salt-bearing, Middle Cambrian/Middle Ordovician salt-free, Late Ordovician salt-bearing, Silurian/Early Devonian salt-free, Middle Devonian salt-bearing, Late Devonian/Permian salt-free), and in South America the Late Carboniferous/Permian salt-bearing stage can be only arbitrarily established, whereas the remaining time from Cambrian to Middle Carboniferous is marked by a salt-free stage. In Africa the Paleozoic epoch was salt-free.

In spite of these differences the history of halite accumulation had many common features on the continents. Two continents are most similar as to the stages of halite sedimentation: Eurasia and North America. On these continents a similar Permian salt stages is distinguished, and the abrupt extension of halite accumulation on both continents is distinctly recorded at the Early/Middle Devonian boundary. The history of halite accumulation from the Middle Devonian to Early Carboniferous did not change greatly. The Late Carboniferous was a salt-free period both in Eurasia and in North America. In fact, stages in the Early Paleozoic were similar. The differences are as follows: Firstly, at the end of the Late Silurian the Přidolian salt stage not recorded in Eurasia is established in North America; whereas in Eurasia the Early Carboniferous episodical salt-bearing stage is established, and not recorded in North America. Secondly, in North America the Middle/Late Ordovician salt stage is established and in Eurasia it is not recorded. Thirdly, in the Cambrian period not one (as in Eurasia), but two stages exist in North America, the Early/Middle Cambrian stage being salt-free. However, these discrepancies are not so important as it seems at first sight, since during the Ordovician, Silurian, and Early Devonian there was either no halite accumulation on both continents, or it was episodical within isolated basins. Moreover, one should bear in mind the poor knowledge of the Ordovician and Silurian halogenic series in Siberia and the possibility of distinguishing the above Late Silurian stage of salt accumulation there.

The history of the Paleozoic halite accumulation in Eurasia and North America, on the one hand, and Australia, on the other, also has some features in common. On all three continents the salt strata are confined to the Middle Devonian. In Eurasia and Australia halite accumulation was documented in the Early Cambrian, and in North America and Australia in the Late Ordovician. The Early/Middle Ordovician, as well as the Early Silurian, were salt-free periods on all the continents. The boundary of the halite accumulation stage which began in the Eifelian age (Middle Devonian) is continuous. The post-Devonian epoch of the Paleozoic in Australia differs greatly from the history of halite sedimentation in Eurasia and North America observed at the same time; in Australia it was a salt-free time.

At present we can describe only schematically the Paleozoic history of salt accumulation in South America. The evidence available shows that salt deposition on this continent, in fact, began only at the end of the Late Carboniferous and, probably, continued in the Permian. Thus, only Late Carboniferous/Permian salt stage is recorded there. It is most likely that in South America halite accumulation could take place in the Cambrian, which is confirmed by the presence of thick anhydritic strata in the Piedmont Andes (Benavides 1968) and by the nondocumented occurrence of Cambrian rock salt in these regions (Lotze 1968). If this assumption proves true, then we shall be able to distinguish the Cambrian salt in South America, a long-term salt-free stage embracing Ordovician, Silurian, Devonian, and most of the Carboniferous. In this case the Paleozoic history of halite accumulation in South America will be, roughly speaking, similar to that of Eurasia and North America, because on all the three continents the Cambrian and Permian (in South America – Late Carboniferous) salt stages could be distinguished.

These data show similar periodicity in the history of the Paleozoic halite accumulation for several continents. During some time intervals halite accumulation was either ubiquitous or was recorded on most of continents which allows us to regard these time intervals as common for the Paleozoic history of the Earth. They are: the Cambrian period when halite accumulation occurred in Eurasia, North America, and Australia and, probably, in South America; the Early Ordovician and Early Silurian, these being salt-free periods on all the continents; the Permian period when halite sedimentation was extensive in Eurasia, North and, probably, South America. The Middle Devonian boundary of halite accumulation is rather conspicuous, being documented in Eurasia, North America, and Australia.

The data available imply that the stages of halogenic process in general distinguished earlier (Fiveg 1962; Strakhov 1962, 1963; Ivanov and Voronova 1972; Lotze 1968) in fact apply only to the history of halite accumulation. Thus, it is extremely significant that some of the above stages of halite accumulation correlate well with the halogenesis epochs established earlier, such as the Permian, Middle/Late Devonian or Early Cambrian. But, in most cases, it is necessary to determine more precisely the earlier halogenesis epochs, based probably on halite accumulation intensity. It means that halite formation took place not only in the Early Cambrian, but throughout the Cambrian period, which constituted a single salt stage. The Late Silurian can only be regarded as an episodical salt stage; it will be shown below that it was not an epoch of intense salt accumulation, as Lotze (1968) supposed. The boundaries of the Middle/Late Devonian stage distinguished by many workers are to be defined more precisely, for salt accumulation did not cease in the Devonian, but continued up to the Visean, so one could call this salt stage the Middle Devonian/Early Carboniferous.

The above evidence throws light on the problem of contemporaneous halite accumulation in different salt basins. Fiveg (1962), who studied Permian deposits, concluded that halite accumulation was not contemporaneous in different basins, and formation of salt strata took place at close intervals of time. The stratigraphic data available allow more or less definite determination of the time of halite accumulation only within an age. It is rather clearly determined within these limits that the beginning and end of salt deposition in various basins, as a rule, do not fall at the same time. Halite accumulation was not strictly isochronous in all places, not only in different basins, but even within one basin. At the same time, it is quite obvious that during salt stages, within different time intervals, halite accumulation was contemporaneous in many salt basins. It may be exemplified by the Kungurian age when salt accumulation took place contemporaneously in the East and Central European, Chu-Sarysu, Midcontinent, and probably North Mexican and Peru-Bolivian Basins. However, contemporaneous salt accumulation does not mean that the beginning and end of the formation of individual salt sequences were also isochronous in all these basins. It can be definitely stated that there was no strict isochroneity, and Fiveg's conclusions remain correct in this respect.

In the Paleozoic potassium accumulation is obviously of episodical character. At present, the Paleozoic history of potassium sedimentation can be subdivided into 11 stages: Early Cambrian, episodical potassium; Middle Cambrian/Early Ludlovian, potassium-free; Late Ludlovian, episodical potassium; Přidolian/Early Devonian, potassium-free; Middle/Late Devonian, episodical potassium; Tournaisian, potassium-free; Visean, episodical potassium; Namurian/Middle Moscovian, potassium-free; Late Moscovian, episodical potassium; Late Carboniferous/Asselian, potassium-free; Sakmarian/Tatarian, episodical potassium stages.

During stages of episodical potassium accumulation the formation of potash salts was, as a rule, confined to a single salt basin. These are the following basins: in the Early Cambrian – East Siberian, in the Late Silurian – Michigan-Pre-Appalachian, in the Eifelian age (Middle Devonian) – first Morsovo and then Tuva Basin, in the Givetian age (Middle Devonian) – West Canadian, in the Frasnian and Famennian ages (Late Devonian) – Upper Devonian Basin of the Russian Platform, in the Visean age (Early Carboniferous) – Maritime, in the Moscovian age (Middle Carboniferous) – two adjacent basins, Paradox and Eagle, in Sakmarian age (Early Permian) – East European, in the Artinskian age (Early Permian) – Supai, in the Kungurian and Ufimian ages (Early and the beginning of the Late Permian) – East European, in the Kazanian age (Late Permian) – Central European. It is only at the very end of the Paleozoic, in the Tatarian age (Late Permian), that potash salts accumulated in three basins: Central European, Alpine, and Midcontinent.

The age and areal distribution of potash salt horizons might point to a peculiar nature of potassium accumulation in the Paleozoic which differed greatly from sulfate or halite accumulation. Potassium sedimentation was never ubiquitous or even extensive on individual continents. Such a sedimentation could take place under favorable climatic and tectonic conditions; thus, potash salts did not accumulate in all salt basins, but only where paleogeography favored the formation of potassium

(Fiveg 1962; Ivanov and Voronova 1972; Strakhov 1962; Valyashko 1962). It is noteworthy that potassium accumulation was observed not only during Paleozoic salt stages, but also during episodical salt stages when some isolated small salt basins existed, e.g., in Late Moscovian or Late Ludlovian time. In the history of the Peleozoic potassium accumulation was very intense at the Early/Middle Devonian boundary, though not so distinct as in the case of sulfate or halite sedimentation when large potash salt deposits began to form. In general, the Middle and Late Devonian, Carboniferous, and Permian were more favorable for potassium accumulation than the older epochs of the Paleozoic, probably except for the Early Cambrian.

Thus, sulfate accumulation during the Paleozoic was continuous and contemporaneous on different continents. Halite accumulation was characterized by a distinct periodicity reflected in a successive alternation of salt and salt-free, or episodical salt stages. Potassium sedimentation had a distinct episodical character expressed, first of all, in accumulation of potash salts in one of the salt basins of a certain age. In spite of these differences in stages of sulfate, halite, and potassium accumulation, a number of major epochs in the Paleozoic history can also be distinguished in evaporite sedimentation in general. There are three more or less definite epochs: Early Cambrian, Middle Cambrian/Early Devonian, and Middle Devonian/Permian. The first and, especially, the third epoch might be regarded as epochs of extensive evaporite accumulation in general, for this time is characterized not only by more intense sulfate accumulation, but also by halite accumulation and potassium sedimentation. The Middle Cambrian/Early Devonian epoch was marked by less extensive sulfate formation, episodical halite accumulation, and extremely rare formation of the potash salts. The Early/Middle Devonian boundary was very important in the entire process of evaporite sedimentation.

CHAPTER III

Areal Extent and Volume of Evaporites. Epochs of Intense Evaporite Accumulation

Introductory Remarks

There is little published evidence on areal extent and volume of Paleozoic evaporites. Such data mostly concern rock salt. As a rule, the volume and areal extent of rock salt have been estimated only from several salt basins. Thus, Fischer (1968) approximately calculated the volume of the Zechstein rock salt in the Central European Basin, Fiveg and Banera (1968) presented estimates for the volume of Kungurian salts in the East European Basin, Pierce and Rich (1962) provided similar data on the Silurian Michigan-Pre-Appalachian Basin, and Zharkov (1969) on the Cambrian East Siberian Basin. Estimates of the volume and areal extent of rock salt in all the Paleozoic basins, or in most of them, as well as the epochs and periods of the Paleozoic for all the continents, were quite rare. The first to be mentioned are: the monograph and paper by Kalinko (1973a,b), a review by Zharkov (1974a, 1977) and a paper by Meyerhoff (1970b). Some investigations were aimed at estimating the total volume of evaporites for different periods and epochs of the Paleozoic. Among these some well-known reviews can be mentioned (Ronov and Khain 1954–1957; Ronov et al. 1974; Ronov 1976; Ronov et al. 1976) in which a given volume of salt formations characterizes roughly the bulk of rock salt and sulfate rocks. However, there are no data on the estimates of areal extent and volume of sulfate rocks, rock and potash salts. At the same time it is this evidence that is indispensable for establishing epochs of intense sulfate, halite, and potassium accumulation, as well as epochs of intense evaporite sedimentation in the Paleozoic.

The volume and areal extent of halogenic rocks were determined [6] in all the sulfate and salt strata using an estimate of their area, average thickness, and percentage of rocks in a sequence. As a result, the total mean thickness of sulfate rocks, rock and potash salts of chloride and sulfate types was measured for a given evaporite basin, and their volume was calculated. As a rule, minimum values were used and, thus, the resultant data on the area and volume are also the minimum possible estimates. It should be remembered that due to the scarce evidence and difficulty of volume calculations the estimate of the volume of halogenic rocks is only approximate within 10% to 25% for different basins. Only sulfate rocks, rock and potash

6 The volume and areal extent of halogenic rocks in the Permian evaporite basins of Eurasia were calculated by Merzlyakov (1976, 1977) and, partly, by the present author

salts by now found in the Paleozoic basins were used for calculations. The assumed volume of halogenic rocks in the poorly studied regions, and areas of their probable distribution were not taken into account, since at present even an approximate estimate of their volume is impossible. Thus, the real volume of Paleozoic evaporites actually exceeds the calculated values. Masses not taken into account can slightly increase a proportion of total volumes of sulfates, rock and potash salts in individual basins, and during different ages, epochs, and periods of the Paleozoic, but they will not change the general pattern of their relative content.

Regardless of the approximate values, they allow comparison of Paleozoic evaporite basins on the basis of their areal extent and volume of the accumulated halogenic rocks, and also establishment of the distribution pattern of the volumes of sulfate rocks, rock salt, and potash salts among major Paleozoic stratigraphic units on different continents and on the Earth in general, using the earlier determined stratigraphic position of sulfate and salt strata, and to distinguish areas where sulfate, halite, and potassium accumulation took place during certain periods, epochs, and ages of the Paleozoic era.

Areal Extent and Volume of Halogenic Rocks in Paleozoic Evaporite Basins

Let us first compare Paleozoic salt and sulfate basins proceeding from the data on the areal extent and volume of the accumulated halogenic rocks (Tables 3 and 4).

It is definitely determined that the Paleozoic evaporite basins differ considerably in the areas where sulfate deposits are developed. Sulfates are most abundant in salt basins, among which the most important are: Cambrian – East Siberian (2×10^6 km^2) and Iran-Pakistan (3×10^6 km^2); Devonian – North Siberian (1.5×10^6 km^2); Permian – Central (1×10^6 km^2) and East European (2×10^6 km^2). In these basins the area of sulfate rocks exceeds 1,000,000 km^2.

The area under sulfates in most salt basins is several hundred thousand square kilometers. They are: Mackenzie, the Ordovician of the Canadian Arctic Archipelago, the Ordovician/Silurian Williston, Canning, Michigan-Pre-Appalachian, the Devonian/Carboniferous Chu-Sarysu, Morsovo, the Upper Devonian of the Russian Platform, West Canadian, Hudson, Sverdrup, the Carboniferous Williston, Maritime, Amazon, Midcontinent, Peru-Bolivian, Moesian. In other salt basins sulfate rocks occupy several tens of thousands of square kilometers. At present there are 9 basins of this type (Tuva, Michigan, Adavale, Mid-Tien Shan, Paradox, Eagle, Supai, Alpine, and the Permian Chu-Sarysu). In only two salt basins (Amadeus and Saltville) does the area of sulfate development not exceed 10,000 km^2, but these basins have not yet been outlined. Thus, in the Paleozoic salt basins the area of the sulfate rocks is mainly over 10,000 km^2, never being less than this value.

On the contrary, within sulfate basins the sulfates usually occupy several tens of thousands or several thousands of square kilometers. Fortynine of 63 sulfate basins have such an area of sulfate accumulation. It is only in 12 sulfate basins that the area exceeds 100,000 km^2, except for the Carboniferous East European Basin where it is 1,000,000 km^2.

Table 3. Area of distribution and volume of sulfate rocks, salt rocks, and potassium-salt rocks within

Basin	Area of basin	Name of salt and sulfate sequence	Sulfate rocks			Salt rocks
			Area km^2	Thickness km	Volume km^3	Area km^2
1	2	3	4	5	6	7
East Siberian	4.5–5 x 10^6	Irkutsk	1.5 x 10^6	0.01	1.5 x 10^4	1 x 10^4
		Usolye	2 x 10^6	0.04	8 x 10^4	1.5–2 x 10^6
		Belsk	1.3 x 10^6	0.025	3.5 x 10^4	1.3 x 10^6
		Angara	1.2 x 10^6	0.05	6 x 10^4	1.2 x 10^6
		Litvintsevo	8 x 10^5	0.01	8 x 10^3	5 x 10^5
		Mayan	6 x 10^5	0.003	1.8 x 10^3	2 x 10^5
		Upper Cambrian	6 x 10^5	0.001	6 x 10^2	2 x 10^5
		Total	2 x 10^6		2 x 10^5	1.5–2 x 10^6
Iran-Pakistan	5 x 10^6	Hormoz and Punjab	3 x 10^6	0.01	3 x 10^4	2 x 10^6
		Mila	3 x 10^4	0.02	6 x 10^2	1 x 10^4
		Total	3 x 10^6		3.06 x 10^4	2 x 10^6
Mackenzie	9 x 10^5	Salina River	6 x 10^5	0.01	6 x 10^3	5 x 10^5
Amadeus	1.5 x 10^5	Chandler	2 x 10^3	0.01	2	1 x 10^3
Canadian Arctic	7 x 10^5	Bay Fjord	1 x 10^5	0.03	3 x 10^3	1 x 10^5
		Baumann Fiord	1 x 10^5	0.05	5 x 10^3	–
		Read Bay	1.5 x 10^5	0.005	7.5 x 10^2	–
		Total	1.5 x 10^5		9.75 x 10^3	1 x 10^5
Williston	3 x 10^5	Stonewall	1 x 10^5	0.002	2 x 10^2	?
		Red River	1.5 x 10^5	0.005	7.5 x 10^2	–
		Stony Mountain	1 x 10^5	0.001	1 x 10^2	–
		Interlake	1 x 10^5	0.001	1 x 10^2	–
		Total	1.5 x 10^5		1.5 x 10^3	?
Canning	5 x 10^5	Lower Carribady	3 x 10^3	0.01	3 x 10^3	2.5 x 10^5
		Upper Carribady	3 x 10^5	0.02	6 x 10^3	2.5 x 10^5
		Stottid	3 x 10^5	0.005	1.5 x 10^3	–
		Total	3 x 10^5		1.05 x 10^4	
Michigan-Pre-Appalachian	5 x 10^5	Salina	3 x 10^5	0.01	3 x 10^3	2.6 x 10^5
North Siberian	2.5 x 10^6	Zubovo	2 x 10^3	0.01	20	5 x 10^2
		Manturovo and Fokin	4.5 x 10^3	0.02	90	5 x 10^2
		Kygyltuus	5 x 10^4	0.005	2.5 x 10^4	1.5 x 10^4
		Nordvik	1.2 x 10^5	0.01	1.2 x 10^3	1.2 x 10^5
		Zubovo	7.5 x 10^4	0.01	7.5 x 10^2	–

Paleozoic salt basins

		Chloride type of potassium salt			Sulfate type of potassium salt			Total volume of sulfate and salt rocks
Thickness km	Volume km^3	Area km^2	Thickness km	Volume km^3	Area km^2	Thickness km	Volume km^3	Volume km^3
8	9	10	11	12	13	14	15	16
0.02	2×10^2	–	–	–	–	–	–	1.52×10^4
0.2	3×10^5	5×10^4	0.0003	15	–	–	–	3.8×10^5
0.1	1.3×10^5	?	?	?	–	–	–	1.65×10^5
0.1	1.2×10^5	4×10^4	–	25	–	–	–	1.8×10^5
0.05	2.5×10^4	–	–	–	–	–	–	3.3×10^4
0.025	5×10^3	–	–	–	–	–	–	6.8×10^3
0.025	5×10^3	–	–	–	–	–	–	5.6×10^3
	5.85×10^5	6×10^4		40				7.85×10^5
300	6×10^5	?	?	?	?	?	?	6.3×10^5
0.1	1×10^3	–	–	–	–	–	–	1.6×10^3
	6.01×10^5							6.32×10^5
0.2	1×10^5	–	–	–	–	–	–	1.06×10^5
0.1	1×10^2	–	–	–	–	–	–	1.02×10^2
0.1	1×10^4	–	–	–	–	–	–	1.3×10^4
–	–	–	–	–	–	–	–	5×10^3
–	–	–	–	–	–	–	–	7.5×10^2
	1×10^4							1.88×10^4
?	?	–	–	–	–	–	–	2×10^2
–	–	–	–	–	–	–	–	7.5×10^2
–	–	–	–	–	–	–	–	1×10^2
–	–	–	–	–	–	–	–	1×10^2
?	?	–	–	–	–	–	–	1.15×10^3
0.3	0.75×10^4	–	–	–	–	–	–	1.05×10^4
0.3	0.75×10^4	–	–	–	–	–	–	1.35×10^4
–	–	–	–	–	–	–	–	1.5×10^3
	1.5×10^4							2.55×10^4
0.1	2.6×10^4	3.6×10^4	0.0005	18	–	–	–	2.9×10^4
0.01	5	?	?	?	–	–	–	25
0.01	5	?	?	?	–	–	–	95
0.1	1.5×10^3	–	–	–	–	–	–	1.75×10^3
0.1	1.2×10^4	–	–	–	–	–	–	1.32×10^4
–	–	–	–	–	–	–	–	7.5×10^2

Table 3 (continued)

Basin	Area of basin	Name of salt and sulfate sequence	Sulfate rocks			Salt rocks
			Area km^2	Thickness km	Volume km^3	Area km^2
1	2	3	4	5	6	7
		Gypsum-bearing sequence of Taimyr	1×10^3	0.01	10	–
		Dezhnev	7×10^4	0.01	7×10^2	–
		Rusanov	5×10^4	0.005	2.5×10^2	–
		Sidin	2×10^5	0.005	1×10^3	–
		Yukta	5×10^3	0.003	15	–
		Atyrkan	1×10^4	0.01	1×10^2	–
		Burkhala	2×10^4	0.005	1×10^2	–
		Sebechan	5×10^4	0.005	2.5×10^2	–
		Vayakh	1×10^4	0.01	1×10^2	–
		Nakokhoz	2×10^5	0.002	4×10^2	–
		Kalargon	1×10^3	0.002	2	–
		Upper Fokin	4×10^3	0.01	40	–
		Namdyr	1×10^5	0.001	1×10^2	–
		Matusevich	2×10^4	0.001	20	–
		Gypsum-bearing sequence of Vrangel Island	2×10^3	0.001	2	–
		Gypsum-bearing sequence of Kyutingda Depression	3×10^4	0.005	1.5×10^2	–
		Tundrin	2×10^3	0.001	2	–
		Kurunguryakh	5×10^4	0.001	50	–
		Sulfate-carbonate sequence of Khatanga Depression	1×10^5	0.001	1×10^2	–
		Total	1.5×10^6		4.7×10^3	1.36×10^5
Tuva	?	Ikhedushiingol	2×10^4	0.01	2×10^2	1×10^4
Chu-Sarysu	3×10^5	Lower salt sequence	5×10^4	0.01	5×10^2	5×10^4
		Upper salt sequence	5×10^3	0.01	50	5×10^3
		Sulfate sequence	1.5×10^5	0.005	7.5×10^2	–
		Sulfate-bearing carbonate-terrigene sequence	1.5×10^5	0.005	7.5×10^2	–
		Sulfate sequence of Ulkunburul, Kashiburul and Tekturmas Mountains	3×10^4	0.01	3×10^2	–
		Kyzylkanat	2×10^5	0.001	2×10^2	–
		Total	2×10^5		2.55×10^3	5×10^4

		Chloride type of potassium salt			Sulfate type of potassium salt			Total volume of sulfate and salt rocks
Thickness km	Volume km^3	Area km^2	Thickness km	Volume km^3	Area km^2	Thickness km	Volume km^3	Volume km^3
8	9	10	11	12	13	14	15	16
–	–	–	–	–	–	–	–	10
–	–	–	–	–	–	–	–	7×10^2
–	–	–	–	–	–	–	–	2.5×10^2
–	–	–	–	–	–	–	–	1×10^3
–	–	–	–	–	–	–	–	15
–	–	–	–	–	–	–	–	1×10^2
–	–	–	–	–	–	–	–	1×10^2
–	–	–	–	–	–	–	–	2.5×10^2
–	–	–	–	–	–	–	–	1×10^2
–	–	–	–	–	–	–	–	4×10^2
–	–	–	–	–	–	–	–	2
–	–	–	–	–	–	–	–	40
–	–	–	–	–	–	–	–	1×10^2
–	–	–	–	–	–	–	–	20
–	–	–	–	–	–	–	–	2
–	–	–	–	–	–	–	–	1.5×10^2
–	–	–	–	–	–	–	–	2
–	–	–	–	–	–	–	–	50
–	–	–	–	–	–	–	–	1×10^2
	1.35×10^4							1.82×10^4
0.1	1×10^3	1×10^3	?	?	–	–	–	1.2×10^3
0.1	5×10^3	–	–	–	–	–	–	5.5×10^3
0.1	5×10^2	–	–	–	–	–	–	5.5×10^2
–	–	–	–	–	–	–	–	7.5×10^2
–	–	–	–	–	–	–	–	7.5×10^2
–	–	–	–	–	–	–	–	3×10^2
–	–	–	–	–	–	–	–	2×10^2
	5.5×10^3							8.05×10^3

Table 3 (continued)

Basin	Area of basin	Name of salt and sulfate sequence	Sulfate rocks			Salt rocks
			Area km^2	Thickness km	Volume km^3	Area km^2
1	2	3	4	5	6	7
Morsovo	1.5×10^6	Morsovo salt sequ.	8×10^4	0.001	80	8×10^4
		Narva	2×10^4	0.001	20	1×10^2
		Morsovo sulfate sequence	8×10^5	0.01	8×10^4	–
		Middle Lopushan	6×10^4	0.005	3×10^2	–
		Total	8.6×10^5		8.04×10^4	8×10^4
Upper Devonian Basin of the Russian Platform	1.5×10^6	Evlanovo-Liven sequence of the Pripyat Depression	2×10^4	0.03	6×10^2	2×10^4
		Dankov-Lebedyan sequence of the Pripyat Depression	3×10^4	0.005	1.5×10^2	2.6×10^4
		Shchigrov sequence of the Dnieper-Donets Depression				
		Lower salt sequence of the Dnieper-Donets Depression				
		Upper salt sequence of the Dnieper-Donets Depression	6×10^4	0.03	1.8×10^3	4×10^4
		Salt sequence of the Pre-Timan Trough	2×10^4	0.01	2×10^2	2×10^3
		Shchigrov sequence of the Pripyat Depression	3×10^4	0.002	60	–
		Voronezh-Evlanovo sequence of the Pripyat Depression	3×10^4	0.002	60	–
		Mezhsolevaya sequence of the Pripyat Depression	3×10^4	0.001	30	–
		Upper Devonian sequence of the Baltic and Moscovian syneclises, and Volga-Ural Region	7×10^5	0.005	3.5×10^4	–
		Ustukhta	6×10^4	0.003	1.8×10^2	–
		Idzhid-Kamensk	1×10^3	0.002	2	–
		Total	8.5×10^5		3.81×10^4	6.8×10^4
West Canadian	1×10^6	Lotsberg	–	–	–	1.4×10^5
		Gold Lake	–	–	–	1.8×10^5

		Chloride type of potassium salt			Sulfate type of potassium salt			Total volume of sulfate and salt rocks
Thickness km	Volume km^3	Area km^2	Thickness km	Volume km^3	Area km^2	Thickness km	Volume km^3	Volume km^3
8	9	10	11	12	13	14	15	16
0.01	8×10^2	2.5×10^2	0.0001	0.02	–	–	–	8.8×10^2
0.01	1	–	–	–	–	–	–	21
–	–	–	–	–	–	–	–	8×10^4
–	–	–	–	–	–	–	–	3×10^2
	8×10^2	2.5×10^2		0.02				8.12×10^4
0.25	5×10^3	?	?	?	–	–	–	5.6×10^3
0.9	2.34×10^4	2×10^3	0.01	20	–	–	–	2.36×10^4
1.0	4×10^4	?	?	?	–	–	–	4.18×10^4
0.5	1×10^3	–	–	–	–	–	–	1.2×10^3
–	–	–	–	–	–	–	–	60
–	–	–	–	–	–	–	–	60
–	–	–	–	–	–	–	–	30
–	–	–	–	–	–	–	–	3.5×10^4
–	–	–	–	–	–	–	–	1.8×10^2
–	–	–	–	–	–	–	–	2
		6.94×10^4		20				1.076×10^6
0.1	1.4×10^4	–	–	–	–	–	–	1.4×10^4
0.03	5.4×10^3	–	–	–	–	–	–	5.4×10^3

Table 3 (continued)

Basin	Area of basin	Name of salt and sulfate sequence	Sulfate rocks			Salt rocks
			Area km^2	Thickness km	Volume km^3	Area km^2
1	2	3	4	5	6	7
		Prairie				
		Black Creek				
		Hubbard	4×10^5	0.03	1.2×10^4	3.5×10^5
		Davidson	3×10^4	0.002	60	2×10^4
		Dinsmore	5×10^3	0.005	25	5×10^3
		Stettler salt sequ.	2×10^4	0.001	20	2×10^4
		Ernestina Lake	1.5×10^5	0.001	1.5×10^2	–
		Chinchaga	8×10^4	0.01	8×10^2	–
		Muskeg	6×10^4	0.05	3×10^3	–
		Beaverhill Lake	1×10^5	0.001	1×10^2	–
		Woodbend and Duperow	2×10^5	0.01	2×10^3	–
		Birdbear and Nisku	5×10^4	0.005	2.5×10^2	–
		Stettler sulfate sequ.	1.5×10^5	0.01	1.5×10^3	–
		Total	6×10^5		2×10^4	3.5×10^5
Hudson	3×10^5	Moose River	8×10^4	0.005	4×10^2	5×10^4
		Upper Kenogami River and Stooping River	2.5×10^5	0.002	5×10^2	–
		Williams Island	1×10^5	0.001	1×10^2	–
		Long Rapids	1×10^5	0.0005	50	–
		Total	2.5×10^5		1.05×10^3	5×10^4
Michigan	2.4×10^5	Lucas	7×10^4	0.02	1.4×10^3	2.5×10^4
Adavale	1×10^5	Boree	1×10^4	0.005	50	8×10^3
Mid-Tien Shan	4×10^4	Salt sequence	3×10^3	0.01	30	3×10^3
		Sulfate-bearing terrigene-carbonate sequence	1×10^4	0.001	10	–
		Chemanda	1×10^4	0.001	10	–
		Aktailyak	5×10^3	0.001	5	–
		Kodzhagul	5×10^3	0.001	5	–
		Total	3×10^4		60	3×10^3
Sverdrup	6×10^5	Otto Fiord	3×10^5	0.05	1.5×10^4	2×10^5
		Borup Fiord	1×10^4	0.01	1×10^2	–
		Antoinetta	1×10^4	0.005	50	–
		Mount Bayley	5×10^4	0.01	5×10^2	–
		Total	3×10^5		1.58×10^4	2×10^5

		Chloride type of potassium salt			Sulfate type of potassium salt			Total volume of sulfate and salt rocks
Thickness km	Volume km^3	Area km^2	Thickness km	Volume km^3	Area km^2	Thickness km	Volume km^3	Volume km^3
8	9	10	11	12	13	14	15	16
0.12	4.2×10^4	7×10^4	0.005	3.5×10^2	–	–	–	5.44×10^4
0.03	6×10^2	–	–	–	–	–	–	6.6×10^2
0.01	50	–	–	–	–	–	–	75
0.01	2×10^2	–	–	–	–	–	–	2.2×10^2
–	–	–	–	–	–	–	–	1.5×10^2
–	–	–	–	–	–	–	–	8×10^2
–	–	–	–	–	–	–	–	3×10^3
–	–	–	–	–	–	–	–	1×10^2
–	–	–	–	–	–	–	–	2×10^3
–	–	–	–	–	–	–	–	2.5×10^2
–	–	–	–	–	–	–	–	1.5×10^3
	6.17×10^4	7×10^4		3.5×10^2				8.21×10^4
0.01	5×10^2	–	–	–	–	–	–	9×10^2
–	–	–	–	–	–	–	–	5×10^2
–	–	–	–	–	–	–	–	1×10^2
–	–	–	–	–	–	–	–	50
	5×10^2							1.55×10^3
0.035	8.7×10^2	–	–	–	–	–	–	2.27×10^3
0.05	4×10^2	–	–	–	–	–	–	4.5×10^2
0.05	1.5×10^2	–	–	–	–	–	–	1.8×10^2
–	–	–	–	–	–	–	–	10
–	–	–	–	–	–	–	–	10
–	–	–	–	–	–	–	–	5
–	–	–	–	–	–	–	–	5
	1.5×10^2							2.1×10^2
0.5	1×10^5	–	–	–	–	–	–	1.15×10^5
–	–	–	–	–	–	–	–	1×10^2
–	–	–	–	–	–	–	–	50
–	–	–	–	–	–	–	–	5×10^2
	1×10^5							1.16×10^5

Table 3 (continued)

Basin	Area of basin	Name of salt and sulfate sequence	Sulfate rocks			Salt rocks
			Area km^2	Thickness km	Volume km^3	Area km^2
1	2	3	4	5	6	7
Williston	5×10^5	Charles and Mission Canyon	2×10^5	0.01	2×10^3	8×10^4
		Kibbey	1×10^5	0.001	1×10^2	–
		Total	2×10^5		2.1×10^3	8×10^4
Maritime	6×10^5	Gautreau	1×10^3	0.01	10	1×10^2
		Windsor	3×10^5	0.02	10	2×10^5
		Total	3×10^5		6.01×10^3	2×10^5
Saltville	?	MacCrady	5×10^3	0.01	50	?
Paradox	5×10^4	Paradox	3.5×10^4	0.02	7×10^2	2.5×10^4
Eagle	2×10^4	Eagle Valley	1×10^4	0.005	50	5×10^2
Amazon	1×10^6	Nova Olinda	4.5×10^5	0.01	4.5×10^3	1×10^5
		Itaituba	4.5×10^5	0.001	4.5×10^2	–
		Total	4.5×10^5		4.95×10^3	1×10^5
Midcontinent	1.3×10^6	Ingleside	5×10^3	0.001	5	5×10^3
		Broom Creek	1.1×10^4	0.001	11	1.1×10^4
		Wellington	1.2×10^5	0.002	2.4×10^2	1.2×10^5
		Owl Canyon salt sequence	1×10^3	0.001	1	?
		Cimarron	1.5×10^5	0.02	3×10^3	1.5×10^5
		Opeche sequence of Denver, Yulesburg, and Williston Basins	5×10^4	0.01	5×10^2	5×10^4
		Salt sequence of Flowerport, Blaine and Yelton Form.	5×10^4	0.005	2.5×10^2	4×10^4
		San Andres-Artesia	2×10^5	0.01	2×10^3	1.5×10^5
		Pine	8×10^4	0.005	4×10^2	6×10^4
		Castile	1×10^5	0.05	5×10^3	1×10^5
		Upper Minnelusa	8×10^4	0.005	4×10^2	–
		Lower Wellington	1.3×10^5	0.01	1.3×10^3	–
		Upper Wellington	1.2×10^5	0.005	6×10^2	–
		Wichita	1×10^5	0.001	1×10^2	–
		Stone Corrall	1×10^5	0.001	1×10^2	–
		Owl Canyon sulfate sequence	1×10^5	0.01	1×10^3	–
		Blaine	1×10^5	0.01	1×10^3	–
		Minnekahta	5×10^4	0.005	2.5×10^2	–

		Chloride type of potassium salt			Sulfate type of potassium salt			Total volume of sulfate and salt rocks
Thickness km	Volume km^3	Area km^2	Thickness km	Volume km^3	Area km^2	Thickness km	Volume km^3	Volume km^3
8	9	10	11	12	13	14	15	16
0.03	2.4×10^3	–	–	–	–	–	–	4.4×10^3
–	–	–	–	–	–	–	–	1×10^2
	2.4×10^3							4.5×10^3
0.1	10	–	–	–	–	–	–	20
0.2	4×10^4	3×10^4	0.001	30	?	?	?	4.6×10^4
	4×10^4	3×10^4		30				4.6×10^4
?	?	–	–	–	–	–	–	50
0.36	9×10^3	1.5×10^4	0.03	4.5×10^2	?	?	?	9.15×10^3
0.3	1.5×10^2	1×10^2	0.0005	0.1	–	–	–	2×10^2
0.2	2×10^4	–	–	–	–	–	–	2.45×10^4
–	–	–	–	–	–	–	–	4.5×10^2
	2×10^4							2.5×10^4
?	?	–	–	–	–	–	–	5
0.02	2.2×10^2	–	–	–	–	–	–	2.31×10^2
0.1	1.1×10^4	–	–	–	–	–	–	1.12×10^4
?	?	–	–	–	–	–	–	1
0.1	1.5×10^4	–	–	–	–	–	–	1.8×10^4
0.02	1×10^3	–	–	–	–	–	–	1.5×10^3
0.05	2×10^3	–	–	–	–	–	–	2.25×10^3
0.1	1.5×10^4	–	–	–	–	–	–	1.7×10^4
0.03	1.8×10^3	–	–	–	–	–	–	2.2×10^3
0.17	1.7×10^4	7×10^4	0.005	3.5×10^2	7×10^4	0.002	1.4×10^2	2.25×10^4
–	–	–	–	–	–	–	–	4×10^2
–	–	–	–	–	–	–	–	1.3×10^3
–	–	–	–	–	–	–	–	6×10^2
–	–	–	–	–	–	–	–	1×10^2
–	–	–	–	–	–	–	–	1×10^2
–	–	–	–	–	–	–	–	1×10^3
–	–	–	–	–	–	–	–	1×10^3
–	–	–	–	–	–	–	–	2.5×10^2

Table 3 (continued)

Basin	Area of basin	Name of salt and sulfate sequence	Sulfate rocks			Salt rocks
			Area km^2	Thickness km	Volume km^3	Area km^2
1	2	3	4	5	6	7
		Artesia	6×10^4	0.02	1.2×10^3	–
		Glendo	1.6×10^5	0.001	1.6×10^2	–
		Day Creek	4×10^4	0.001	40	–
		Dewey Lake	1×10^5	0.001	1×10^2	–
		Total	1×10^6		1.77×10^4	6×10^5
Supai	2.5×10^5	Middle Supai	5×10^3	0.001	5	5×10^3
		Upper Supai	6×10^3	0.001	6	6×10^3
		Sulfate-bearing terrigene-carbonate sequence	5×10^3	0.005	25	–
		Total	1.1×10^4		36	6×10^3
North Mexican	5×10^5	Salt sequence	?	?	?	?
Peru-Bolivian	1.25×10^6	Mitu	5×10^5	0.01	5×10^3	2×10^5
		Chuquichambi	3×10^5	0.01	3×10^3	2×10^5
		Capacabana	8×10^5	0.005	4×10^3	–
		Total	8×10^5		1.2×10^4	5×10^5
Central European	1.2×10^6	Upper Rotliegendes	–	–	–	8×10^4
		Werra salt sequence	–	–	8.75×10^3	–
		Stassfurt	–	–	4.7×10^3	–
		Leine	–	–	4.6×10^3	–
		Aller	–	–	6.48×10^2	–
		Ten Boer	2.5×10^5	0.01	2.5×10^4	–
		Werra sulfate sequ.	1.93×10^4	0.22	4.25×10^3	–
		Hartlepool	1×10^4	0.01	1×10^2	–
		Sandwich	8×10^3	0.002	16	–
		Fleswik	8×10^3	0.002	16	–
		Suduvsk	9×10^4	0.03	2.7×10^3	–
		Total	1×10^6		5.07×10^4	7×10^5
Alpine	1 x 10	Salt sequence of the East Alps and the Spišsko-Gemerskoe rudogoři	3×10^4	0.01	3×10^2	2.6×10^4
		Sulfate sequence of the East Alps	1.5×10^4	0.001	15	–
		Sulfate sequence of the Spišsko-Gemerskoe rudogoři	5×10^3	0.005	25	–

		Chloride type of potassium salt			Sulfate type of potassium salt			Total volume of sulfate and salt rocks
Thickness km	Volume km^3	Area km^2	Thickness km	Volume km^3	Area km^2	Thickness km	Volume km^3	Volume km^3
8	9	10	11	12	13	14	15	16
–	–	–	–	–	–	–	–	1.2×10^3
–	–	–	–	–	–	–	–	1.6×10^2
–	–	–	–	–	–	–	–	40
–	–	–	–	–	–	–	–	1×10^2
	6.3×10^4	7×10^4		3.5×10^2	7×10^4		1.4×10^2	8.12×10^4
0.01	50	–	–	–	–	–	–	55
0.05	3×10^2	1×10^3	0.0001	0.1	–	–	–	3.06×10^2
–	–	–	–	–	–	–	–	25
	3.5×10^2	1×10^3		0.1				3.86×10^2
?	?	–	–	–	–	–	–	?
0.1	3×10^4	–	–	–	–	–	–	3.5×10^4
0.1	2×10^4	–	–	–	–	–	–	2.3×10^4
–	–	–	–	–	–	–	–	4×10^3
	5×10^4							6.2×10^4
0.1	8×10^3	–	–	–	–	–	–	8×10^3
–	6×10^3	–	–	20	–	–	–	1.5×10^4
–	9.1×10^4	–	–	1.4×10^3	–	–	52	9.72×10^4
–	3.4×10^4	–	–	3.54×10^2	–	–	3.82×10^2	3.93×10^4
–	6.7×10^3	–	–	3	–	–	2	7.35×10^3
–	–	–	–	–	–	–	–	2.5×10^4
–	–	–	–	–	–	–	–	4.25×10^3
–	–	–	–	–	–	–	–	1×10^2
–	–	–	–	–	–	–	–	16
–	–	–	–	–	–	–	–	16
–	–	–	–	–	–	–	–	2.7×10^3
	1.46×10^5	4×10^5		1.7×10^3	2×10^5		3.8×10^2	1.99×10^5
0.2	5.2×10^3	?	?	?	–	–	–	5.5×10^3
–	–	–	–	–	–	–	–	15
–	–	–	–	–	–	–	–	25

Table 3 (continued)

Basin	Area of basin	Name of salt and sulfate sequence	Sulfate rocks			Salt rocks
			Area km^2	Thickness km	Volume km^3	Area km^2
1	2	3	4	5	6	7
		Sulfate sequence of the Tribeč Mountains	5×10^3	0.003	15	–
		Sulfate sequence of the Low Tatra Mountains	5×10^3	0.003	15	–
		Total	3×10^4		3.7×10^2	2.6×10^4
Moesian	1×10^5	Salt sequence	3×10^3	0.01	30	3×10^3
		Sulfate-bearing red bed	8×10^5	0.001	8×10^2	–
		Sulfate-bearing terrigene sequence	5×10^4	0.002	1×10^2	–
		Sulfate-carbonate sequence of the Varna basin	1×10^3	0.01	10	–
		Total	8×10^5		9.4×10^2	3×10^3
East European	2.5×10^6	Nikitovo	4.3×10^3	–	58	2×10^4
		Slavyansk	8.4×10^3	–	1.4×10^2	3×10^4
		Kramatorsk	6.9×10^3	–	60	1×10^4
		Upper Kuloi	8×10^4	0.02	1.6×10^3	8×10^4
		Beresniki	1×10^4	0.06	6×10^2	9×10^3
		Salt sequence of the Verkhnepechora Depression	7×10^3	0.04	2.8×10^2	6.6×10^3
		Iren sequence of the Chusovaya and the Yuryuzan-Sylva Depressions	1.5×10^3	0.04	60	1.2×10^3
		Salt sequence of the Pre-Caspian Depression, Volga-Ural Region and Cis-Uralian Trough	8×10^5	–	3.3×10^4	6×10^5
		Salt sequence of East Pre-Caspian Depression	5×10^4	0.001	50	5×10^4
		Hydrochemical sequence	8×10^4	0.01	8×10^2	8×10^4
		Salt-bearing red beds	?	?	?	5×10^4
		Upper Kartamysh	2.7×10^3	–	19	–

		Chloride type of potassium salt			Sulfate type of potassium salt			Total volume of sulfate and salt rocks
Thickness km	Volume km^3	Area km^2	Thickness km	Volume km^3	Area km^2	Thickness km	Volume km^3	Volume km^3
8	9	10	11	12	13	14	15	16
–	–	–	–	–	–	–	–	15
–	–	–	–	–	–	–	–	15
	5.2×10^3							5.57×10^3
0.3	9×10^2	?	?	?	–	–	–	9.3×10^2
–	–	–	–	–	–	–	–	8×10^2
–	–	–	–	–	–	–	–	1×10^2
–	–	–	–	–	–	–	–	10
	9×10^2							1.84×10^3
0.08	1.6×10^3	–	–	–	–	–	–	1.66×10^3
0.1	3×10^3	–	–	–	–	–	–	3.14×10^3
0.1	1×10^3	1×10^3	–	5	8×10^2	–	2.35	1.07×10^3
0.003	2.4×10^2	–	–	–	–	–	–	1.84×10^3
0.5	4.5×10^3	2.5×10^3	–	1.19×10^2	–	–	–	5.22×10^3
0.15	1×10^3	1.5×10^3	–	4	–	–	–	1.28×10^3
0.03	36	–	–	–	–	–	–	96
–	9.9×10^5	3×10^5	–	1.45×10^4	2×10^5	–	5.5×10^2	1.038×10^6
0.01	5×10^2	–	–	–	–	–	–	5.5×10^2
0.1	8×10^3	–	–	–	–	–	–	8.8×10^3
0.01	5×10^2	–	–	–	–	–	–	5×10^2
–	–	–	–	–	–	–	–	19

Table 3 (continued)

Basin	Area of basin	Name of salt and sulfate sequence	Sulfate rocks			Salt rocks
			Area km^2	Thickness km	Volume km^3	Area km^2
1	2	3	4	5	6	7
		Sulfate-carbonate sequence	3×10^5	–	1.74×10^4	–
		Sulfate-bearing red beds	5×10^5	0.003	1.5×10^3	–
		Total	2×10^6		5.56×10^4	5.56×10^4
Chu-Sarysu	2.5×10^5	Zhidelisai-Kingir	3×10^4	–	1.38×10^2	1.15×10^4
		Tuzkol	5×10^4	–	2.8×10^2	2×10^4
		Total	8×10^4		4.18×10^2	3.15×10^4
Total			15.7×10^6		5.77×10^5	7.45×10^6

Similar data show that in the Paleozoic the most extensive sulfate accumulation occurred within salt basins, and outside the basins sulfate sedimentation was more limited. This conclusion is of great importance for the establishment of a more precise paragenetic relationship between sulfate and salt accumulation, allowing the assumption that halite sedimentation in the ancient epoch was mostly observed in regions with conditions favorable for extensive (not less than 10,000 km² and mostly greater than 100,000 km²) sulfate deposition. Thus, it can be assumed that within sulfate basins with a present area of sulfate rocks over 100,000 km² either salt accumulation took place in the past but at present no salt strata are preserved, or these basins remain salt basins but rock salt horizons cannot be distinguished because of the poor knowledge of the basins. These are: Tarim, Lena-Yenisei (Ordovician/Silurian), Canadian Arctic (Devonian), South Peruvian, Parnaiba, East Greenland and, probably, Arabian.

The areal extent of rocks enables subdivision of all the Paleozoic evaporite basins into four groups on the basis of sulfate accumulation area: (1) vast (more than 1,000,000 km²), (2) large (hundreds of thousands of square kilometers), (3) medium (tens of thousands of square kilometers), and (4) small (about 10,000 km²).

At present six vast Paleozoic basins are known, five of them being salt basins and one a sulfate basin. They existed in the Cambrian (East Siberian and Iran-Pakistan), Devonian (North Siberian), Carboniferous (East European), and Permian (Central and East European).

There are 28 large basins (17 salt and 11 sulfate basins). They formed in the Cambrian (Mackenzie and Tarim), Ordovician and Silurian (Williston, Michigan-Pre-Appalachian, Canadian Arctic, Canning, Lena-Yenisei), Devonian (Morsovo, Upper Devonian of the Russian Platform, Moesian-Wallachian, Chu-Sarysu, West Canadian,

		Chloride type of potassium salt			Sulfate type of potassium salt			Total volume of sulfate and salt rocks
Thickness km	Volume km^3	Area km^2	Thickness km	Volume km^3	Area km^2	Thickness km	Volume km^3	Volume km^3
8	9	10	11	12	13	14	15	16
–	–	–	–	–	–	–	–	1.74×10^4
–	–	–	–	–	–	–	–	1.5×10^3
	1.01×10^6	3.05×10^5		1.46×10^4	2.01×10^5		5.52×10^2	1.08×10^6
–	2.8×10^3	–	–	–	–	–	–	2.94×10^3
–	3×10^3	–	–	–	–	–	–	3.28×10^3
	5.8×10^3							6.22×10^3
	2.944×10^6	9.91×10^5		1.76×10^4	4.71×10^5		1.07×10^3	3.54×10^6

Hudson, Canadian Arctic, Illinois), Carboniferous (Sverdrup, Williston, Maritime, Amazon, Reggan, Tindouf), and Permian (Midcontinent, Peru-Bolivian, Moesian, South Peruvian, Parnaiba, East Greenland, and Arabian).

There are 42 medium basins (9 salt basins and 33 sulfate basins). They show the following distribution through geological time: Cambrian – 1, Ordovician and Silurian – 8, Devonian – 7, Carboniferous – 18, Permian – 8.

There are 18 small basins, 16 of them being sulfate basins. They existed mostly in the Carboniferous (10 basins) and Permian (4 basins). However, two basins are distinguished in the Cambrian and two in the Devonian.

It is quite obvious that vast sulfate basins formed during the Paleozoic, but most of them existed in the Cambrian, Devonian, and Permian. It should be also noted that medium and small sulfate basins were most extensively developed in the Carboniferous.

The Paleozoic salt basins can be subdivided on the basis of rock salt areal extent into: vast (more than 1,000,000 km^2), large (hundreds of thousands of square kilometers), medium (tens of thousands of square kilometers), and small (less than 10,000 km^2). There are two vast salt basins at present, both of Cambrian age: East Siberian and Iran-Pakistan. There are 13 large salt basins whose distribution through geological periods is as follows: Cambrian – 1 (Mackenzie), Ordovician – 2 (Canadian Arctic and Canning), Silurian – 1 (Michigan-Pre-Appalachian), Devonian – 2 (North Siberian and West Canadian), Carboniferous – 3 (Sverdrup, Maritime, and Amazon), Permian – 4 (Midcontinent, Peru-Bolivain, Central and East European). There are ten medium basins of rock salt accumulation: six in the Devonian (Tuva, Chu-Sarysu, Morsovo, Upper Devonian of the Russian Platform, Hudson, Michigan), two in the Carboniferous (Williston and Paradox), two in the Permian (Alpine and

Table 4. Area of distribution and volume of sulfate rocks within Paleozoic sulfate basins

Basin	Area of basin km^2	Name of sulfate sequence	Area km^2	Thickness km	Volume km^3
1	2	3	4	5	6
Tarim	4×10^5	Sulfate sequence	1×10^5	0.0005	50
Anti-Atlas	1×10^4	Sulfate sequence	?	?	?
Michigan	8×10^4	Munising	5×10^3	0.001	5
Cis-Andean	1×10^5	Limbo	5×10^4	0.03	1.5×10^3
Anadarko	?	West Spring Creek	?	?	?
South Illinois	1.5×10^5	Joachim	6×10^4	0.001	60
Georgina	2.7×10^5	Toko	5×10^4	0.001	50
Lena-Yenisei	2×10^6	Sulfate-carbonate sequence of Norilsk Region	8×10^4	0.01	8×10^2
		Sulfate-dolomite sequence of Tunguska syneclise	2×10^5	0.001	2×10^2
		Irbukla-Kochak	1×10^5	0.001	1×10^2
		Stan	1×10^5	0.0003	10
		Kharyalakh	1×10^5	0.001	1×10^2
		Tochilnin	5×10^4	0.0001	5
		Bratsk	1×10^5	0.0005	50
		Utokan	1.5×10^5	0.0005	75
		Kongda	1.5×10^5	0.0005	75
		Kholyukhan	3.5×10^5	0.001	3.5×10^2
		Total	7.3×10^5		1.7×10^3
Baltic	1×10^5	Tallin	1×10^3	0.0001	0.1
		Itfer	1×10^3	0.0001	0.1
		Ievsk	1×10^3	0.00001	0.01
		Paprenyai-Pagegyai	1×10^4	0.0001	1
		Total	1×10^4		1.2
Severnaya Zamlya	1×10^5	Komsomolsk	5×10^4	0.001	50
		Gypsum-bearing sequence	5×10^4	0.005	2.5×10^2
		Total	5×10^4		3×10^2
Moose River	1.5×10^5	Dolomite sequence	5×10^4	0.001	50
		Kenogami River	2.5×10^4	0.0005	7.5
		Moose River	1.8×10^4	0.01	1.8×10^2
		Williams Island	1.5×10^4	0.001	15
		Long Rapids	?	?	?
		Total	5×10^4		2.53×10^2
Pechora	2.5×10^5	Kosju-Adak	2×10^4	0.001	20
		Filippyelsk	2×10^4	0.001	20
		Total	2×10^4		40

Table 4 (continued)

Basin	Area of basin km^2	Name of sulfate sequence	Area km^2	Thickness km	Volume km^3
1	2	3	4	5	6
Dniester-Prut	1×10^5	Pugai	2×10^4	0.001	40
Carnarvon	5.7×10^4	Dirk Hartox	3×10^4	0.001	30
Minusinsk	2.5×10^4	Abakan-Askyz	1×10^4	0.001	10
		Beysk	1.5×10^4	0.005	75
		Total	1.5×10^4		85
Kuznetsk	1.5×10^4	Podonin	1×10^4	0.001	10
Teniz	1.2×10^5	Sulfate sequence	5×10^4	0.001	50
Turgai	1×10^4	Terrigene-carbonate sequence	1×10^3	0.001	1
Moesian-Wallachian	2×10^5	Carbonate-sulfate sequence	1×10^5	0.01	1×10^3
Tindouf	2×10^5	Sulfate-bearing red beds	5×10^4	0.001	50
Canadian Arctic	7×10^5	Bird Fiord	2×10^5	0.01	2×10^3
Illinois		Jeffersonville	1×10^5	0.001	1×10^2
Central Iowa	1×10^4	Kenwood	1×10^3	0.001	1
Cedar Valley			1×10^3	0.001	1
		Total	1×10^3		2
Pechora-Novaya Zemlya	1.5×10^5	Sulfate-carbonate sequence of Pechora Depression	5×10^4	0.005	2.5×10^2
		Gypsum-bearing sequence of Novaya Zemlya	1×10^3	0.01	10
		Total	5.1×10^4		2.6×10^2
Spitsbergen		Lower gypsum-bearing sequence	1×10^4	0.005	50
East European	2×10^6	Ozersk	8×10^5	0.01	8×10^3
		Oka-Serpukhov	1×10^6	0.005	5×10^3
		Kashira-Myachkov	1×10^6	0.005	5×10^3
		Kasimov	8×10^5	0.005	4×10^3
		Klyazma	1.5×10^6	0.003	4.5×10^3
		Noginsk	6×10^5	0.001	6×10^2
		Total	1.5×10^6		2.71×10^4

Table 4 (continued)

Basin	Area of basin km^2	Name of sulfate sequence	Area km^2	Thickness km	Volume km^3
1	2	3	4	5	6
East Uralian		Gypsum-bearing sequence of Magnitogorsk Region	1 x 10^3	0.01	10
		Sulfate-bearing terrigene-carbonate sequence of Bagaryak	1 x 10^3	0.001	1
		Sulfate-bearing terrigene sequence	1 x 10^3	0.001	1
		Total	3 x 10^3		12
Teniz	1.2 x 10^5	Kirey	1 x 10^3	0.001	1
Tyup	2.5 x 10^3	Tyup and Chaarkuduk	2 x 10^3	0.001	2
Chimkent	5 x 10^3	Gypsum-bearing sequence	2.5 x 10^3	0.01	25
Aksu	2 x 10^4	Kurukusum	1 x 10^3	0.001	1
Achikkul	3 x 10^4	Arktag	2 x 10^4	0.001	20
Lhasa	2 x 10^4	Pando	1 x 10^4	0.001	10
Fitzroy	5 x 10^3	Anderson	2 x 10^3	0.005	10
Rhadames	6 x 10^4	Sulfate sequence	3 x 10^4	0.001	30
Illisie	5 x 10^4	El-Adeb-Larach	2 x 10^4	0.005	1 x 10^2
		Tigentourin	2 x 10^4	0.005	1 x 10^2
		Total	2 x 10^4		2 x 10^2
Ahnet	6 x 10^4	Sulfate sequence	1 x 10^3	0.001	1
Reggane	1.2 x 10^5	Gypsum-bearing sequence	1 x 10^5	0.01	1 x 10^3
		Sulfate-bearing red beds	1 x 10^5	0.001	1 x 10^2
		Total	1 x 10^5		1.1 x 10^3
Tindouf	2 x 10^5	Ouarkziz	1 x 10^5	0.02	2 x 10^3
Illinois	2 x 10^5	St. Louis	5 x 10^4	0.001	50
South Iowa	2 x 10^4	St. Louis	1 x 10^4	0.001	10
		Pella	1 x 10^4	0.001	10
		Total	1 x 10^4		20
Michigan	8 x 10^4	Michigan	2.5 x 10^3	0.001	2.5
		Saginaw	2.5 x 10^3	0.001	2.5
		Total	2.5 x 10^3		5

Table 4 (continued)

Basin	Area of basin km^2	Name of sulfate sequence	Area km^2	Thickness km	Volume km^3
1	2	3	4	5	6
Orogrande	8×10^4	Magdalena	4×10^4	0.02	8×10^2
San Juan	4×10^4	Paradox	2×10^4	0.001	20
East Wyoming	1.2×10^5	Minnelusa	6×10^4	0.001	60
Venezuela	1×10^5	Palmarito	5×10^4	0.001	50
South Peruvian	2×10^5	Tarma	1×10^5	0.001	1×10^2
Northumberland	3×10^4	Gypsum-bearing calcare-ous sandstones	1×10^4	0.001	10
North Ireland	7×10^4	Roscunish	1×10^4	0.005	50
Central England	1×10^4	Anhydrite sequence	5×10^3	0.001	5
Dobruja	5×10^4	Sulfate-carbonate sequence	3×10^4	0.005	1.5×10^2
Rio Blanco	3.2×10^5	Patquia	6×10^4	0.001	60
Parnaiba	6×10^5	Pedra-do-Fogo	3×10^5	0.01	3×10^3
North Italy	1.5×10^5	Collio	1.5×10^3	0.001	1.5
		Groden	5×10^4	0.001	50
		Lower Bellerophon	2×10^4	0.02	4×10^2
		Upper Bellerophon	2×10^4	0.05	1×10^3
		Total	7×10^4		1.45×10^3
Dinarids	9×10^4	Sulfate-carbonate sequence	6×10^4	0.01	6×10^2
		Sulfate-bearing red beds	2×10^4	0.001	20
		Total	8×10^4		6.2×10^2
Mecsek	1×10^4	Gypsum-bearing terrigene sequence	5×10^3	0.001	5
		The first anhydrite sequ.	1×10^4	0.01	1×10^2
		The second anhydrite sequ.	1×10^4	0.01	1×10^2
		Total	1×10^4		2×10^2
Rakhov	3.5×10^3	Gypsum-bearing sequence	3×10^3	0.001	3
Karasu-Ishsai	2×10^3	Karasu	1×10^3	0.001	1
Darvaza	2.5×10^3	Shakarsen	2×10^3	0.01	20
Spitsbergen	3.5×10^4	Upper gypsum-bearing sequence	1×10^4	0.01	1×10^2
East Greenland	3×10^5	Gypsum-bearing sequence	2×10^5	0.03	6×10^3
Arabian	3×10^6	Sulfate sequence	3.5×10^5	0.005	1.75×10^3
Total			1.8×10^6		5.28×10^4

Chu-Sarysu). There are six small basins: one in the Cambrian (Amadeus) and one in the Devonian (Adavale), two in the Carboniferous (Mid-Tien Shan and Eagle) and two in the Permian (Supai and Moesian). As mentioned, in three basins a salt accumulation area is unknown.

The above data show that some extensive salt areas existed in the Paleozoic. It is interesting to find out what part of the Peleozoic evaporite basins was taken up by salt accumulation if we consider the correlation between the total area of basins and that of sulfate and salt accumulation in these basins. The figures for certain basins are given below.

In the Cambrian East Siberian sedimentary basin (about 4.5–5 $\times$ 10^6 km^2) rock salt accounts for 33%–40% of the basins; and about 85%–90% of sulfate sedimentation. In other basins studied the amount of rock salt is as follows: in the Iran-Pakistan Basin – 40%–45% of the total basin area and 65 % of sulfate accumulation area, in Mackenzie – 55% and 83%, in Amedeus – 1% and 50%, in the Ordovician Basin of the Canadian Arctic Archipelago – 14% and 66%, in Canning – 50% and 83%, in Michigan-Pre-Appalachian – 52% and 87%, in North Siberian – 6% and 9%, in Chu-Sarysu (Devonian/Carboniferous) – 20% and 25%, in Morsovo – 8% and 9%, in Upper Devonian of the Russian Platform – 7% and 8%, in West Canadian – 35% and 58%, in Hudson – 16% and 20%, in Michigan – 10% and 36%, in Adavale – 8% and 80%, in Mid-Tien Shan – 8% and 10%, in Sverdrup – 33% and 66%, in Williston (Carboniferous) – 16% and 40%, in Maritime – 33% and 66%, in Paradox – 50% and 71%, in Eagle – 3% and 5%, in Amazon – 10% and 22%, in Midcontinent – 50% and 60%, in Supai – 2% and 55%, in Peru-Bolivian – 40% and 62%, in Central European – 58% and 70%, in Alpine – 26% and 86%, in Moesian – 3% and 4%, in East European – 35% and 36%, and in Chu-Sarysu (Permian) – 12% and 39%.

The above values imply that many Paleozoic salt basins occupied vast seas where salt accumulation covered one third and even half of the entire sedimentary basin. They may be exemplified by the Cambrian East Siberian, Iran-Pakistan and Mackenzie Basins, the Ordovician Canning Basin, the Silurian Michigan-Pre-Appalachian Basin, the Devonian West Canadian, the Carboniferous Sverdrup, Maritime, Paradox, the Permian Midcontinent, Peru-Bolivian, Central and East European, Alpine. In the Recent we cannot find any equivalents of such salt-producing seas. This confirms the conclusion drawn by some investigators (Yanshin 1961; Fiveg 1962; Strakhov 1962) that many old salt basins differing from recent ones in their size are extensive sea basins.

An interesting regularity issues from the rock salt area and its correlation with that of sedimentary basins, i.e., the greatest number of the large salt-producing seas existed only during some periods of the Paleozoic. For example, in the Cambrian in most basins, in spite of their considerable size, salt deposition accounted for 40%–50% of their area, and rock salt accumulation areas were either very large, or large; and in the Permian, when salt accumulation areas were mostly large and often constituted 35% to 60% of a basin area. It is to the Cambrian and Permian periods that all the vast salt basins and five of 13 large ones are confined. Both in the Ordovician and Silurian there was only one basin where salt accumulation area covered 50% of the territory. It is noteworthy that the Canning Basin is arbitrarily regarded as the Ordovician. In the Devonian and Carboniferous sedimentary basins the area of salt deposition was limited, usually not exceeding 10%–20% of the total basin. It is only in one Devonian basin – West Canadian – that rock salt covered 35% of the basin. There were three basins of this type in the Carboniferous (Sverdrup, Maritime, and Paradox), but it is only in the first two that the salt accumulation area was large.

The relationship between a salt accumulation area and that of sulfate sedimentation suggests that in all the Cambrian and in most Permian basins salt strata occupied not less than 50% of a sulfate-formation area. In fact, the same is true for three salt basins of Ordovician and Silurian age. In Devonian and Carboniferous basins the picture is quite different. Salt accumulation in many basins of this age was observed in 10%–20% of sulfate sedimentation area. This probably means that, beginning with the Devonian, paleogeographic and paleotectonic conditions responsible for salt accumulation have changed in comparison with the older epochs of the Paleozoic which affected the relationships between sulfate and halite sedimentation. Similar changes in salt accumulation conditions probably took place in the Permian period when proportions of areas of halite and sulfate accumulation were similar to those of the Early Paleozoic salt basins.

The above data show that salt accumulation conditions in various periods of the Paleozoic differed greatly. It is obvious that the Cambrian and Permian were two distinct periods in the Paleozoic when very large and large salt basins predominated, and within these basins salt accumulation was very extensive. Salt deposition often constituted 50% of sulfate sedimentation area there.

Let us analyze the data on the potash salt areal extent in the Paleozoic salt basins. They are at present found in 13 of 34 Paleozoic basins. The most extensive potassium accumulation areas occupy Permian salt basins. In two of them (Central and East European) this area might be regarded as large, since it attains 4×10^5 and 3.5×10^5 km^2, respectively. Additionally, in another Permian basin (Midcontinent) a potash salt areal extent is medium, but close to large (7×10^4 km^2). Such great areas of the potash salt development have not been recorded from older salt basins. It should be noted that in all the above Permian basins potash salts of sulfate type have been found and they are widespread. Thus, in the Central and East European Basins they attain 2×10^5 km^2, and in the Midcontinent Basin 7×10^4 km^2. Sulfate potash salts have not been found in other basins. It is known that salt accumulation areas in the above Permian salt basins were large, however potassium sedimentation was recorded in these basins: in the Central European Basin it accounts for 42% of a salt zone and in the Midcontinent Basin for 12%. The Devonian and Carboniferous salt basins, as a rule, had medium and small areas of potassium accumulation. It is noteworthy that potassium sedimentation in basins of this age was observed not only in large, but in medium and small salt basins as well; their area was limited and usually did not exceed 20%–25% of salt accumulation area. Only in the Paradox Basin did potassium sedimentation area constitute 60% of a salt accumulation area. Conditions of potassium deposition in the Devonian and Carboniferous might have differed greatly from those in the Permian. The data available show that in the Silurian and Cambrian salt basins an area of potassium accumulation was medium and occupied a small part of the salt-producing area (4% and 14% in the East Siberian and Michigan-Pre-Appalachian Basins, respectively). Only Permian salt basins had a considerable area of potassium accumulation in the Paleozoic.

Paleozoic evaporite basins are also different with respect to the volume of halogenic rocks accumulation. Thus, the entire bulk of halogenic sediments of the Permian East European Basin is 1.08 x 10^6 km^3. The volume of evaporites in the East Siberian and Iran-Pakistan Cambrian Basins is 7.85 x 10^5 km^3 and 6.32 x 10^5 km^3, respectively. Then come the following basins: Central European (1.99 x 10^5 km^3), Sverdrup (1.16 x 10^5 km^3), Mackenzie (1.02 x 10^5 km^3), Upper Devonian of the Russian Platform (1.076 x 10^5 km^3), West Canadian (8.81 x 10^4 km^3), Morsovo (8.12 x 10^4 km^3), Midcontinent (8.12 x 10^4 km^3), Peru-Bolivian (6.2 x 10^4 km^3), Maritime (4.6 x 10^4 km^3), Michigan-Pre-Appalachian (2.9 x 10^4 km^3), Canning (2.55 x 10^4 km^3), Amazon (2.5 x 10^4 km^3), Ordovician of the Canadian Arctic (1.88 x 10^4 km^3), North Siberian (1.82 x 10^4 km^3), East European Carboniferous sulfate basin (2.71 x 10^4 km^3). In other Paleozoic basins the volume of halogenic rocks does not exceed several thousands of cubic kilometers.

These estimates are significant. First, they imply the accumulation of the greatest volumes of evaporites in the very large and large Paleozoic basins. Second, the bulk of halogenic rocks in the Paleozoic appeared to form within a limited number of basins, in fact in three salt basins, two Cambrian (East Siberian and Iran-Pakistan) and one Permian (East European). Almost 73% of the total volume of all the Paleozoic evaporites is concentrated in these basins. If we assume for the Cambrian and Permian the existence of such salt basins as the Central European, Mackenzie, Midcontinent, and Peru-Bolivian which, according to their volume of halogenic rocks, hold the fourth, sixth, tenth, and eleventh place, respectively, and contain 12.5% of evaporites (4.44 x 10^5 km^3), then the total volume of Cambrian and Permian halogenic deposits will make up 85.5% of the bulk of Paleozoic evaporites accumulated. Thus, it is possible to conclude that Cambrian and Permian evaporite basins had the largest volume of halogenic rocks among all the Paleozoic basins.

During other periods of the Paleozoic, basins with such great volumes of evaporites were quite rare. One basin of this type is known in the Carboniferous (Sverdrup) and one in the Devonian (Upper Devonian of the Russian Platform). As a rule, in the Ordovician, Silurian, Devonian, and Carboniferous the volume of halogenic rocks formed in evaporite basins did not exceed several tens of thousands of cubic kilometers.

A similar phenomenon is observed when volumes of sulfate rocks, rock and potash salts in the basins are compared. Thus, most sulfate rocks are confined to two Cambrian (East Siberian and Iran-Pakistan) and four Permian salt basins (Midcontinent, Peru-Bolivian, Central, and East European) where about 3.7 x 10^5 km^3 (64%) of the Paleozoic sulfates occur. This regularity becomes more obvious if it is determined by using volumes of rock and potash salts. It is only in four Cambrian and Permian salt basins (East Siberian, Iran-Pakistan, Central, and East European) that a little less than 80% of the entire Paleozoic rock salt and more than 92% of potash salts of chloride type occur. Sulfate potash salts are almost exclusively confined to Permian basins.

Thus, evaporite basins with the largest distribution area and volume of halogenic rocks existed mainly in the Cambrian and Permian periods. The conditions then existing were favorable for the formation of vast and large salt-producing seas where big masses of halogenic rocks accumulated. Thus, the Cambrian and Permian periods might be easily determined as epochs of considerable evaporite accumulation in general and extensive halite accumulation in particular. The Permian period was an epoch especially favorable for potassium deposition.

Areal Extent of Evaporites During Various Periods, Epochs, and Ages of the Paleozoic

The areas under sulfate, halite, and potassium accumulation on different continents and on the Earth in general during various time intervals of the Paleozoic are given in Tables 5 and 6.

Let us first discuss the areal extent of sulfate rocks (Table 5). The total area occupied by sulfate rocks changes greatly during the Paleozoic history. There were periods when the entire area under sulfates was very extensive. This applies to the Permian and Cambrian periods; sulfate rocks of this age occupy 6,616,000 and 5,652,000 km^2, respectively, i.e., they cover 4% to 5% of the entire continental area. Devonian and Carboniferous sulfate rocks also occupy an extensive area; they account for about 3% of the area under continents. In contrast, the Ordovician and Silurian sulfates occupied a much smaller area, usually about 1% of the area under recent continents. At the same time, the above facts reflect only the most general features of the spatial distribution of sulfate rocks in general for any period in the Paleozoic. The picture becomes clear if the dimensions of the area occupied by sulfate rocks in certain epochs and ages of the Paleozoic history are discussed. These data are rather significant, and allow us to establish those time intervals in the Paleozoic when the area of sulfate accumulation was largest.

Thus, in the Early Cambrian epoch sulfate sediments accumulated over an area exceeding 5,000,000 km^2; they covered almost 4% of the entire continental area. Then, from the Middle Cambrian to the end of Early Devonian a sulfate accumulation area during each age (Cambrian epochs) did not exceed 1% of the entire continental area. In Eifelian time (Middle Devonian) a sulfate sedimentation area expanded considerably, and attained 1.5%. The same area with minor deviations is also determined for the rest of the Devonian period. The areal pattern of Carboniferous sulfate rocks, as well as that of the beginning of the Permian period, is different. At that time the entire area of sulfate accumulation underwent great changes, sometimes it decreased, as in the case of the Namurian, Bashkirian, or Asselian ages (about 0.4%), and sometimes increased greatly, as in the Visean or Gzhelian ages when sulfate rocks covered 1.6% and 1.8% of the entire area of the present continents. Beginning with the Kungurian age to the end of the Paleozoic a new, considerable expansion of sulfate accumulation area was observed. It attained 1.9%, 1.7%, 2.4%, and 2.0% in the Kungurian, Ufimian, Kazanian, and Tatarian, respectively.

Estimates of sulfate sedimentation area for the Paleozoic allow us to distinguish three major stages discussed generally in the preceding chapter: Early Cambrian, Middle Cambrian/Early Devonian, and Middle Devonian/Permian. The first stage was characterized by an extremely extensive total area of sulfate sedimentation; the second – by a minimum area of sulfate accumulation; whereas during the third sulfate were again widespread. The latter is subdivided into three substages: Middle/Late Devonian, Carboniferous/Early Permian (or rather Tournaisian/Artinskian), and Early/Late Permian (Kungurian/Tatarian). The Middle/Late Devonian and Kungurian/Tatarian substages are characterized by an almost persistent area of sulfate sedimentation, and during Tournaisian/Artinskian the total area of sulfate accumulation changed from one age to the other. In general, the largest areas of sulfate accumulation in the Paleozoic were recorded in the Early Cambrian epoch and second half of Permian, from the Kungurian to Tatarian ages.

Table 5. Area of distribution of sulfate rocks over continents in different periods and ages of the Paleozoic (in 10^6 km^2 and in % of the continent's area)

Age	Eurasia			North America	South America	Africa	Australia	Total
	Europe	Asia	Total					
1	2	3	4	5	6	7	8	9
Permian	4.0 (40.0)	0.433 (0.99)	4.433 (8.3)	1.022 (4.2)	1.16 (6.3)	–	–	6.616 (4.9)
Late Permian	2.5 (25.0)	0.433 (0.99)	2.933 (5.5)	0.46 (1.9)	–	–	–	3.393 (2.5)
Tatarian	2.442 (24.4)	–	2.442 (4.6)	0.3 (1.2)	–	–	–	2.742 (2.0)
Kazanian	2.349 (23.5)	0.43 (0.99)	2.779 (5.2)	0.46 (1.9)	–	–	–	3.239 (2.4)
Ufimian	1.33 (13.3)	0.433 (0.99)	1.763 (3.3)	0.46 (1.9)	–	–	–	2.223 (1.7)
Early Permian	1.4 (14.0)	0.033 (0.08)	1.433 (2.7)	0.84 (3.5)	1.16 (6.3)	–	–	2.593 (1.9)
Kungurian	1.248 (12.5)	0.033 (0.08)	1.281 (2.4)	0.42 (1.7)	0.86 (4.7)	–	–	2.562 (1.9)
Artinskinian	0.36 (3.6)	–	0.36 (0.7)	0.262 (1.0)	0.8 (4.4)	–	–	1.422 (1.1)
Sakmarian	0.402 (4.02)	–	0.402 (0.7)	0.247 (1.0)	0.3 (1.6)	–	–	0.949 (0.7)
Asselian	0.323 (3.23)	–	0.323 (0.6)	0.05 (0.2)	–	–	–	0.373 (0.3)
Carboniferous	1.616 (16.2)	0.42 (0.97)	2.036 (3.8)	0.855 (3.5)	0.6 (3.3)	0.251 (0.8)	0.002 (0.03)	3.744 (2.8)
Late Carboniferous	1.5 (15.0)	0.205 (0.47)	1.705 (3.2)	0.11 (0.4)	0.6 (3.3)	0.02 (0.1)	–	2.435 (1.8)
Orenburgian	0.6 (6.0)	0.205 (0.47)	0.805 (1.5)	0.11 (0.4)	0.6 (3.3)	–	–	1.515 (1.1)
Gzhelian	1.5 (15.0)	0.2 (0.45)	1.7 (3.2)	0.11 (0.4)	0.6 (3.3)	0.02 (0.1)	–	2.43 (1.8)
Middle Carboniferous	1.01 (10.1)	0.242 (0.56)	1.252 (2.3)	0.45 (1.9)	–	0.02 (0.1)	–	1.722 (1.3)
Moskovian	1.0 (10.0)	0.224 (0.52)	1.224 (2.3)	0.15 (0.6)	–	–	–	1.374 (1.0)
Bashkirian	0.01 (0.1)	0.242 (0.56)	0.252 (0.4)	0.3 (1.2)	–	0.02 (0.1)	–	0.572 (0.4)
Early Carboniferous	1.106 (11.06)	0.385 (0.86)	1.491 (2.8)	0.675 (2.8)	–	0.231 (0.8)	0.002 (0.03)	2.399 (1.8)

Table 5 (continued)

Age	Eurasia			North America	South America	Africa	Austra-lia	Total
	Europe	Asia	Total					
1	2	3	4	5	6	7	8	9
Namurian	0.05 (0.5)	0.173 (0.4)	0.223 (0.4)	0.01 (0.05)	–	0.13 (0.4)	0.002 (0.03)	0.365 (0.3)
Visean	1.106 (11.06)	0.273 (0.7)	1.379 (2.6)	0.565 (2.3)	–	0.201 (0.7)	0.002 (0.03)	2.147 (1.6)
Tournaisian	0.83 (8.3)	0.088 (0.2)	0.918 (1.7)	0.21 (0.9)	–	–	–	1.128 (0.9)
Devonian	0.991 (9.9)	1.646 (3.78)	2.637 (4.9)	1.269 (5.2)	–	0.05 (0.2)	0.01 (0.1)	3.966 (2.96)
Late Devonian	0.871 (8.7)	0.431 (0.95)	1.284 (2.4)	0.465 (1.9)	–	0.05 (0.2)	–	1.799 (1.3)
Fammenian	0.811 (8.1)	0.288 (0.66)	1.099 (2.0)	0.25 (0.1)	–	–	–	1.349 (1.0)
Frasnian	0.84 (8.4)	0.25 (0.58)	1.09 (2.0)	0.465 (1.9)	–	0.05 (0.2)	–	1.605 (1.2)
Middle Devonian	0.98 (9.8)	0.7 (1.6)	1.68 (3.1)	1.169 (4.8)	–	0.05 (0.2)	0.01 (0.1)	1.909 (1.4)
Givetian	0.1 (1.0)	0.45 (1.04)	0.55 (1.1)	0.766 (3.2)	–	0.05 (0.2)	–	1.366 (1.0)
Eifelian	0.98 (9.8)	0.417 (0.96)	1.397 (2.6)	0.699 (2.9)	–	–	0.01 (0.1)	2.106 (1.6)
Early Devonian	–	0.821 (1.9)	0.821 (1.5)	0.25 (0.1)	–	–	–	1.071 (0.8)
Silurian	0.05 (0.5)	0.5 (1.15)	0.55 (1.1)	0.575 (2.4)	–	–	0.3 (3.9)	1.425 (1.1)
Late Silurian	0.05 (0.5)	0.5 (1.15)	0.55 (1.1)	0.475 (1.96)	–	–	0.3 (3.9)	1.325 (0.99)
Přidolian	0.05 (0.5)	0.35 (0.8)	0.4 (0.7)	0.325 (1.3)	–	–	0.3 (3.9)	1.025 (0.8)
Ludlovian	0.05 (0.5)	0.5 (1.15)	0.55 (1.1)	0.475 (1.96)	–	–	0.3 (3.9)	1.325 (0.99)
Early Silurian	0.05 (0.5)	0.15 (0.35)	0.2 (0.4)	0.1 (0.4)	–	–	0.3 (3.9)	0.6 (0.4)
Wenlockian	0.03 (0.3)	–	0.03 (0.1)	0.1 (0.4)	–	–	0.3 (3.9)	0.43 (0.3)
Llandoverian	0.02 (0.2)	0.15 (0.35)	0.17 (0.4)	0.1 (0.4)	–	–	–	0.27 (0.2)

Table 5 (continued)

Age	Eurasia			North America	South America	Africa	Australia	Total
	Europe	Asia	Total					
1	2	3	4	5	6	7	8	9
Ordovician	0.001 (0.01)	0.78 (1.8)	0.781 (1.5)	0.36 (1.5)	–	–	0.35 (4.6)	1.491 (1.1)
Late Ordovician	–	0.43 (0.99)	0.43 (0.8)	0.28 (1.2)	–	–	0.3 (3.9)	1.01 (0.7)
Ashgillian	–	0.43 (0.99)	0.43 (0.8)	0.15 (0.6)	–	–	0.3 (3.9)	0.88 (0.6)
Late Caradoc	–	0.43 (0.99)	0.43 (0.8)	0.2 (0.8)	–	–	0.3 (3.9)	0.93 (0.7)
Middle Ordovician	0.001 (0.01)	0.53 (1.22)	0.531 (0.99)	0.16 (0.7)	–	–	0.05 (0.7)	0.741 (0.5)
Early and Middle Caradoc	0.001 (0.01)	0.53 (1.22)	0.531 (0.99)	0.1 (0.4)	–	–	0.05 (0.7)	0.681 (0.5)
Llandeilo	0.001 (0.01)	0.53 (1.22)	0.531 (0.99)	0.16 (0.7)	–	–	0.05 (0.7)	1.041 (0.8)
Llanvirn	–	0.53 (1.22)	0.53 (0.99)	0.66 (0.2)	–	–	0.05 (0.7)	0.64 (0.5)
Early Ordovician	–	0.48 (1.1)	0.48 (0.9)	0.1 (0.4)	–	–	0.05 (0.7)	0.585 (0.4)
Arenigian	–	0.48 (1.1)	0.48 (0.9)	0.1 (0.4)	–	–	0.05 (0.7)	0.585 (0.4)
Late Tremadoc	–	0.38 (0.88)	0.38 (0.7)	–	–	–	–	0.38 (0.3)
Cambrian	–	5.1 (11.8)	5.1 (9.6)	0.6 (2.4)	0.05 (0.3)	?	0.002 (0.03)	5.652 (4.2)
Late Cambrian	–	0.6 (1.38)	0.6 (1.1)	0.6 (2.4)	–	–	–	1.2 (0.9)
Middle Cambrian	–	0.83 (1.9)	0.83 (1.6)	–	–	–	–	0.83 (0.6)
Early Cambrian	–	5.1 (11.8)	5.1 (9.6)	–	0.05 (0.3)	?	0.002 (0.03)	5.152 (3.8)

Once again, just as in case of halogenesis stages, the variations in sulfate sedimentation area during the Paleozoic era were established using data on separate continents. These variations demonstrate a pattern for the Earth in general. As to the continents, it might be essentially different. It is only in Eurasia and North America that the pattern established for the Earth can be traced in general. Thus, on both continents the Devonian period and second half of the Permian (Kungurian to Tatarian ages) are easily recognized when their sulfate sedimentation area was rather great, while in the Ordovician, the area was minimum. However, in the Silurian, overall regularities outlined for the Earth are observed only in Eurasia where a sulfate sedimentation area was minimum, whereas in North America, in the Late Silurian, a very extensive sulfate accumulation area was recorded. As to South America, Africa, and Australia, certain stages can be distinguished when a sulfate accumulation area was maximum. For instance, in South America this was the time comprising the Artinskian and the Kungurian ages, and in Australia, presumably, the Late Ordovician and Silurian.

Let us discuss halite accumulation areas for different time intervals of the Paleozoic (Table 6). The Early Cambrian epoch and Kungurian age (Early Permian) are distinguished in the Paleozoic by considering the total area under rock salt on the Earth. In the Early Cambrian the area occupied by rock salt was more than 3,500,000 km^2, i.e., about 2.6% of the entire continental area. Such extensive areas of halite accumulation have not been recorded for any subsequent epochs of the Paleozoic. Even in the Kungurian the area of rock salt deposition was only 1,419,000 km^2, i.e., about one-third of that of the Early Cambrian. During the remaining ages of the Paleozoic, salt deposits covered less than 1% of the Earth's area.

On some continents variations in halite sedimentation areas in Paleozoic history slightly differ. For instance, in Eurasia the Early Cambrian epoch is clearly distinguished, for it is there that all the areas of salt accumulation of that time were located; they constituted more than 6.5% of the Eurasian continental area. However, the Kungurian age cannot be distinguished in Eurasia on the basis of a halite sedimentation area. The Kazanian and Tatarian rock salt occurs on larger territories within the continent. Thus, a stage corresponding to the second half of the Permian period (Kungurian/Tatarian) can be distinguished in Eurasia; then a halite accumulation area was great and covered 1.4% to 1.7% of the continental area. Either in the Early Cambrian epoch, or in the Kungurian age in North America the area of rock salt was not large. Maximum areas there were in the Late Cambrian, Late Silurian, Middle Devonina, and Visean age (Early Carboniferous). In South America and in Australia a vast area of rock salt formation is distinguished only for one stage, the Kungurian age and Middle and Late Ordovician, respectively.

The greatest areas of potash salts both of chloride and sulfate type were recorded for the second half of the Permian period, the Kungurian, Kazanian, and Tatarian ages. The same is true for Eurasia and North America. For the latter the Givetian age is also characterized by a potassium sedimentation area commensurable with that of Permian time.

Table 6. Area of distribution of salt rocks and potassium salt rocks over continents in different periods and ages of the Paleozoic (in 10^6 km^2 and in % of the continents area)

Age	Salt rock							Potassium salt rock						
								Chloride type					Sulfate type	
	Eurasia			North America	South America	Australia	Total	Eurasia			North America	Total	Eurasia (Europe)	North America
	Europe	Asia	Total					Europe	Asia	Total				
1	2	3	4	5	6	7	8	9	10	11	12	13	14	15
Permian	1.456 (14.6)	0.032 (0.07)	1.488 (2.78)	0.606 (2.5)	0.5 (2.7)	–	2.594 (1.9)	0.705 (7.05)	–	0.705 (1.3)	0.071 (0.3)	0.776 (0.6)	0.401 (4.0)	0.07 (0.3)
Late Permian	0.918 (9.2)	0.03 (0.07)	0.921 (1.72)	0.25 (0.8)	–	–	1.171 (0.9)	0.4 (4.0)	–	0.4 (0.7)	0.07 (0.3)	0.47 (0.4)	0.2 (2.0)	0.07 (0.3)
Tatarian	0.779 (7.8)	–	0.779 (1.46)	0.1 (0.4)	–	–	0.879 (0.6)	0.4 (4.0)	–	0.4 (0.7)	0.07 (0.3)	0.47 (0.4)	0.2 (2.0)	0.07 (0.3)
Kazanian	0.783 (7.8)	0.02 (0.05)	0.803 (1.5)	0.21 (0.8)	–	–	1.013 (0.8)	0.4 (4.0)	–	0.4 (0.7)	–	0.4 (0.3)	0.2 (2.0)	–
Ufimian	0.059 (0.6)	0.03 (0.07)	0.089 (0.17)	0.15 (0.6)	–	–	0.239 (0.2)	–	–	–	–	–	–	–
Early Permian	0.807 (8.1)	0.032 (0.07)	0.839 (1.57)	0.37 (1.5)	0.5 (2.7)	–	1.709 (1.3)	0.305 (3.05)	–	0.305 (0.6)	0.001 (0.004)	0.306 (0.2)	0.201 (2.0)	–
Kungurian	0.697 (6.97)	0.032 (0.07)	0.729 (1.36)	0.19 (0.8)	0.5 (2.7)	–	1.419 (1.1)	0.305 (3.05)	–	0.305 (0.6)	–	0.305 (0.2)	0.2 (2.0)	–
Artinskinian	–	–	–	0.156 (0.6)	–	–	0.156 (0.1)	–	–	–	0.001 (0.004)	0.001 (0.001)	–	–
Sakmarian	0.09 (0.9)	–	0.09 (0.17)	0.142 (0.6)	–	–	0.232 (0.2)	0.001 (0.01)	–	0.001 (0.004)	–	0.001 (0.001)	0.0008 (0.008)	–
Asselian	0.03 (0.3)	–	0.03 (0.06)	–	–	–	0.03 (0.02)	–	–	–	–	–	–	–

Carboniferous	–	0.006 (0.01)	0.006 (0.01)	0.506 (2.0)	0.1 (0.5)	–	0.612 (0.5)	–	–	–	0.046 (0.2)	0.046 (0.03)	–	?
Late Carboniferous	–	–	–	–	0.1 (0.5)	–	0.1 (0.07)	–	–	–	–	–	–	–
Orenburgian	–	–	–	–	0.1 (0.5)	–	0.1 (0.07)	–	–	–	–	–	–	–
Gzhelian	–	–	–	–	–	–	–	–	–	–	–	–	–	–
Middle Carboniferous	–	–	–	0.226 (0.9)	–	–	0.226 (0.2)	–	–	–	0.016 (0.07)	0.016 (0.01)	–	?
Bashkirian	–	–	–	0.2 (0.8)	–	–	0.2 (0.2)	–	–	–	–	–	–	–
Early Carboniferous	–	0.006 (0.01)	0.006 (0.01)	0.28 (1.2)	–	–	0.286 (0.2)	–	–	–	0.03 (0.1)	0.03 (0.02)	–	?
Namurian	–	–	–	–	–	–	–	–	–	–	–	–	–	–
Visean	–	0.003 (0.007)	0.003 (0.06)	0.28 (1.2)	–	–	0.283 (0.2)	–	–	–	0.03 (0.1)	0.03 (0.02)	–	?
Tournaisian	–	0.003 (0.007)	0.003 (0.06)	0.08 (0.3)	–	–	0.083 (0.06)	–	–	–	–	–	–	–
Devonian	0.148 (1.5)	0.196 (0.45)	0.344 (0.6)	0.425 (1.7)	–	0.008 (0.1)	0.777 (0.6)	0.002 (0.02)	0.001 (0.002)	0.003 (0.006)	0.07 (0.3)	0.073 (0.05)	–	–
Late Devonian	0.068 (0.7)	0.065 (0.15)	0.133 (0.25)	0.025 (0.1)	–	–	0.158 (0.1)	0.002 (0.02)	?	0.002 (0.004)	–	0.002 (0.001)	–	–
Fammenian	0.068 (0.7)	0.065 (0.15)	0.133 (0.25)	0.02 (0.1)	–	–	0.153 (0.1)	0.002 (0.02)	?	0.002 (0.004)	–	0.002 (0.001)	–	–
Frasnian	0.06 (0.6)	0.016 (0.04)	0.076 (0.14)	0.025 (0.1)	–	–	0.101 (0.7)	?	–	?	–	?	–	–
Middle Devonian	0.08 (0.8)	0.131 (0.3)	0.211 (0.4)	0.425 (1.7)	–	0.008 (0.1)	0.644 (0.5)	0.0002 (0.002)	0.001 (0.002)	0.001 (0.004)	0.07 (0.3)	0.071 (0.05)	–	–

Table 6 (continued)

Age	Salt rock							Potassium salt rock						
								Chloride type					Sulfate type	
	Eurasia			North America	South America	Austra-lia	Total	Eurasia			North America	Total	Eurasia (Europe)	North America
	Europe	Asia	Total					Europe	Asia	Total				
1	2	3	4	5	6	7	8	9	10	11	12	13	14	15
Givetian	–	0.131 (0.3)	0.131 (0.25)	0.35 (1.4)	–	–	0.166 (0.1)	–	0.001 (0.002)	0.001 (0.004)	0.07 (0.3)	0.071 (0.05)	–	–
Eifelian	0.08 (0.8)	0.131 (0.3)	0.211 (0.4)	0.255 (1.1)	–	0.008 (0.1)	0.474 (0.3)	0.0002 (0.002)	0.001 (0.002)	0.001 (0.004)	–	0.001 (0.001)	–	–
Early Devonian	–	0.0005 (0.001)	0.0005 (0.001)	–	–	–	0.0005 (0.0004)	–	?	?	–	?	–	–
Silurian	–	–	–	0.26 (1.1)	–	–	0.26 (0.2)	–	–	–	0.036 (0.15)	0.036 (0.03)	–	–
Late Silurian	–	–	–	0.26 (1.1)	–	–	0.26 (0.2)	–	–	–	(0.036 (0.15)	0.036 (0.03)	–	–
Přidolian	–	–	–	0.26 (1.1)	–	–	0.26 (0.2)	–	–	–	–	–	–	–
Ludlovian	–	–	–	0.26 (1.1)	–	–	0.26 (0.2)	–	–	–	0.036 (0.15)	0.036 (0.03)	–	–
Early Silurian	–	–	–	–	–	–	–	–	–	–	–	–	–	–
Wenlockian	–	–	–	–	–	–	–	–	–	–	–	–	–	–
Llandoverian	–	–	–	–	–	–	–	–	–	–	–	–	–	–
Ordovician	–	–	–	0.1 (0.4)	–	0.25 (3.3)	0.35 (0.3)	–	–	–	–	–	–	–

Late Ordovician	–	–	–	–	–	0.25 (3.3)	0.25 (0.2)	–	–	–	–	–	–	–
Ashgillian	–	–	–	–	–	0.25 (3.3)	0.25 (0.2)	–	–	–	–	–	–	–
Late Caradoc	–	–	–	–	–	0.25 (3.3)	0.25 (0.2)	–	–	–	–	–	–	–
Middle Ordovician	–	–	–	0.1 (0.4)	–	0.25 (3.3)	0.35 (0.3)	–	–	–	–	–	–	–
Early and Middle Caradoc	–	–	–	0.1 (0.4)	–	–	0.1 (0.07)	–	–	–	–	–	–	–
Llandeilo	–	–	–	0.1 (0.4)	–	–	0.1 (0.07)	–	–	–	–	–	–	–
Llanvirn	–	–	–	0.1 (0.4)	–	–	0.1 (0.07)	–	–	–	–	–	–	–
Early Ordovician	–	–	–	–	–	–	–	–	–	–	–	–	–	–
Arenigian	–	–	–	–	–	–	–	–	–	–	–	–	–	–
Late Tremadoc	–	–	–	–	–	–	–	–	–	–	–	–	–	–
Cambrian	–	3.5 (8.1)	3.5 (6.55)	0.5 (2.0)	–	0.001 (0.01)	4.511 (3.4)	–	0.061 (0.14)	0.061 (0.1)	–	0.061 (0.05)	–	–
Late Cambrian	–	0.2 (0.46)	0.2 (0.37)	0.5 (2.0)	–	–	0.502 (0.4)	–	–	–	–	–	–	–
Middle Cambrian	–	0.51 (1.17)	0.51 (0.95)	–	–	–	0.51 (0.4)	–	–	–	–	–	–	–
Early Cambrian	–	3.5 (8.1)	3.5 (6.55)	–	–	0.001 (0.01)	3.501 (2.6)	–	0.061 (0.14)	0.061 (0.1)	–	0.061 (0.05)	–	–

The Volume of Evaporite Sedimentation in Different Periods, Epochs, and Ages of the Paleozoic

Volumes of halogenic rocks are given in Tables 7–9. Table 7 shows the volume of all the evaporites and their distribution among major stratigraphic units of the Paleozoic on the continents and on the Earth in general, while Tables 8 and 9 show the distribution pattern of volumes of sulfate rocks, rock and potash salts of chlorite and sulfate types.

The total volume of Paleozoic evaporites is estimated at about 3.6 x 10^6 km^3 (Table 7). This value correlates well with the total volume of rock salt, gypsum, and anhydrite, calculated by A.B. Ronov, V.E. Khain, and K.S. Seslavinsky for the Paleozoic. It attained 3.83 x 10^6 km^3 (Ronov 1976). However, the distribution of this volume among the epochs of the Paleozoic is usually inconsistent with our data. Thus, the volume of the Middle Cambrian evaporites is almost one order higher than we have calculated. The mass of the Late Cambrian halogenic rocks is more than five times reduced, that of the Silurian almost four times increased, Carboniferous two times increased, Middle and Late Carboniferous more than five times reduced, Late Permian 1.5 times increased. Thus, the values obtained by these investigators should be used very carefully when the volume relations of the evaporite deposits in different epochs of the Paleozoic are established. At the same time, Ronov (1976) and the present author gave almost the same estimates of evaporite volume for the epochs of Early Cambrian and Early Permian. Their mass is so great that all the above discrepancies do not affect the total volume of Paleozoic halogenic rocks.

During the Paleozoic the greatest volume of evaporites formed in the Early Cambrian and Kungurian age (Early Permian) (see Table 7). In the Early Cambrian 38.2% of evaporites formed, and in the Kungurian 32.1%. Thus, during these two stages more than 70% of the total mass of Paleozoic evaporites accumulated. The remaining 30% are distributed irregularly among all the other ages and epochs of the Paleozoic. Among them the Middle Cambrian epoch and the Eifelian age (Middle Devonian) are distinguished when 3.2% of the total volume of Paleozoic evaporites formed. In the Fammenian age, Bashkirian age (Middle Carboniferous), Kazanian, and Tatarian ages (Late Permian) 2%, 3.2%, 2.8%, and 2% formed, respectively, It is quite evident that 17.6% of the volume of the halogenic rocks of the Paleozoic age were accumulated during these ages. The remaining evaporites accounting for 12.4% formed in the Ordovician, Silurian, much of the Carboniferous, and in the first third of the Permian period (Asselian, Sakmarian, and Artinskian ages). Such an age distribution pattern of evaporite volume allows us to distinguish only the Early Cambrian epoch and Kungurian age (Permian) as stages of extensive evaporite sedimentation. However, it should be remembered that these stages are incommensurable in their duration and, hence, in the intensity of halogenic sedimentation. The total volume of Early Cambrian evaporites accumulated during at least 40–50,000,000 years, whereas the volume of Kungurian halogenic deposits formed during much shorter time period, 6–7,000,000 years, i.e., in the Kungurian age a vast mass of evaporites was extremely rapidly accumulated. The other epochs and ages of the Paleozoic era are far inferior to the Early Cambrian and Kungurian and cannot be regarded as very large, proceeding from the volume of salt deposits.

In the preceding chapter the history of the Paleozoic evaporite sedimentation was subdivided into three epochs: Early Cambrian, Middle Cambrian/Early Devonian, and Middle Devonian/Permian. The first and third periods were distinguished as the epochs of considerable evaporite accumulation in general, whereas the second was outlined as an epoch with a limited accumulation of salts. The volumes, on the one hand, confirm these stages, and on the other hand make their determination more precise. Thus, the Early Cambrian and Middle Cambrian/Early Devonian epochs are characterized, respectively, by maximum and minimum volumes of salts accumulated. The Middle Devonian/Permian epoch can be rather easily subdivided into three parts using the distribution of evaporite volumes: (1) Middle/Late Devonian when volumes of halogenic sediments slightly increased, (2) Carboniferous/Early Permian (Tournaisian/Artinskian), when the volume of evaporites in general was extremely limited, (3) Early/Late Permian from Kungurian age of vast halogenesis to Kazanian and Tatarian ages, also marked by a considerable volume of salts.

As a rule, a distribution of evaporites on different continents does not follow the pattern established for the Earth in general. Almost the total volume of Early Cambrian halogenic rocks is confined to Asia, whereas Kungurian rocks occur mainly in Europe. The same preferential concentration of evaporites on one continent, and within this continent even in one basin, is also characteristic of all the other ages of the Paleozoic. Thus, it follows in Paleozoic history that distribution of evaporites was more or less regular among continents, but was concentrated mostly in one of them, within a limited number of basins which were major regions of halogenic sedimentation.

Let us now find out whether the distribution pattern of volumes among the epochs and ages of the Paleozoic established for evaporites in general is in accord with sulfate rocks, rock and potash salts.

The volume of sulfate accumulation allows us to establish the Early Cambrian epoch when one third of Paleozoic sulfate rocks formed (Table 8). The Middle Cambrian/Early Devonian epoch is also clearly outlined, characterized by an extremely small mass of sulfate accumulated. During the interval Middle Devonian/Permian time the volume of sulfates slightly increased, and it is only in the Eifelian age and the Kungurian age (Early Permian) that it was considerable, never reaching however the values observed in the Early Cambrian. Thus, a distribution pattern of sulfate rock volume corresponds, on the one hand, to the above stages of sulfate accumulation and, on the other hand, to the stages of evaporite sedimentation in the Paleozoic in general. The volume of sulfate rocks during the Kungurian age is not as important as that of evaporites. This can be accounted for by the fact that most Kungurian evaporites are not composed of sulfate rocks, but of rock salts.

Let us analyze the data given in Table 9 showing the distribution of rock and potash salt volumes among different time units of the Paleozoic. At present it can be assumed that the total volume of rock salt accumulated during the Paleozoic era is about 3×10^6 km^3. Earlier we estimated this volume at 2.8×10^6 km^3 (Zharkov 1974a, 1977). This increase is due to new salt strata found recently both in newly discovered salt basins (Ordovician of the Canadian Arctic Archipelago, Hudson, Mid-Tien Shan, Sverdrup, Alpine, and Moesian) and in some basins known before (Canning, Maritime, Peru-Bolivian, etc.).

Table 7. Volume of salt rocks and sulfate rocks over continents in different periods and ages of the Paleozoic (in km^3 and in % of volume of evaporite rocks on continents)

Age	Eurasia			North America	South America	Africa	Australia	Total
	Europe	Asia	Total					
1	2	3	4	5	6	7	8	9
Permian	1.289×10^6 (85.5)	8×10^3 (0.5)	1.297×10^6 (43.8)	8.52×10^4 (16.9)	6.5×10^4 (70.2)	–	–	1.447×10^6 (40.2)
Late Permian	1.87×10^5 (12.4)	5.9×10^3 (0.4)	1.93×10^5 (6.5)	3.57×10^4 (7.1)	–	–	–	2.29×10^5 (6.4)
Tatarian	5.6×10^4 (3.7)	–	5.6×10^4 (1.9)	2.41×10^4 (2.8)	–	–	–	8.01×10^4 (2.2)
Kazanian	1.29×10^5 (8.5)	1.9×10^3 (0.1)	1.31×10^5 (4.4)	4.7×10^3 (0.9)	–	–	–	1.36×10^5 (3.8)
Ufimian	2.5×10^3 (0.2)	4×10^3 (0.3)	6.5×10^3 (0.2)	6.9×10^3 (3.4)	–	–	–	1.34×10^4 (0.4)
Early Permian	1.102×10^6 (73.1)	2.1×10^3 (0.1)	1.104×10^6 (37.3)	4.95×10^4 (9.8)	6.6×10^4 (70.2)	–	–	1.218×10^6 (33.8)
Kungurian	1.077×10^6 (71.5)	2.1×10^3 (0.1)	0.079×10^6 (36.4)	1.55×10^4 (3.1)	6.1×10^4 (65.8)	–	–	1.155×10^6 (32.1)
Artinskinian	6.4×10^3 (0.4)	–	6.4×10^3 (0.2)	1.83×10^4 (3.6)	2.03×10^3 (2.2)	–	–	2.67×10^4 (0.7)
Sakmarian	1.04×10^4 (0.7)	–	1.04×10^4 (0.4)	1.5×10^4 (3.0)	2×10^3 (2.2)	–	–	2.74×10^4 (0.8)
Asselian	9.1×10^3 (0.5)	–	8.1×10^3 (0.3)	7×10^2 (0.1)	–	–	–	8.8×10^3 (0.2)

Carboniferous	2.77×10^4 (1.8)	3.15×10^3 (0.2)	3.08×10^4 (1.0)	1.77×10^5 (34.9)	2.6×10^4 (28.2)	3.31×10^3 (98.5)	10 (0.04)	2.37×10^5 (6.6)
Late Carboniferous	9.1×10^3 (0.6)	1.05×10^2 (0.09)	9.2×10^3 (0.3)	8.8×10^2 (0.2)	2.6×10^4 (28.2)	1×10^2 (3.0)	–	3.62×10^4 (1.0)
Orenburgian	6×10^2 (0.04)	5 (+)	6.1×10^2 (0.02)	4.25×10^2 (0.1)	2.5×10^2 (27.1)	–	–	2.6×10^4 (0.7)
Gzhelian	8.5×10^3 (0.6)	1×10^2 (0.03)	8.6×10^3 (0.3)	4.55×10^2 (0.1)	9.75×10^2 (1.1)	1×10^2 (3.0)	–	1.01×10^4 (0.3)
Middle Carboniferous	5.05×10^3 (0.3)	1.46×10^2 (0.05)	5.2×10^3 (0.2)	1.25×10^5 (24.6)	–	1×10^2 (3.0)	–	1.3×10^5 (3.7)
Moscovian	5×10^3 (0.3)	1.16×10^2 (0.04)	5.12×10^3 (0.2)	1.05×10^4 (2.0)	–	–	–	1.56×10^4 (0.5)
Bashkirian	50 (+)	30 (+)	80 (+)	1.15×10^5 (22.5)	–	1×10^2 (3.0)	–	1.15×10^5 (3.2)
Early Carboniferous	1.35×10^4 (0.9)	2.89×10^3 (0.2)	1.64×10^4 (0.5)	5.08×10^4 (10.1)	–	3.11×10^3 (92.5)	10 (0.04)	7.03×10^4 (1.9)
Namurian	1.25×10^2 (0.01)	4.6×10^2 (0.03)	5.85×10^2 (0.02)	1×10^2 (0.02)	–	1.1×10^2 (3.3)	5 (0.02)	8×10^2 (0.02)
Visean	5.28×10^3 (0.3)	1.65×10^3 (0.12)	6.93×10^3 (0.2)	4.93×10^4 (9.8)	–	3×10^3 (89.2)	5 (0.02)	5.92×10^4 (1.6)
Tournaisian	8.1×10^3 (0.6)	7.8×10^2 (0.05)	8.88×10^3 (0.3)	1.42×10^3 (0.3)	–	–	–	1.02×10^4 (0.3)
Devonian	1.9×10^5 (12.6)	2.58×10^4 (1.8)	2.16×10^3 (7.2)	8.86×10^4 (17.6)	–	50 (1.5)	4.5×10^2 (1.7)	3.05×10^5 (8.5)
Late Devonian	1.08×10^5 (7.1)	7.91×10^3 (0.5)	1.16×10^5 (3.9)	4.92×10^3 (1.0)	–	25 (0.75)	–	1.21×10^5 (3.4)
Fammenian	6.58×10^4 (4.4)	3.58×10^3 (0.2)	6.94×10^4 (2.3)	1.75×10^3 (0.3)	–	–	–	7.12×10^4 (2.0)

Table 7 (continued)

Age	Eurasia			North America	South America	Africa	Australia	Total
	Europe	Asia	Total					
1	2	3	4	5	6	7	8	9
Frasnian	4.17 x 10^4 (2.7)	4.33 x 10^3 (0.3)	4.6 x 10^4 (1.6)	3.17 x 10^3 (0.7)	–	–	–	4.92 x 10^4 (1.4)
Middle Devonian	8.22 x 10^4 (5.5)	1.64 x 10^4 (1.2)	9.86 x 10^4 (3.3)	8.33 x 10^4 (16.5)	–	25 (0.75)	4.5 x 10^2 (1.7)	1.82 x 10^5 (5.1)
Givetian	5 x 10^2 (0.03)	8.33 x 10^3 (0.6)	8.83 x 10^3 (0.3)	5.85 x 10^4 (11.6)	–	–	–	6.73 x 10^4 (1.9)
Eifelian	8.17 x 10^4 (5.5)	8.04 x 10^3 (0.6)	8.97 x 10^4 (3.0)	2.48 x 10^4 (4.9)	–	–	4.5 x 10^2 (1.7)	1.15 x 10^5 (3.2)
Early Devonian	–	1.49 x 10^3 (0.1)	1.49 x 10^3 (0.05)	5 x 10^2 (0.1)	–	–	–	1.99 x 10^3 (0.06)
Silurian	81 (+)	7.5 x 10^2 (0.05)	8.31 x 10^2 (0.03)	2.99 x 10^4 (5.9)	–	–	1.53 x 10^3 (5.8)	3.23 x 10^4 (0.9)
Late Silurian	40.5 (+)	6.75 x 10^2 (0.04)	7.15 x 10^2 (0.02)	2.98 x 10^4 (5.9)	–	–	1 x 10^3 (3.8)	3.15 x 10^4 (0.9)
Přidolian	10 (+)	3 x 10^2 (0.02)	3.1 x 10^2 (0.01)	2.15 x 10^4 (4.3)	–	–	5 x 10^2 (1.9)	2.23 x 10^4 (0.6)
Ludlovian	30.5 (+)	3.75 x 10^2 (0.02)	4.06 x 10^2 (0.01)	8.27 x 10^3 (1.6)	–	–	5 x 10^2 (1.9)	9.18 x 10^3 (0.3)
Early Silurian	40.5 (+)	75 (0.01)	1.15 x 10^2 (+)	1 x 10^2 (+)	–	–	5.3 x 10^2 (2.0)	7.45 x 10^2 (0.02)
Wenlockian	20.5 (+)	–	20.5 (+)	50 (+)	–	–	5.3 x 10^2 (2.0)	6 x 10^2 (0.02)
Llandoverian	20 (+)	75 (0.01)	95 (+)	50 (+)	–	–	–	1.45 x 10^2 (+)

Ordovician	0.3 (+)	1.29×10^{3} (0.1)	1.29×10^{3} (0.04)	1.94×10^{4} (3.8)	–	–	2.41×10^{4} (92.1)	4.5×10^{4} (1.2)
Late Ordovician	–	3.5×10^{2} (0.03)	3.5×10^{2} (0.01)	1.1×10^{3} (0.2)	–	–	2.4×10^{4} (91.9)	2.55×10^{4} (0.7)
Ashgillian	–	1.7×10^{2} (0.01)	1.7×10^{2} (+)	3.25×10^{2} (0.1)	–	–	1.35×10^{4} (51.6)	1.4×10^{4} (0.4)
Late Caradoc	–	1.8×10^{2} (0.02)	1.8×10^{2} (+)	7.75×10^{2} (0.1)	–	–	1.05×10^{4} (40.3)	1.15×10^{4} (0.3)
Middle Ordovician	0.3 (+)	4.5×10^{2} (0.04)	4.5×10^{2} (0.01)	1.33×10^{4} (2.6)	–	–	25 (0.1)	1.38×10^{4} (0.4)
Early and Middle Caradoc	0.2 (+)	1.6×10^{2} (0.02)	1.6×10^{2} (+)	4.8×10^{3} (0.9)	–	–	–	4.96×10^{3} (0.14)
Llandeilo	0.1 (+)	1.5×10^{2} (0.01)	1.5×10^{2} (+)	5.1×10^{3} (0.9)	–	–	–	5.25×10^{3} (0.14)
Llanvirn	–	1.4×10^{2} (0.01)	1.4×10^{2} (+)	3.43×10^{3} (0.7)	–	–	–	3.6×10^{3} (0.12)
Early Ordovician	–	4.9×10^{2} (0.04)	4.9×10^{2} (0.02)	5×10^{3} (1.0)	–	–	25 (0.1)	5.52×10^{3} (0.14)
Arenigian	–	2.5×10^{2} (0.02)	2.5×10^{2} (0.01)	5×10^{3} (1.0)	–	–	–	5.25×10^{3} (0.14)
Late Tremadoc	–	2.4×10^{2} (0.02)	2.4×10^{2} (0.01)	–	–	–	–	2.4×10^{2} (+)
Cambrian	–	1.417×10^{6} (97.35)	1.417×10^{6} (47.9)	1.06×10^{5} (20.9)	1.5×10^{3} (1.6)	?	1.02×10^{2} (0.4)	1.525×10^{6} (42.6)
Late Cambrian	–	5.6×10^{3} (0.4)	5.6×10^{3} (0.2)	1.06×10^{5} (20.9)	–	–	–	1.12×10^{5} (3.2)
Middle Cambrian	–	4.14×10^{4} (2.8)	4.14×10^{4} (1.4)	–	–	–	–	4.14×10^{4} (1.2)
Early Cambrian	–	1.37×10^{6} (94.15)	1.37×10^{6} (46.3)	–	1.5×10^{3} (1.6)	?	1.02×10^{2} (0.4)	1.372×10^{6} (38.2)
Total	1.507×10^{6}	1.456×10^{6}	2.963×10^{6}	5.06×10^{5}	9.25×10^{4}	3.36×10^{3}	2.62×10^{4}	3.592×10^{6}

Table 8. Volume of sulfate rocks over continents in different periods and ages of the Paleozoic (in km^3 and in % of volume of sulfate rocks on continents)

Age	Eurasia			North America	South America	Africa	Australia	Total
	Europe	Asia	Total					
1	2	3	4	5	6	7	8	9
Permian	1.1×10^5 (42.9)	2.19×10^3 (0.9)	1.12×10^5 (22.3)	2.12×10^4 (23.5)	1.5×10^4 (66.5)	–	–	1.48×10^5 (23.5)
Late Permian	3.2×10^4 (12.4)	2.08×10^3 (0.8)	3.4×10^4 (6.8)	1.14×10^4 (12.6)	–	–	–	4.54×10^4 (7.2)
Tatarian	1.09×10^4 (4.2)	–	1.09×10^4 (2.2)	6.6×10^3 (7.3)	–	–	–	1.75×10^4 (2.8)
Kazanian	1.96×10^4 (7.6)	8.9×10^2 (0.3)	2.05×10^4 (4.1)	2.9×10^3 (3.2)	–	–	–	2.34×10^4 (3.7)
Ufimian	1.5×10^3 (0.6)	1.19×10^3 (0.5)	2.69×10^3 (0.5)	1.9×10^3 (2.1)	–	–	–	4.59×10^3 (0.7)
Early Permian	7.8×10^4 (30.2)	1.1×10^2 (0.05)	7.81×10^4 (15.5)	9.8×10^3 (10.9)	1.5×10^4 (66.5)	–	–	1.03×10^5 (16.3)
Kungurian	5.89×10^4 (22.8)	1.1×10^2 (0.05)	5.9×10^4 (11.7)	2.5×10^3 (2.8)	1.1×10^4 (48.8)	–	–	7.25×10^4 (11.5)
Artinskinian	6.4×10^3 (2.5)	–	6.4×10^3 (1.3)	3.1×10^3 (3.4)	2.03×10^3 (9.0)	–	–	1.15×10^4 (1.8)
Sakmarian	9.2×10^3 (3.6)	–	9.2×10^3 (1.8)	3.5×10^3 (3.9)	2×10^3 (8.7)	–	–	1.47×10^4 (2.3)
Asselian	3.5×10^3 (1.4)	–	3.5×10^3 (0.7)	7×10^3 (0.8)	–	–	–	4.2×10^3 (0.7)
Carboniferous	2.77×10^4 (10.7)	2.5×10^3 (1.0)	3.02×10^4 (6.0)	2.5×10^4 (27.6)	6.01×10^3 (26.8)	3.31×10^3 (98.5)	10 (0.1)	6.45×10^4 (10.2)

Late Carboniferous	9.1×10^3 (3.6)	1.05×10^2 (0.05)	9.2×10^3 (1.8)	8.8×10^2 (1.0)	6.01×10^3 (26.8)	1×10^2 (3.0)	–	1.62×10^4 (2.6)
Orenburgian	6×10^2 (0.2)	5 (0.002)	6.1×10^2 (0.1)	5.25×10^2 (0.5)	5.03×10^3 (22.5)	–	–	6.07×10^3 (1.0)
Gzhelian	8.5×10^3 (3.4)	1×10^2 (0.04)	8.6×10^3 (1.7)	4.55×10^2 (0.5)	9.75×10^2 (4.3)	1×10^2 (3.0)	–	1.01×10^4 (1.6)
Middle Carboniferous	5.05×10^3 (1.9)	1.46×10^2 (0.06)	5.2×10^3 (1.1)	1.58×10^4 (17.4)	–	1×10^2 (3.0)	–	2.11×10^4 (3.3)
Moscovian	5×10^3 (1.9)	1.16×10^2 (0.05)	5.12×10^3 (1.0)	8.23×10^2 (0.9)	–	–	–	5.94×10^3 (0.9)
Bashkirian	50 (0.02)	30 (0.01)	80 (0.1)	1.5×10^4 (16.5)	–	1×10^2 (3.0)	–	1.52×10^4 (2.4)
Early Carboniferous	1.35×10^4 (5.2)	2.24×10^3 (0.9)	1.57×10^4 (3.1)	8.33×10^3 (9.2)	–	3.11×10^3 (92.5)	10 (0.1)	2.72×10^4 (4.3)
Namurian	1.25×10^2 (0.05)	4.6×10^2 (0.2)	5.85×10^2 (0.1)	1×10^2 (0.1)	–	1.1×10^2 (3.3)	5 (0.05)	8×10^2 (0.2)
Visean	5.28×10^3 (2.0)	1.5×10^3 (0.6)	6.78×10^3 (1.4)	7.22×10^3 (8.0)	–	3×10^3 (89.2)	5 (0.05)	1.7×10^4 (2.6)
Tournaisian	8.1×10^3 (3.1)	2.8×10^2 (0.1)	8.38×10^3 (1.6)	1.01×10^3 (1.1)	–	–	–	9.39×10^3 (1.5)
Devonian	1.2×10^5 (46.3)	6.26×10^3 (2.6)	1.26×10^5 (25.2)	2.47×10^4 (27.6)	–	50 (1.5)	50 (0.5)	1.51×10^5 (24.0)
Late Devonian	3.81×10^4 (14.8)	1.41×10^3 (0.6)	3.95×10^4 (7.9)	4.07×10^3 (4.6)	–	25 (0.7)	–	4.36×10^4 (6.9)
Fammenian	2.14×10^4 (8.3)	5.8×10^2 (0.2)	2.2×10^4 (4.3)	1.55×10^3 (1.8)	–	–	–	2.36×10^4 (3.7)
Frasnian	1.67×10^4 (6.5)	8.3×10^2 (0.4)	1.75×10^4 (3.6)	2.52×10^3 (2.8)	–	–	–	2×10^4 (3.2)

Table 8 (continued)

Age	Eurasia			North America	South America	Africa	Australia	Total
	Europe	Asia	Total					
1	2	3	4	5	6	7	8	9
Middle Devonian	8.14×10^4 (31.5)	3.37×10^3 (1.4)	8.48×10^4 (17.0)	2.01×10^4 (22.4)	–	25 (0.8)	50 (0.5)	1.05×10^5 (16.8)
Givetian	5×10^2 (0.2)	1.83×10^3 (0.75)	2.33×10^3 (0.5)	1.61×10^4 (17.9)	–	–	–	1.84×10^4 (2.9)
Eifelian	8.09×10^4 (31.3)	1.54×10^3 (0.6)	8.25×10^4 (16.5)	4×10^3 (4.5)	–	–	50 (0.5)	8.66×10^4 (13.9)
Early Devonian	–	1.48×10^3 (0.6)	1.48×10^3 (0.3)	5×10^2 (0.6)	–	–	–	1.98×10^3 (0.3)
Silurian	81 (0.04)	7.5×10^2 (0.3)	8.31×10^2 (0.17)	3.85×10^3 (4.3)	–	–	1.53×10^3 (14.4)	6.21×10^3 (1.0)
Late Silurian	40.5 (0.02)	6.75×10^2 (0.3)	7.15×10^2 (0.15)	3.75×10^3 (4.2)	–	–	1×10^3 (9.4)	5.47×10^3 (0.9)
Přidolian	10 (0.005)	3×10^2 (0.12)	3.1×10^2 (0.07)	1.5×10^3 (1.6)	–	–	5×10^2 (4.7)	2.31×10^3 (0.4)
Ludlovian	30.5 (0.01)	3.75×10^2 (0.15)	4.06×10^2 (0.08)	2.25×10^3 (2.6)	–	–	4×10^2 (4.7)	3.16×10^3 (0.5)
Early Silurian	40.5 (0.02)	75 (0.003)	1.15×10^2 (0.02)	1×10^2 (0.1)	–	–	5.3×10^2 (5.0)	7.45×10^2 (0.1)
Wenlockian	20.5 (0.01)	–	20.5 (0.005)	50 (0.05)	–	–	5.3×10^2 (5.0)	6×10^2 (0.1)
Llandoverian	20 (0.01)	75 (0.003)	95 (0.02)	50 (0.05)	–	–	–	1.45×10^2 (0.02)
Ordovician	0.3 (+)	1.29×10^3 (0.5)	1.29×10^3 (0.3)	9.4×10^3 (10.4)	–	–	9.05×10^3 (85.4)	2×10^4 (3.2)

Late Ordovician	–	3.5×10^{2} (0.14)	3.5×10^{2} (0.07)	1.1×10^{3} (1.2)	–	–	9×10^{3} (85.0)	1.05×10^{4} (1.7)
Ashgillian	–	1.7×10^{2} (0.07)	1.7×10^{2} (0.03)	3.25×10^{2} (0.4)	–	–	6×10^{3} (56.6)	6.5×10^{3} (1.1)
Late Caradoc	–	1.8×10^{2} (0.07)	1.8×10^{2} (0.04)	7.75×10^{2} (0.8)	–	–	3×10^{3} (28.4)	4×10^{3} (0.6)
Middle Ordovician	0.3 (+)	4.5×10^{2} (0.2)	4.5×10^{2} (0.09)	3.3×10^{3} (3.6)	–	–	25 (0.2)	3.78×10^{3} (0.6)
Early and Middle Caradoc	0.2 (+)	1.6×10^{2} (0.07)	1.6×10^{2} (0.03)	1.5×10^{3} (1.6)	–	–	–	1.66×10^{3} (0.3)
Llandeilo	0.1 (+)	1.5×10^{2} (0.06)	1.5×10^{2} (0.03)	1.8×10^{3} (9.0)	–	–	–	1.95×10^{3} (0.3)
Llanvirn	–	1.4×10^{2} (0.06)	1.4×10^{2} (0.03)	30 (0.003)	–	–	–	1.7×10^{2} (0.03)
Early Ordovician	–	4.9×10^{2} (0.2)	4.9×10^{2} (0.1)	5×10^{3} (5.6)	–	–	25 (0.2)	5.52×10^{3} (0.9)
Arenigian	–	2.5×10^{2} (0.1)	2.5×10^{2} (0.05)	5×10^{3} (5.6)	–	–	–	5.25×10^{3} (0.8)
Late Tremadoc	–	2.4×10^{2} (0.1)	2.4×10^{2} (0.05)	–	–	–	–	2.4×10^{2} (0.05)
Cambrian	–	2.31×10^{5} (94.7)	2.31×10^{5} (46.0)	6×10^{3} (6.6)	1.5×10^{3} (6.7)	?	2 (0.02)	2.39×10^{5} (38.1)
Late Cambrian	–	6×10^{2} (0.3)	6×10^{2} (0.1)	6×10^{3} (6.6)	–	–	–	6.6×10^{3} (1.1)
Middle Cambrian	–	1.04×10^{4} (4.4)	1.04×10^{4} (2.1)	–	–	–	–	1.04×10^{4} (1.7)
Early Cambrian	–	2.2×10^{5} (90.0)	2.2×10^{5} (43.9)	–	1.5×10^{3} (6.7)	?	2.0 (0.02)	2.22×10^{5} (35.3)
Total	2.58×10^{5}	2.44×10^{5}	5.02×10^{5}	9.02×10^{4}	2.25×10^{4}	3.36×10^{3}	1.06×10^{4}	6.29×10^{5}

Table 9. Volume of salt rocks and potassium salt rocks over continents in different periods and ages

Age	Salt rock					
	Eurasia			North America	South America	Australia
	Europe	Asia	Total			
1	2	3	4	5	6	7
Permian	1.162×10^6 (94.3)	5.8×10^3 (0.5)	1.168×10^6 (47.7)	6.35×10^4 (15.3)	5×10^4 (71.4)	–
Late Permian	1.53×10^5 (13.5)	3.8×10^3 (0.3)	1.57×10^5 (6.4)	2.38×10^4 (5.7)	–	–
Tatarian	4.44×10^4 (3.7)	–	4.44×10^4 (1.8)	1.7×10^4 (4.2)	–	–
Kasanian	1.08×10^5 (8.8)	1×10^3 (0.1)	1.09×10^5 (4.4)	1.8×10^3	–	–
Ufimian	1×10^3 (0.01	2.8×10^3 (0.2)	3.8×10^3 (0.2)	5×10^3 (0.1)	–	–
Early Permian	1.009×10^6 (81.8)	2×10^3 (0.2)	1.011×10^6 (41.3)	3.97×10^4 (9.6)	5×10^4 (71.4)	–
Kungurian	1.003×10^6 (81.3)	2×10^3 (0.2)	1.005×10^6 (41.05)	1.3×10^4 (3.2)	5×10^4 (71.4)	–
Artinskinian	–	–	–	1.52×10^4 (3.8)	–	–
Sakmarian	1.24×10^3 (0.1)	–	1.24×10^3 (0.05)	1.15×10^4 (2.6)	–	–
Asselian	4.6×10^3 (0.4)		4.6×10^3 (0.2)	–	–	–
Carboniferous	–	6.5×10^2 (0.05)	6.5×10^2 (0.05)	1.51×10^5 (36.6)	2×10^4 (28.6)	–
Late Carboniferous	–	–	–	–	2×10^4 (28.6)	–
Orenburgian	–	–	–	–	2×10^4 (28.6)	–
Gzhelian	–	–	–	–	–	–
Middle Carboniferous	–	–	–	1.092×10^5 (26.4)	–	–
Moscovian	–	–	–	9.2×10^3 (2.2)	–	–
Bashkirian	–	–	–	1×10^5 (24.2)	–	–
Early Carboniferous	–	6.5×10^2 (0.05)	6.5×10^2 (0.05)	4.24×10^4 (10.2)	–	–
Namurian	–	–	–	–	–	–

of the Paleozoic (in km^3 and in % of volume of salt rocks on continents)

	Potassium salt rock						
	Chloride type					Sulfate type	
Total	Eurasia			North America	Total	Eurasia (Europe)	North America
	Europe	Asia	Total				
8	9	10	11	12	13	14	15
1.281 x 10^6 (43.5)	1.64 x 10^4 (100.0)	–	1.64 x 10^4 (99.4)	3.5 x 10^2 (29.2)	1.67 x 10^4 (95.3)	9.34 x 10^2 (100.0)	1.4 x 10^2 (100.0)
1.81 x 10^5 (6.1)	1.78 x 10^3 (10.9)	–	1.78 x 10^3 (11.0)	3.5 x 10^2 (29.2)	2.13 x 10^3 (12.2)	3.82 x 10^2 (40.8)	1.4 x 10^2 (100.0)
6.14 x 10^4 (2.1)	3.57 x 10^2 (2.2)	–	3.57 x 10^2 (2.2)	3.5 x 10^2 (29.2)	7.07 x 10^2 (4.1)	3.3 x 10^2 (35.4)	1.4 x 10^2 (100.0)
1.11 x 10^5 (3.7)	1.42 x 10^3 (8.7)	–	1.42 x 10^3 (8.8)	–	1.41 x 10^3 (8.1)	52.0 (5.4)	–
8.8 x 10^3	–	–	–	–	–	–	–
1.1 x 10^6 (37.4)	1.46 x 10^4 (89.1)	–	1.46 x 10^4 (89.0)	0.1 (+)	1.46 x 10^4 (83.1)	5.52 x 10^2 (59.2)	–
1.068 x 10^6 (36.4)	1.46 x 10^4 (89.1)	–	1.46 x 10^4 (89.0)	–	1.46 x 10^4 (83.1)	5.5 x 10^2 (58.9)	–
1.52 x 10^4 (0.5)	–	–	–	0.1 (+)	0.1 (+)	–	–
1.17 x 10^4 (0.4)	5.0 (0.05)	–	5.0 (0.05)	–	5.0 (0.03)	2.35 (0.3)	–
4.6 x 10^3 (0.1)	–	–	–	–	–	–	–
1.71 x 10^5 (5.8)	–	–	–	4.8 x 10^2 (40.1)	4.8 x 10^2 (2.5)	–	?
1 x 10^4 (0.7)	–	–	–	–	–	–	–
2 x 10^4 (0.7)	–	–	–	–	–	–	–
–	–	–	–	–	–	–	–
1.092 x 10^5 (3.6)	–	–	–	4.5 x 10^2 (37.6)	4.5 x 10^2 (2.3)	–	?
1.092 x 10^5 (0.3)	–	–	–	4.5 x 10^2 (37.6)	4.5 x 10^2 (2.3)	–	?
1 x 10^5 (3.3)	–	–	–	–	–	–	–
4.31 x 10^4 (1.4)	–	–	–	30.0 (2.5)	30.0 (0.2)	–	?
–	–	–	–	–	–	–	–

Table 9 (continued)

Age	Salt rock					
	Eurasia			North America	South America	Australia
	Europe	Asia	Total			
1	2	3	4	5	6	7
Visean	–	1.5×10^2 (0.01)	1.5×10^2 (+)	4.2×10^4 (10.1)	–	–
Tournaisian	–	5×10^2 (0.04)	5×10^2 (+)	4.1×10^2 (0.1)	–	–
Devonian	7.02×10^4 (5.7)	1.95×10^4 (1.6)	8.97×10^4 (3.7)	6.37×10^4 (15.4)	–	4×10^2 (2.6)
Late Devonian	6.94×10^4 (5.6)	6.5×10^3 (0.5)	7.59×10^4 (3.1)	8.5×10^2 (0.2)	–	–
Fammenian	4.44×10^4 (3.6)	3×10^3 (0.2)	4.74×10^4 (1.9)	2×10^2 (0.04)	–	–
Frasnian	2.5×10^4 (2.0)	3.5×10^3 (0.2)	2.85×10^4 (1.2)	6.5×10^2 (0.16)	–	–
Middle Devonian	8×10^2 (0.1)	1.3×10^4 (1.1)	1.38×10^4 (0.6)	6.28×10^4 (15.2)	–	4×10^2 (2.6)
Givetian	–	6.5×10^3 (0.5)	6.5×10^3 (0.3)	4.2×10^4 (10.1)	–	–
Eifelian	8×10^2 (0.1)	6.5×10^3 (0.5)	7.3×10^3 (0.3)	2.08×10^4 (5.1)	–	4×10^2 (2.6)
Early Devonian	–	5 (+)	5 (+)	–	–	–
Silurian	–	–	–	2.6×10^4 (6.3)	–	–
Late Silurian	–	–	–	2.6×10^4 (6.3)	–	–
Přidolian	–	–	–	2×10^4 (4.8)	–	–
Ludlovian	–	–	–	6×10^3 (1.5)	–	–
Early Silurian	–	–	–	–	–	–
Wenlockian	–	–	–	–	–	–
Llandoverian	–	–	–	–	–	–
Ordovician	–	–	–	1×10^4 (2.4)	–	1.5×10^4 (96.8)
Late Ordovician	–	–	–	–	–	1.5×10^4 (96.8)

Total	Potassium salt rock						
	Chloride type					Sulfate type	
	Eurasia			North America	Total	Eurasia (Europe)	North America
	Europe	Asia	Total				
8	9	10	11	12	13	14	15
4.22×10^4 (1.4)	–	–	–	30.0 (2.5)	30.0 (0.2)	–	?
9.1×10^2 (0.03)	–	–	–	–	–	–	–
1.54×10^5 (5.2)	20.0 (+)	?	20.0 (0.2)	3.5×10^2 (29.2)	3.7×10^2 (1.9)	–	–
7.68×10^4 (2.6)	20 (+)	?	20 (0.2)	–	20 (0.1)	–	–
4.76×10^4 (1.6)	20 (+)	?	20 (0.2)	–	20 (0.1)	–	–
2.92×10^4 (1.0)	?	–	?	–	?	–	–
7.7×10^4 (2.6)	0.02 (+)	?	0.02 (+)	3.5×10^2 (29.2)	3.5×10^2 (1.8)	–	–
4.85×10^4 (1.6)	?	?	?	3.5×10^2 (29.2)	3.5×10^2 (1.8)	–	–
2.85×10^4 (1.0)	0.02 (+)	?	0.02 (+)	–	0.02 (+)	–	–
5 (+)	–	?	?	–	?	–	–
2.6×10^4 (0.9)	–	–	–	18 (1.5)	18 (0.1)	–	–
2.6×10^4 (0.9)	–	–	–	18 (1.5)	18 (0.1)	–	–
2×10^4 (0.7)	–	–	–	–	–	–	–
6×10^3 (0.2)	–	–	–	18 (1.5)	18 (0.1)	–	–
–	–	–	–	–	–	–	–
–	–	–	–	–	–	–	–
–	–	–	–	–	–	–	–
2.5×10^4 (0.8)	–	–	–	–	–	–	–
1.5×10^4 (0.5)	–	–	–	–	–	–	–

Table 9 (continued)

Age	Salt rock					
	Eurasia			North America	South America	Australia
	Europe	Asia	Total			
1	2	3	4	5	6	7
Ashgillian	–	–	–	–	–	7.5×10^3 (48.4)
Late Caradoc	–	–	–	–	–	7.5×10^3 (48.4)
Middle Ordovician	–	–	–	1×10^4 (2.4)	–	–
Early and Middle Caradoc	–	–	–	3.3×10^3 (0.8)	–	–
Llandeilo	–	–	–	3.3×10^3 (0.8)	–	–
Llanvirn	–	–	–	3.4×10^3 (0.8)	–	–
Early Ordovician	–	–	–	–	–	–
Arenigian	–	–	–	–	–	–
Late Tremadoc	–	–	–	–	–	–
Cambrian	–	1.186×10^6 (97.85)	1.86×10^6 (48.55)	1×10^5 (24.0)	–	1×10^2 (0.6)
Late Cambrian	–	5×10^3 (0.45)	5×10^3 (0.2)	1×10^5 (24.0)	–	–
Middle Cambrian	–	3.1×10^4 (2.5)	3.1×10^4 (1.3)	–	–	–
Early Cambrian	–	1.15×10^6 (94.9)	1.15×10^6 (47.05)	–	–	1×10^2 (0.6)
Total	1.32×10^6	1.212×10^6	2.444×10^6	4.15×10^5	7×10^4	1.55×10^4

The volume of the Paleozoic rock salt shows the following distribution pattern among the periods: Cambrian 1.286×10^6 km^3, Ordovician 2.5×10^4 km^3, Silurian 2.6×10^4 km^3, Devonian 1.54×10^5 km^3, Carboniferous 1.72×10^5 km^3, Permian 1.281×10^6 km^3. Let us compare this with Kalinko's data (1973a,b). He calculated volumes for almost all the salt basins and for each of the Paleozoic periods. However, his estimates are much higher. This difference is due to the fact that M.K. Kalinko calculated volumes of salt strata in general, and not only volumes of rock salts. This follows from his data on the thicknesses. Kalinko (1973a, Table 2, pp 10–14) points

	Potassium salt rock						
	Chloride type					Sulfate type	
Total	Eurasia			North America	Total	Eurasia (Europe)	North America
	Europe	Asia	Total				
8	9	10	11	12	13	14	15
7.5×10^3 (0.25)	–	–	–	–	–	–	–
7.5×10^3 (0.25)	–	–	–	–	–	–	–
1×10^4 (0.3)	–	–	–	–	–	–	–
3.3×10^3 (0.1)	–	–	–	–	–	–	–
3.3×10^3 (0.1)	–	–	–	–	–	–	–
3.4×10^3 (0.1)	–	–	–	–	–	–	–
–	–	–	–	–	–	–	–
–	–	–	–	–	–	–	–
–	–	–	–	–	–	–	–
1.286×10^6 (43.8)	–	40 (100)	40 (0.4)	–	40 (0.4)	–	–
1.05×10^5 (3.6)	–	–	–	–	–	–	–
3.1×10^4 (0.1)	–	–	–	–	–	–	–
1.15×10^6 (39.1)	–	40 (100)	40 (0.4)	–	40 (0.4)	–	–
2.944×10^6	1.64×10^4	40	1.64×10^4	1.2×10^3	1.76×10^4	9.34×10^2	1.4×10^2

out that it was the thicknesses of strata that were used in calculations, and not thicknesses of rock salt; but later, on his diagrams (Kalinko 1973a, Fig. 2, p. 15; Kalinko 1973b, Fig. 2, p 98) only rock salt volumes are plotted. In fact, these diagrams do not show the distribution of the total amount of rock salt, but that of all evaporites, i.e., halite together with sulfate and, probably, carbonate and other rocks. This is confirmed by thicknesses presented. Thus, the estimates for rock salt volume in the East Siberian Cambrian Basin assumed a thickness of 2.1 km for the area of $1.125. \times 10^6$ km^2. Undoubtedly, this is an average thickness of the Cambrian halogenic series as a whole, probably even overestimated for the area discussed. We have already shown (Zharkov 1969) that the total thickness of rock salt in this basin at present is about 500 m. Even in the case of maximum estimates it would not exceed 1000 m, i.e., it will be one half of that proposed by M.K. Kalinko. The same applies to the thickness of Silurian salt deposits in the Michigan-Pre-Appalachian Basin where M.K. Kalinko estimated it in the deepest (Huron) zone as equal to 1.5 km; in fact, Pierce and Rich (1962) showed that the average thickness of rock salt there is not more than 100 m. It may be also exemplified by Devonian salt deposits of the Pripyat Trough: an assumed thickness of these beds (1.9 km) corresponds to the entire section and should not be used only for rock salt.

Thus, M.K. Kalinko's values mostly reflect total volumes of all the rocks of salt strata; therefore they are much higher than our values. Only for the Cambrian salt are the volumes similar. But this was quite incidental, for the volume of the Cambrian rock salt given by M.K. Kalinko was obtained for the East Siberian Basin only, and the volume of salt deposits in the Iran-Pakistan Basin is greatly lowered and taken for both Proterozoic and Cambrian formations.

Both M.K. Kalinko and the author had great problems when Ordovician rock salt was studied. Originally its volume was not determined (Kalinko 1973a; Zharkov 1974a, 1977), but later an approximate estimate for rock salt of this age was given by Kalinko (1973b). Here we also thought it possible to assign, though arbitrarily, salt strata of the Canning Basin to the Ordovician.

The data presented by M.K. Kalinko are highly important. They showed for the first time the real correlation of volumes of all the salt deposits in the Paleozoic periods. It should be emphasized that on the basis of M.K. Kalinko's evidence two periods of the vast development of salt strata are distinguished in the Paleozoic, viz. Cambrian and Permian. All the other periods are characterized by relatively small salt volumes compared to those given above. This phenomenon is well in accord with the age distribution pattern for the Paleozoic rock salts.

The Cambrian and Permian periods of vast halite accumulation are clearly distinguished in the Paleozoic. The volumes or rock salt accumulated in the Devonian and Carboniferous periods were almost one order less rock salt volume in the latter, exceeding that in the Devonian owing to salt strata found in the newly discovered salt basins. The Ordovician and Silurian are marked by comparatively small halite accumulation volumes.

The most characteristic features of rock salt volume distribution among the ages and epochs of the Paleozoic are given below. In the Cambrian period most of the rock salt was formed during the Early Cambrian, and thus it is only this time that might be regarded as a period of vast halite accumulation, proceeding from the volume calculated. All the Silurian rock salts accumulated at the end of the Ludlovian and Přidolian ages. The salt mass of the Devonian period shows an almost regular distribution among the ages of the Middle and Late Devonian. The bulk of rock salt of the Carboniferous period formed in the Bashkirian age. In the Permian period the greatest volume of rock salt formed in the Kungurian which is distinguished as an epoch of intense salt accumulation; by the volume of rock salt deposited it is comparable to the Early Cambrian epoch. These data show once again that the bulk of rock salt in the Paleozoic accumulated, like evaporites, during two stages, Early

Cambrian and Kungurian, when 39.1% and 36.4% of the total salt volume formed, respectively. It was during these stages that 75.5% of the total volume of the Paleozoic rock salt formed. Thus, the present assumptions that the Late Silurian, Middle and Late Devonian, and Late Permian were the epochs of considerable salt accumulation in the Paleozoic (Strakhov 1962. Lotze 1968; Ivanov and Voronova 1972) should be revised. All these epochs are much inferior to the Early Cambrian and Kungurian ages which alone can be actually regarded as very large ones on the basis of the extent of halite sedimentation.

The history of the Paleozoic salt accumulation might be subdivided into three major epochs on the basis of distribution of rock salt volume: Early Paleozoic (or Cambrian), Middle Paleozoic comprising the Ordovician, Silurian, and Early Devonian, and the Upper Paleozoic (Middle Devonian/Permian). These epochs correlate well with those outlined in the preceding chapter.

No coincidence is observed in the distribution of rock salt volumes on the continents and on the Earth in general. The total volume of the Early and Middle Cambrian salt occurs within Eurasia. On the other continents no large masses of halite accumulated during this epoch, though the conditions favored its deposition. Almost the total volume of the Late Cambrian rock salt is confined to North America, whereas on other continents it is either small, or might be only tentatively assumed. The bulk of Middle Ordovician rock salt mass and that of Ludlovian and Přidolian ages (Late Silurian) was recorded only from North America, of the Eifelian from North America and Europe, Givetian mostly from North America, Frasnian and Fammenian mostly from Europe, Visean, Bashkirian and Moscovian from North America, Orenburgian from South America, Asselian from Europe, Sakmarian and Artinskian mostly from North America, Kungurian from Europe, North and South Americas, Ufimian, Kazanian and Tatarian from Europe and North America. Thus, it turns out once again that distribution of rock salt mass was not regular among all the continents, and concentrated mainly on one of them, within a limited number of salt basins which were then the major regions of halite accumulation.

One regularity becomes obvious in the established distribution of potash salts of chloride type: their bulk mass (83.1%) formed in the Kungurian age (Early Permian), and 12.2% accumulated during Kazanian and Tatarian ages (Late Permian). During all the other potassium stages (Early Cambrian, Late Ludlovian, Middle/Late Devonian, Visean, and Late Moscovian) the data available show only 4.7% of potash chlorides. Potash salts of sulfate type accumulated mainly during the Permian period and predominantly in the East and Central European salt basins.

General Conclusions

Conclusions concerning the epochs of intense sulfate, halite, and potassium sedimentation are as follows:

1. Sulfate sedimentation was recorded throughout the entire Paleozoic era. However, an areal extent of sulfate sediments and their volume suggests that Early Cambrian and Middle Devonian/Permian epochs of considerable sulfate formation

can be distinguished in Paleozoic history. This conclusion is confirmed by following data: just during these epochs the greatest number of evaporite basins with very large and large areas and volumes of sulfate sedimentation existed; the entire area of sulfate sediment accumulation on the continents considerably expanded; and the main volume of sulfate rocks was formed.

2. The history of the Paleozoic halite accumulation was characterized not only by distinct stages, but also by the epochs of vast salt accumulation. Such epochs were the Early Cambrian and the second half of the Permian period (Kungurian/Tatarian time); it was only during that time that there were some enormous salt-producing seas with an extensive salt accumulation area where most of Paleozoic rock salts accumulated. However, a really intense halite accumulation in all the Paleozoic history was recorded only in the Kungurian age (Early Permian), when more than one third of the total mass of the Paleozoic salts deposited.

3. Maximum area and extremely vast volumes of potassium accumulation were observed during the second half of the Permian period, beginning with the Kungurian age up to the end of the Paleozoic. However, the data available imply that the Kungurian age (Early Permian) might be regarded as the epoch of the most intense potassium sedimentation, since at that time the bulk of the discovered potash salts of chloride and sulfate types formed.

4. The Early Cambrian and Kungurian age (Early Permian) on the whole might be regarded as the major epochs of halogenesis, since they are clear-cut in the Paleozoic both by their areal extent and the volume of all evaporite sediments accumulated.

CHAPTER IV

Paleogeography of Continents and Paleoclimatic Zonation of Evaporite Sedimentation

Introductory Remarks

One of the most important problems of halogenesis which attracts widespread attention is connected with paleogeography and paleoclimate of the old evaporite basins.

The old evaporite deposits are known now to have a distribution on the continents different from recent sediments in the areas of halogenic sedimentation. Recent evaporites are accumulated mainly in arid and semi-arid climatic zones, i.e., in regions of arid climate and minimum net atmospheric precipitation. It was proved by many authors (Ivanov 1953; Strakhov 1960, 1962, 1963; Fiveg 1962; Lotze 1968; Valyashko 1969, and others). Recent evaporites are associated with two zones located within tropical and subtropical belts. They are situated north and south of the equatorial belt and almost parallel to it (Lotze 1968). Old evaporite formations including salt-bearing ones, unlike the recent evaporites, are found in quite different parts of the continents outside the present tropical and subtropical belts. The Paleozoic salt- and sulfate-bearing beds often occur farther north behind the Polar Circle. For example, the Cambrian salt deposits in the East Siberian Basin reach 70° N and those in the Mackenzie Basin probably occur even farther than 70° N. The Silurian halogenic beds are known from Severnaya Zemlya and within the Canadian Arctic Archipelago, i.e., approximately at 70° and even 80° N. The Devonian evaporites were found in Taimyr, Severnaya Zemlya, Vrangel Island, and Canadian Arctic Islands at 70° to 80°N. The Carboniferous evaporite strata also occur up to 80° N, they were reported from Taimyr, Spitsbergen, and the Sverdrup Basin. The Permian sulfate-bearing deposits are known from eastern Greenland, Spitsbergen, and the Canadian Arctic Archipelago, i.e., between 70° and 80° N.

Such an unusual distribution of Paleozoic evaporites poses the following questions: is it possible to use evaporitic sediments as paleoclimatic indicators to restore arid areas within tropic and subtropic zones for the past geological epochs? If it is possible then why are Paleozoic evaporites widespread far north? Is it connected with the fact that in Paleozoic epochs of evaporite sedimentation climate on the Earth was hot and dry, and conditions were favorable for evaporite accumulation to have existed even in the Polar Circle? Was the climatic zonation in Paleozoic time similar to the recent one but is present-day distribution of Paleozoic evaporites inconsistent with the past climatic zonation due to polar wandering or a different position of the equator, or due to continental drift according to the new global tectonics? Finally, may a climatic zonation of evaporite sedimentation in the Paleozoic have been in general different from that of the Recent?

Answers to all these questions can be obtained if we use all the results available on the regularities for the spatial distribution of Paleozoic evaporite basins and do not fail to take into account paleomagnetic data and paleotectonic reconstructions with analysis of relative displacement of continents and, probably, some platform blocks. In this basis we can propose a more substantiated zonation of the evaporite sedimentation for certain periods of the Paleozoic and establish regularities more close to the present ones which are responsible for migration of zones with time.

The second group of questions which arise when we consider the history of Paleozoic sedimentation is connected with a paleogeographic position of salt basins in different epochs of the Paleozoic with respect to the then continents and seas. To obtain an answer one should also make wide paleogeographic reconstruction for the entire Earth, taking into account all the recent data of new global tectonics and paleomagnetism.

To discuss these problems we have constructed lithologo-paleogeographic maps for certain periods of a considerable evaporite sedimentation, namely: one for the Early Paleozoic (Early Cambrian) and four for the Late Paleozoic (Middle and Late Devonian, Early and Late Permian). There are two versions: (1) without regard for the continental drift and (2) taking into account the continental drift on the basis of paleomagnetic data.

The first maps were constructed without regard for the displacement of continents. They show the distribution on continents of different deposits, namely: red and gray terrigenous rocks, carbonate and terrigenous-carbonate sediments, salt and sulfate formations and coal-bearing beds. Land areas, volcanic zones, and belts were also recognized. The old land areas existing on the present continents are specified. Additionally, the possible connections between the land masses and marine basins mainly across the Atlantic, Indian, and Arctic Oceans for different continents are shown on the basis of the distribution of similar lithologo-paleogeographic zones in terms of new global tectonics. Thus we obtain artificially enlarged land areas and marine basins at the expense of ocean basins, whatever their real size should be if we do not take into account the displacement of continents or changes in the dimensions of the Earth.

Our lithologo-paleogeographic maps without regard for the continental drift were compiled mainly using the *Atlas of the Lithologo-Paleogeographic Maps of the USSR* (1969); *Atlas of the Lithologo-Paleogeographic Maps of the Russian Platform and its Geosynclinal Margin* (1960); *Paleogeographic Atlas of China* (1962); *Lithofacies Maps: An Atlas of the United States and Canada* (Sloss et al. 1960); Paleogeographic Maps of Asia in: *The Geological Development of the Japanese Islands* (1968); *Geological Atlas of Poland* (1968); Lithofacies Maps of Western Canada (Geological History . . . 1964); paleotectonic maps of the *Permian System in the United States* (McKee et al. 1967), as well as the papers of the International Symposium on the Devonian System (Int. Symp. 1967). In addition, we used paleogeographic maps and data published by Ronov and Khain (1954–1956), Richter-Bernburg (1957, 1972), Harrington (1962), Stöcklin (1962, 1968a,b), Strakhov (1962), Termier and Termier (1964), Ustritsky (1967, 1972), Brown et al. (1968), Jung (1968a,b), Katzung (1968, 1975), Sinitsyn (1970), Dutro and Saldukas (1973), Kholodiv (1973), Falke (1974), Ronov et al. (1974), Flügel (1975), Watson and Swanson (1975), Ziegler (1975), Ronov et al. (1976), and many others.

Schemes taking into account the continental drift were constructed on maps in the Mollweide projection. This projection was chosen primarily because it clearly shows a near-equatorial zone and both northern and southern near-polar areas may be represented. A reconstruction of continents is based mainly on the data of Creer (1973) and for the Early and Late Permian of Creer (1973) and Robinson (1973). We also took into account data of Kravchinsky (1970, 1973), Khramov et al. (1974), Bukha et al. (1976), Zonenshein (1976). Continents were plotted on maps in the Mollweide projection according to their position after Creer and Robinson with outlines of litho-paleogeographic zones taken from the earlier schemes without regard for the continental drift. Thus our schemes show not only the location of continents for one or another epoch in the Paleozoic, but ancient dry land and ancient seas in their position with respect to the equator and poles of a certain epoch.

Paleoclimatic reconstructions usually show a position for the ancient near-equatorial warm, near-polar temperate, and Arctic climatic zones. Equatorial and subequatorial belts as well as tropical and subtropical belts are often assigned to warm zones. However, works of Strakhov (1960, 1962, 1963, 1971) showed the importance of arid zones which used to be located in the tropical and subtropical belt south and north of the equator with an equatorial tropical humid zone between them. It is also important to single out a zone of temperate humid climate. Principles for the reconstruction of the past climatic zones were worked out in detail by Strakhov (1960). He was right in saying that "in any paleoclimatic reconstruction preference should be given to lithological and not to paleontological factors" (p. 161). Moreover, it is quite necessary to use all the data available for rocks as climatic indicators; among them N.M. Strakhov mentioned iron, manganese ores, kaolin, coal, carbonate-free redstones; evaporites (gypsum, anhydrite, rock and potash salts), carbonate redstones, and glacial deposits as indicators of humid, arid, and near-polar cold zones, respectively. Strakhov emphasized the importance of a comprehensive approach for paleoclimatic reconstructions. "We cannot restrict ourselves to a restoration of one particular zone, for instance an arid zone, and to collecting rock material typical of arid conditions, as in the case of F. Lotze when he reconstructed 'salt belts' " (p. 161).

Recently many works were published dealing with paleoclimatic zonation for different periods of the Paleozoic and distribution of rocks – climate indicators – such as evaporites, redstones, glacial deposits, coal etc. There are many different viewpoints: some believe that the position of continents has not changed and conclude polar wandering and displacement of the equator; others prove continental drift on the basis of paleomagnetic data and new global tectonics; still others confirm the unchanged position for the equator, poles, and continents throughout the entire Phanerozoic. So we should consider all the versions of a climatic zonation to choose the most valid one.

Paleogeography of Continents and Paleoclimatic Zonation for Evaporite Sedimentation in the Early Cambrian

Reconstruction Without Regard for Continental Drift

Proceeding from the present location of continents we can obtain an asymmetric distribution of land masses (Fig. 26) for the Early Cambrian. They were located mainly in the southern hemisphere and occupied most of South and East Africa, a considerable part of South America and Antarctica as well as south-western Australia. These land masses formed a large continent, Gondwanaland. There was no continent comparable to Gondwanaland in the northern hemisphere of the Earth. There were relatively small land masses such as America with the Canadian Shield and a part of Greenland; Fennoscandia, which covered the area of the Baltic and Ukrainian Shields, the Voronezh Mass and adjacent regions of the East European Platform with, possibly, northern West Siberia; Katasia in south-easternmost and Beringia (Beringland) in north-easternmost Asia. Smaller land masses occupied southern Europe, Central Asia, Kazakhstan, southern Siberia within the Yana-Kolyma region, China, and some other areas.

Marine basins are recognized on distribution of sedimentary strata in the following regions: along the periphery of North America in the Rockies, Appalachians, and Canadian Arctic Islands; along the northern and north-eastern margins of Greenland; in western South America in the Cordilleras and in the Amazon Basin and, possibly, in the Parana Basin; in north-western and north Africa; in Europe within the Portugal-South Spanish and Spanish-Sardinian Basins, Central European Basin, stretching from England to the west to Czechoslovakia and Poland to the east; Norwegian, Baltic, German-Polish and Caucasus basins; in Asia where Early Cambrian deposits are widespread – in the Iran-Pakistan Basin, the Aral-Turgai Trough, within the Tien Shan, in the Karatau and Central Kazakhstan, in the Sayan-Altai fold area, on the Siberian platform, in the Transbaikal area, in the Mongol-Amur and Sikhote-Alin Basins, as well as in China and South Eastern Asia within the North Chinese, South Chinese and Indo-Chinese Basins; in central and south-eastern Australia within the Ord Basin, Bonaparte Bay, Daly River, Amadeus, Georgina, Adelaide, Kanmentu and maybe Canning, in eastern Australia and in Tasmania; in West Antarctica within the Transantarctic and Welsworth Mountains.

The distribution of marine basins points to their relation to the northern hemisphere, namely, to North America, South Africa and Eurasia. They occupied a vast area in Eurasia and more than half of it was covered by sea in the Early Cambrian. Another peculiar feature of the Early Cambrian is evident from the scheme drawn without regard to continental drift. It shows that marine basins in the southern hemisphere as a rule were located along the margins of Gondwanaland and were typical marginal seas. A similar basin seems to have occurred in North America. By contrast, in Eurasia most of the marine basins were situated on the continent and were mainly inland seas.

Ocean basins as shown on the paleogeographic scheme without regard to continental drift existed in place of the present Pacific and Arctic Oceans. There was probably another ocean basin on the site of the present North Atlantic between the

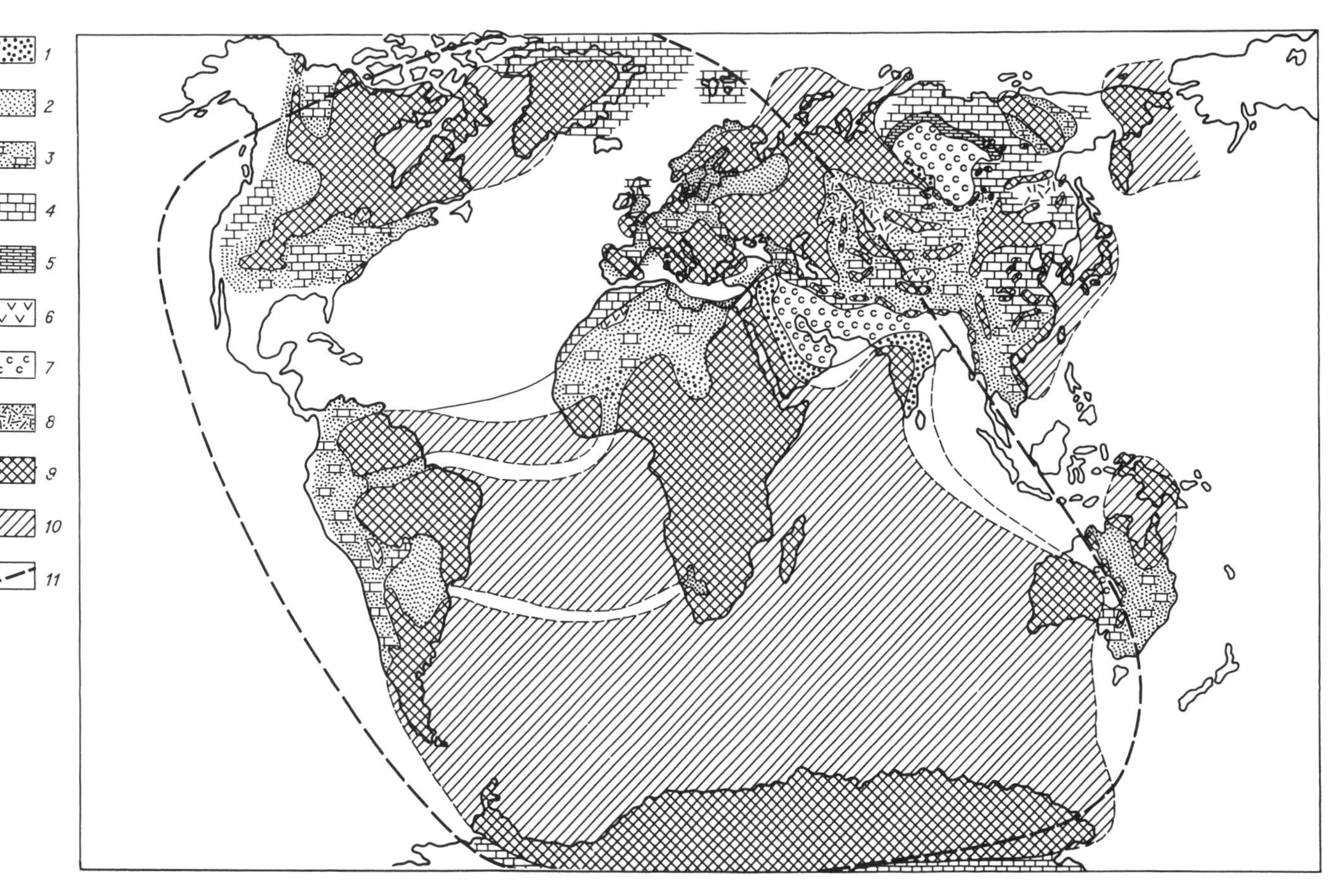

Fig. 26. Litho-paleogeographic map. Early Cambrian epoch.

1–8 areas of occurrence of terrigenous red beds *(1)*, terrigenous gray-colored beds *(2)*, terrigenous-carbonate gray-colored beds *(3)*, carbonate *(4)*, carbonate biogenic *(5)*, sulfate *(6)*, salt *(7)*, volcanic and sedimentary-volcanic rocks *(8)*, *9* land masses on recent continents; *10* inferred areas of land in oceans; *11* a position of the paleoequator

American continent on the one side and Fennoscandia and Gondwanaland on the other side. It is not valid to postulate the existence of an ocean basin in place of the Indian Ocean.

Gray-colored terrigenous or carbonate-terrigenous sediments accumulated in most Early Cambrian marine basins under conditions of normal salinity and in fresh water near land masses. The terrigenous gray-colored deposits were widespread mainly in near-shore areas of marginal seas on American and African continents in seas off West Gondwanaland and within inland seas of Eurasia (Baltic, Central European, Caucasus, West Chinese, Indo-Chinese, and others). A gradual increase in portions of carbonate rocks and locally a complete replacement of carbonates by terrigenous rocks is traced from the near-shore toward the center of the seas and oceanward. Gray terrigenous strata are characterized by a considerable amount of glauconite.

Carbonate strata during the Early Cambrian deposited mainly within the following eleven regions: (1) almost over the entire northern and eastern margins of Gondwanaland and in Spitsbergen, (2) in the Central Michigan-Pre-Appalachian Basin and in south-western America, (3) in the outer Cordillera Basin in western America, (4) in the northern Mackenzie Basin, (5) in the north-eastern North Africa Basin far from Gondwanaland and on the site of the Spanish-Sardinian Basin, (6) in the Central Asian Basin, (7) over most of the Sayan-Altai fold area, (8) in the north and north-eastern Siberian platform, Taimyr and Baikal-Patom area, (9) in the North Chinese, partly South Chinese and Mongol-Okhotsk Basins, (10) in the Adelaide Basin in southern Australia, and (11) in West Antarctica. Thus the areas of carbonate accumulation were both central and outer parts of marginal seas, and many inner marine basins, biogenic buildups, archeocyathid algal bioherms, biostromes, carbonate banks and masses were rather common in some zones of carbonate sedimentation. They were found in the Sayan-Altai fold area, in the Siberian platform (Strakhov 1962; Khomentovsky and Repina 1965; Zhuravleva 1968), in north-western Africa (Termier and Termier 1964), in North America, South Australia, the Adelaide Basin (Daily 1956), and in West Antarctica (Hill 1964).

In the Early Cambrian evaporites accumulated in three salt basins: the East Siberian, Iran-Pakistan, and Amadeus Basins and within three sulfate basins: the Tarim, Anti-Atlas, and Cis-Andean. The Amadeus Basin could reach the Canning Basin and stretches even farther into the Indian Ocean. It is quite probable that the Iran-Pakistan salt basin advanced into the Hindustan Peninsula and was within the same evaporite zone as the Amadeus Basin of Australia, i.e., on the margin of Gondwanaland.

Thick Early Cambrian terrigenous red beds occur in the near-shore parts of evaporite sedimentary basins. They were reported from the Iran-Pakistan Basin, in the south-western East Siberian and Amadeus Basins. Relatively thin terrigenous red beds were found in North Africa (Reggane, Ahnet, Tindouf Basins) and within the Appalachians in eastern North America. Red terrigenous sediments as a rule are not common outside salt accumulation areas. However, variegated argillaceous-carbonate deposits were widespread in Early Cambrian marine basins. They were found in the northern and eastern Siberian platform, in the Kolyma region, North and South Chinese Basins and locally in the Sayan-Altai fold area, in the Pamirs.

Now let us see what the spatial distribution of different types of sediments was with respect to the present equator and present arrangement of continents on the globe. Much thought should be given to such sediments as evaporites, redstones, and biogenic carbonate buildups, which are considered to be climate indicators.

No regularity in association of gray terrigenous deposits with certain zones can be found for the Early Cambrian. They are scattered both in marginal and in inland seas from 40° to 70° S. One should very carefully postulate that gray terrigenous deposits are less common at high latitudes.

As to carbonate sequences, one may say that they occur mainly in the northern hemisphere, and are found between 20° and 80° N. It is within this zone that major areas of carbonate sedimentation are known in north-western Africa, North America, Greenland, and Eurasia. In the southern hemisphere the relation of carbonate strata to certain zones is not so strict. However, if we take into account the occurrence of carbonate in Adelaide, Australia, and in West Antarctica then their distribution in the southern hemisphere will be roughly limited by latitudes from 30° to 90° S. Thus the present arrangement of continents enables the assumption for the Early Cambrian of two belts of carbonate accumulation roughly parallel to the recent equator which embraced wide zones near the poles.

Such a distribution of Early Cambrian carbonates on continents hardly fits the present day zonation of carbonate sedimentation. First, it would not be correct to compare carbonate distribution pattern in ancient epicontinental basins with zonation of carbonate sedimentation in oceans where it is concentrated in the recent epoch. Second, the distribution of recent carbonates in the Atlantic and Pacific Oceans varies peculiarly with depth and direction of warm currents (Lisitsyn and Petelin 1970; Emelyanov et al. 1971; Lisitsyn 1974; Strakhov 1976). Generally speaking, an intense precipitation of carbonate material takes place within a wide latitudinal zone from 70° N to 60° S. In the Atlantic and Pacific Oceans this zone is shifted northward (70° N–55° S) and southward (10° N–60° S), respectively. Within this zone the most intense carbonate sedimentation (in absolute values) is related to two rather narrow belts north and south of the equator. Some workers (Briden and Irving 1968; Fairbridge 1968) point to 15° and 25° N as well as between 15° and 25° S, others (Strakhov 1976) direct attention to a place much farther from the equator, approximately at latitudes from 25° to 40° S. A "carbonate-free equatorial belt" (Fairbridge 1968, p. 301) running from 15° S to 15° N is suggested between these belts of considerable carbonate accumulation (coral reefs exclusive). At high latitudes (above 40°) of both the northern and southern hemispheres carbonate is being accumulated in small amounts.

If we attempt to compare present zonation of carbonate accumulation in oceans with that in epicontinental basins in the Early Cambrian we shall see that the latter occupy a much wider area than recent ones on the scheme, which does not invoke continental drift, i.e., they run up to the poles. A carbonate-free equatorial belt for the Early Cambrian stretches from 30° N to 20° S. It is twice as wide as the recent one. However, Early Cambrian zones of carbonate accumulation, despite some significant but quite explicable differences from the recent carbonate distribution, may be grouped into latitudinal belts with respect to the modern equator.

Less certain is a correlation for distribution of Early Carbonate and recent carbonate biogenic buildups. The recent reef formation takes place in a near-equatorial zone between latitudes 30° N and 30° S; coral reefs are most numerous at 10° and 20° south and north of the equator (Wells 1957; Opdyke 1966; Briden and Irving 1968; Fairbridge 1968). Early Cambrian biogenic carbonate buildups are known mainly outside this zone and so a wide near-equatorial belt is carbonate-free. They occur over the entire Early Cambrian belts both north of 20° N and south of 30° S and very often at high latitudes. For example, in East Siberia almost at 70° N and in Antarctica, where biogenic carbonates were reported south of 85° S. Despite such a scatter within the Early Cambrian, carbonate belts in the northern hemisphere are restricted by a zone between 30° and 70° S. It is there where all the carbonate buildups are known in the Siberian platform, Sayan-Altai fold area, and in other regions of Eurasia as well as in north-western Africa and North America. A distribution of Early Cambrian bioherms differs greatly from recent ones. However, Early Cambrian carbonate belts having been much wider than recent ones, zones of bioherms should also be wider. Proportionally, Early Cambrian and recent belts are very similar. A belt of Early Cambrian carbonate bioherms in the northern hemisphere roughly parallels the recent equator. Such a parallel pattern is not observed in the southern hemisphere.

Lotze (1968) was the first to attempt to establish an areal distribution of Cambrian halogenic deposits with regard to the present arrangement of continents. He placed Cambrian salt deposits of the Iran-Pakistan and East Siberian Basins and those of the Mackenzie Basin into a single evaporite belt. The belt runs almost longitudinally from the Persian Gulf through East Siberia and the Arctic Ocean into north-western Canada. Lotze suggested that such a strike of the Cambrian evaporite belt was due to a different position of the equator in Cambrian time. Zharkov (1970) and Ivanov and Voronova (1972) proposed another explanation for salt basin distribution in the Cambrian of Asia. They put the East Siberian and Iran-Pakistan salt basins into northern and southern arid zones respectively with the equator in between. Meyerhoff (1970b) on the contrary believes that Cambrian evaporite deposits in the northern hemisphere are associated with a belt symmetric with respect to recent arid and semi-arid climatic zones. A considerable shift of Cambrian evaporites northward within East Siberia was caused, in his opinion, by overall warming of climate on the Earth, and due to this fact arid zones reached 70° to 80° N and temperate zones were in polar regions; secondly, by a peculiar displacement of warm currents and dry wind belts toward the Arctic Ocean and interiors of Eurasia; these displacements take place at the present time, and, according to Meyerhoff, existed continuously for at least during 1 b.y. Meyerhoff and Meyerhoff consider that due to the position of the rotational axis of the Earth distribution of continents and ocean basins was persistent for 1.6 b.y. and thus a correlation should exist between areal distribution of ancient evaporite belts and recent climatic zones.

Now we shall analyze the evidence for distribution of salt and other halogenic beds of Early Cambrian age on continents and try to group, if possible, areas of Early Cambrian evaporite sedimentation into a belt running along the present equator or present arid climatic zones, i.e., along the present tropical and sub-tropical belts.

The Early Cambrian East Siberian salt basin lies at present between latitude 50° N and 70° N. Evaporite formations of the same age are unknown at these latitudes either in Eurasia or in North America. However, one may refer to the Mackenzie Basin in North America containing Late Cambrian halogenic beds. The Mackenzie Basin area appeared to lie within the arid zone in Early Cambrian time. Such an assumption is confirmed by the presence of Late Precambrian evaporite deposits and so we can conclude that the Mackenzie Basin was in the arid zone in the Late Precambrian and throughout the Cambrian, including the Early Cambrian. In this case it is possible to suggest that the arid zone of sedimentation reached in the Early Cambrian epoch as far as 70° N in North America and that the evaporite belt may be extended up to this latitude. In southern Eurasia the Iran-Pakistan salt basin is situated now between latitude 15° and 40° N. Gypsum and anhydrite exposures are known at about the same latitudes in north-western Africa in the Anti-Atlas, while in North America halogenic formations have not been found at these latitudes. However, it is possible to suggest the presence of Cambrian salt deposits for the Central and South Appalachians (Rodgers 1970). They could lie at about 40° N. In Eurasia Early Cambrian evaporites were reported from the Tarim Basin in China, situated between the East Siberian and Iran-Pakistan salt basins (Some . . . 1965; Meyerhoff and Meyerhoff 1972). These data show that Cambrian evaporites occur at present in the northern hemisphere between latitude 15° and 70° N. At these latitudes they are known in Eurasia, north-western Africa, and in North America, i.e., on different continents. This allows us to place the areas of evaporite sedimentation into a single belt parallel to the modern equator. This belt is rather wide, probably due to an overall warming of climate on the Earth, as A. Meyerhoff suggests.

In the southern hemisphere only one salt basin of Early Cambrian age is known now in the Amadeus Basin, Australia. It is situated between latitude 20° and 30° N. If we assume that salt strata of the Canning Basin are of Cambrian age, then a zone of evaporites in Australia may go as far as latitude 15° S. In Africa no evaporite deposits are known at the same latitudes. However, in South America they have been found at the foothills of the Andes. Both areas on far-spaced continents can be placed into an evaporite belt parallel to the modern equator.

Thus in reconstructions which do not invoke continental drift it is possible to place Early Cambrian evaporite deposits into two evaporite belts at a certain distance from the modern equator and in general parallel to it. The northern evaporite and southern evaporite belts will be between latitudes 15° and 70° N and between 15° and 30° S, respectively, with a near-equatorial area in between. In other words, even by the recent arrangement of continents, zones of Early Cambrian evaporite sedimentation can be restored to form latitudinal belts parallel to the present climatic zones.

It is to be noted that reconstructions proposed by Lotze (1968) for the Early Cambrian evaporite belts look less well substantiated. His Cambrian evaporite belt was established only on the basis of spatial distribution of evaporites in the northern hemisphere, mainly in Eurasia and North America, and there is no place for sulfate deposits. Moreover, it is impossible to stretch it into the southern hemisphere because it is impossible to place into it either Cambrian evaporites of Australia or South America, or to suggest that evaporites of Australia occupy another evaporite belt without considerable displacement of these continents.

There is another possibility for grouping Cambrian halogenic beds into evaporite belts. Thus, salt strata of Australia and the Iran-Pakistan Basin or sulfate beds of north-western Africa and South America, i.e., all the evaporite known from basins around Gondwanaland, may be placed into the southern belt. In this case the northern evaporite belt will combine salt deposits of the East Siberian and Mackenzie Basins. However, these belts are not latitudinal with respect to the modern equator or arid zones. Their present position can be explained only under the assumption that the equator was in another place during the Cambrian.

The observable facts show that if we do not take into account paleomagnetic data or continental drift there are two possibilities for the existence of evaporite belts in the Early Cambrian.

The first is in accord with the recent climate zonation and present position of the equator. Unlike the present belts those of the Early Cambrian are wider, the northern belt being much wider than the southern one. To explain this it is necessary to assume that in the Early Cambrian, climate on the Earth was much warmer than at present and the arid zone ran as far as 70° N. The smaller width of the southern belt can be accounted for by the present arrangement of continents because, owing to the longitudinal position of southern continents and their small size, the arid zone of the southern hemisphere both at present and in the past has been, according to Meyerhoff (1970a,b), always more narrow than that of the northern hemisphere. A considerable width of the northern evaporite belt of the Early Cambrian is consistent with the above-mentioned great width of Early Cambrian belts of carbonate sedimentation and carbonate biogenic buildups. Moreover, it is possible to see the association of the Early Cambrian belt of carbonate bioherms in the northern hemisphere just with the evaporite belt, i.e., predominantly with warm climatic zones. It is also to be noted that on the evidence of spatial distribution both of carbonates and evaporites we identify a near-equatorial climatic area for the Early Cambrian coinciding with that of the Recent. All these consistencies may confirm the first version of evaporite belts for the Early Cambrian epoch. This version is well in accord with Meyerhoff's ideas (1970a,b; Meyerhoff and Meyerhoff 1972). However, we cannot agree with A. Meyerhoff as to paleogeographic conditions of salt accumulation in northern basins including the East Siberian Basin. In his reconstructions warm saline water of the Atlantic, brought by currents along the Lomonosov Ridge in the central Arctic Ocean, entered the East Siberian and all Cambrian, Ordovician, Silurian, and Devonian basins of Central and North Siberia.

The second version for the Early Cambrian evaporite belt position is inconsistent with recent global climate zonation. For its explanation we should use the polar wandering and different position of the equator in the Early Cambrian. Figure 26 shows a proposed equator location for that time. By such reconstructions both evaporite and carbonate zones, biogenic included, fall into the near-equatorial area. Belts of evaporite and carbonate sedimentation are becoming narrow while their position relative to the present equator remains almost the same. This allows us to say that the second version for the Early Cambrian evaporite belts is also in accord with observable facts. It is more accurate because paleomagnetic data were taken into account.

The paucity of evidence on Early Cambrian terrigenous red beds does not enable us to establish their extent. They occur essentially in areas of evaporite sedimentation and apparently within the evaporite belts of that time. So the same two versions may be used for the distribution of red beds. There is no evidence on the position of variegated argillaceous-carbonate and carbonate deposits either with respect to the modern or the ancient equator.

Reconstruction Which Invokes Continental Drift

Now let us consider the Early Cambrian paleogeography which invokes continental drift, as well as new global tectonics and paleomagnetics data. Again we shall try to show the distribution of different deposits over the globe with respect to reconstructed continents and the ancient equator [7].

The location of land masses and marine basins by such a reconstruction is shown on a litho-paleogeographic scheme (Fig. 27). It differs from the scheme which does not invoke continental drift, as follows. Firstly, configuration of ocean basins changed due to a different arrangement of continents, and basins increased in size. The Northern Ocean is much larger and embraces most of the northern hemisphere. The Southern Ocean occupies a certain position about the southern pole between Gondwanaland, on the one hand, and America, Europe, and Asia or Laurasia as a whole, on the other hand. There is no ocean basin in place of the Atlantic, i.e., between America and Europe. Secondly, land masses occupied a different position. Most of them shifted into the southern hemisphere. In America, Europe, and Asia and also in Antarctica and Australia they lie in a near-equatorial zone. Thirdly, many marine basins of Europe and Asia, which earlier were considered as inner basins, became marginal seas due to a junction of the Arabian and Hindustan Peninsulas with remaining land masses of Gondwanaland, and spread within the Tethys. In general, in these reconstructions all the marine basins around Gondwanaland, including the Iran-Pakistan Basin, are marginal basins. Inland basins persist only in Central Asia, as well as those between America and Europe, while on the scheme which does not invoke continental drift they follow the margins of the Atlantic.

The relation of different types of deposit to marine basins being the same, we shall not dwell upon the subject. Let us discuss a spatial distribution of gray terrigenous, carbonate, evaporitic, and red beds of the Early Cambrian with respect to a proposed equator.

The situation is the same: no regularity exists as to the distribution of gray terrigenous deposits. They were found both about the poles and at the equator. Carbonate beds show quite different distribution. They are widespread near the equator between 40° S and 30° N. Ten of the eleven areas of Early Cambrian carbonate sedimentation lie within this band, namely: (1) Greenland-Spitsbergen, (2) Michigan-Pre-Appalachian, (3) Cordilleran, (4) Mackenzie, (5) Central Asian, (6) Sayan-Altai, (7) East Siberian, Taimyr, Baikal-Patom, (8) North Chinese, South Chinese, and Mongol-Okhotsk, (9) Adelaide, and (10) Antarctic basins. Only one area of carbonate sedimentation in north-western Africa stands apart; it lies near the southern pole. Most

---

7 It has been already mentioned that we used the reconstructions of Creer (1973), where the southern pole is located in the Eastern Atlantic, not far off north-western Africa. If we take it for the northern pole (Khramov 1967; Khramov et al. 1974; Bukha et al. 1976) then we get a mirror reflection of a reconstruction

of the Early Cambrian carbonate strata on the scheme which invokes continental drift is located not at the equator but far from it about 10° and 30° both to south and north of it. Carbonate bioherms occur mainly in areas close to the equatorial belt.

All the Early Cambrian salt basins are situated between 40° S and 30° N. As a rule they lie at some distance from the equator, either north or south of it. On the reconstruction which invokes continental drift the East Siberian and Iran-Pakistan Basins are in the southern hemisphere, i.e., in a single evaporite belt, and not in different ones as was suggested earlier (Zharkov 1970; Ivanov and Voronova 1972). An assumption as to the link between the Iran-Pakistan Basin and a Cambrian basin of Australia has not been proved; these basins lie far from each other and in different hemispheres. The Anti-Atlas and Cis-Andean sulfate basins also do not fall into an evaporite belt. The former is situated near the southern polar circle and the latter lies between 40° and 50° S. All the thick red beds of Early Cambrian age as well as variegated argillaceous-carbonate and carbonate deposits occur in a near-equatorial area.

Such a distribution pattern of carbonate, evaporite (except for sulfate basins) and red beds of the Early Cambrian is well in accord with that of the Recent. In all probability the Early Cambrian belt between 40° S and 30° N encompassed equatorial, sub-equatorial, tropical, and sub-tropical climatic zones of this time. Opdyke (1966), Briden and Irving (1968) arrived at almost the same conclusion about distribution of deposits (including Cambrian evaporites) indicating tropical and sub-tropical climate. N. Opdyke, on the basis of paleomagnetic data, determined a latitudinal interval for an Early Cambrian warm climatic zone between 0° and 35° south and north of the equator, and according to Irving and Briden (1962) it is between 20° S and 20° N. This evidence enables us to place areas of evaporite sedimentation in a belt lying between the tropics if continental drift is invoked.

Conclusions Concerning Paleoclimate and Paleogeography of Evaporite Basins in the Early Cambrian

There are three possibilites for the position of evaporite belts on the Earth in the Early Cambrian. Two possibilites do not invoke continental drift and the third does. Comparing these three versions it is difficult to give preference to one of them because they all need strong evidence. The first version is invalid without an assumption about an overall warming of climate on the Earth in the Early Cambrian. To explain the second possibility considerable polar wandering should have taken place and a different position for the equator is to be postulated. The third version is valid only when continental drift is involved. A position of evaporite belts in all three versions under certain assumptions is consistent with the modern climatic zonation of evaporite sedimentation. Only after the discussion about climatic zonation of the evaporite sedimentation for other epochs of the Paleozoic can one make a final choice. It is possible to say with some assurance that evaporite sedimentation in the Early Cambrian took place in arid zones of tropical and sub-tropical belts.

Now let us consider an arrangement of Early Cambrian salt belts on continents and in oceans and links existing at that time between marine basins and saliniferous zones. First it becomes evident from litho-paleogeographic maps both accounting and not accounting for continental drift that Early Cambrian salt basins were all

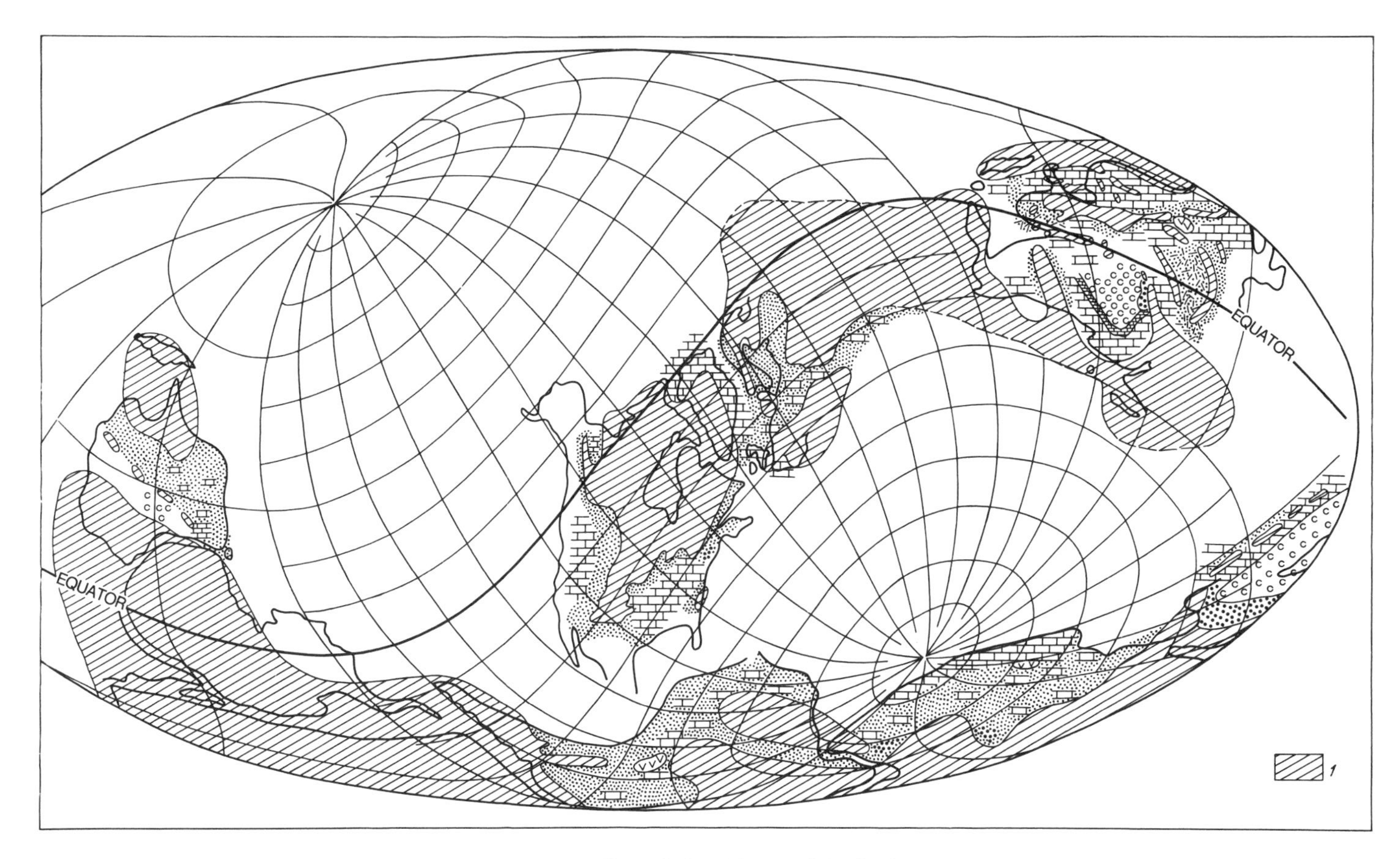

Fig. 27. Litho-paleogeographic map which invokes continental drift. Early Cambrian epoch. Mollweide projection. *1* land masses. For explanation see Fig. 26

epicontinental seas. They either occupied continental margins, and in this case they were typical marginal seas, or when they were situated on continents they formed inland marine basins. A paleogeography which does not invoke continental drift shows two big Early Cambrian salt basins of inland type, i.e., the Iran-Pakistan and East Siberian, while the Amadeus Basin in Australia may be tentatively assigned to the marginal type. On the contrary, when continental drift is involved, only the East Siberian Basin will be of the inland type and the Iran-Pakistan and Amadeus Basins become marginal seas.

The second peculiarity of an Early Cambrian paleogeography seen in some reconstructions is the presence of broad connections between salt basins and seas of normal salinity through intermediate basins which predominantly were on archipelagoes (Zharkov 1971b, 1976). A marine water supply into salt areas of the Iran--Pakistan Basin was undoubtedly from the Tethys or Southern Ocean. It is quite probable that water of normal salinity from the Tethys was brought also into the East Siberian Basin. Paleo-biogeographic data also shows evidence of wide connections between marine and ocean basins in the Early Cambrian (Termier and Termier 1964; Repina 1968; Zhuravleva 1968). Despite great disparities as to recognition of biogeographic realms and provinces using archaeocyathes and trilobites, some conclusions made by the above authors seem to be rather important. Thus, archaeociathes enable us to establish a permanent connection between the Sayan-Altai fold area and the Siberian platform which confirms a supply of marine water and hence migration of fauna between marine basins of these regions. When trilobites are used this connection is not so evident. However, Repina (1968) suggests such a link for the Aldanian, Botomian, and especially Lena age.

Thus, the Early Cambrian salt basins were big seas situated along margins or in the interior of continents and broad connections existed with marine basins and oceans of normal salinity. They undoubtedly were situated in arid zones of tropical and subtropical belts of the Earth, however it is impossible to show an unambiguous position for these evaporite belts because of the different interpretation of the observed facts.

Paleogeography of Continents and Paleoclimatic Zonation of Evaporite Sedimentation for the Middle and Late Devonian

Reconstruction Which Does Not Invoke Continental Drift

The Middle and Late Devonian was another epoch of intense evaporite sedimentation in Paleozoic history. We shall try to show paleogeographic environments of sedimentation on continents for this time and identify a position of evaporite basins with respect to climatic zones.

First of all we shall discuss a Middle Devonian litho-paleogeographic map which does not invoke continental drift (Fig. 28). The figure shows that the Middle Devonian paleogeography has greatly differed as compared to the Early Cambrian. Land masses, ocean and marine basins occupy a new position on the Earth. Great changes

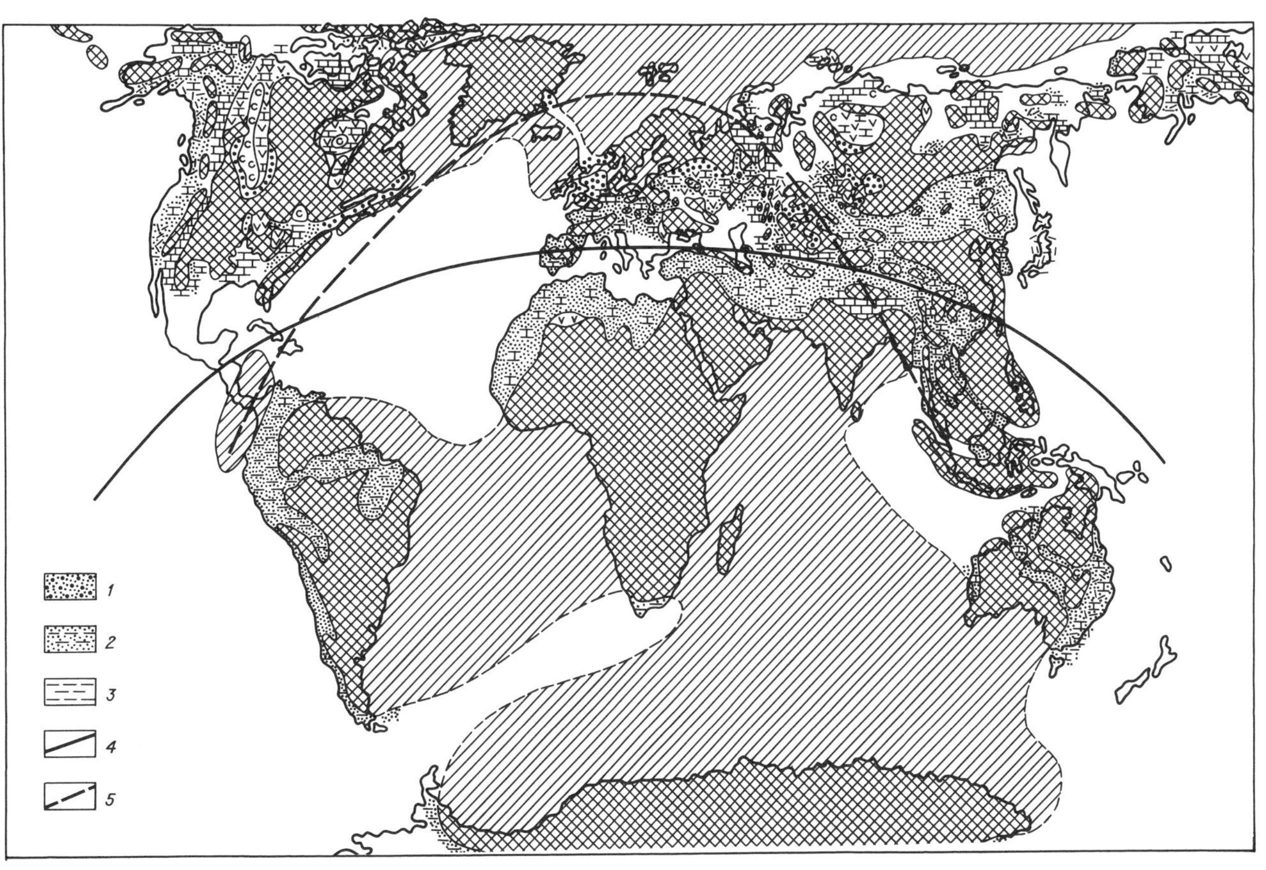

Fig. 28. Litho-paleogeographic map. Middle Devonian epoch.

1–3 areas of occurrence of variegated terrigenous *(1)*, sandy-clay *(2)*, and clay *(3)* deposits; *4* position of the paleoequator after Sinitsyn; *5* position of the paleoequator after Strakhov. For other explanations see Fig. 26

took place in the northern hemisphere. Here one can clearly recognize a big continent, Laurasia, encompassing most of North America, Greenland, north-western and northern Europe, namely the Baltic Shield and adjacent areas and apparently the present Arctic Ocean. In the southern hemisphere Gondwanaland is another big continent comprising southern and eastern Africa, eastern South America, Australia, eastern Antarctica, and the Arabian and Hindustan Peninsulas.

A wide tectonically complex zone comprising a large number of epicontinental marine basins and smaller land masses is visible between these two supercontinents. The southern part of this zone stretching from north-western Africa and Europe through Asia Minor and Central Asia into Indo-Malaysia is usually called Tethys. The north-eastern part of the zone embracing eastern Europe and large areas of northern, north-eastern, and eastern Asia were occupied by numerous subcontinents and islands separated by seaways. Such a distribution of seaways and land masses on the Eurasian continent made Sinitsyn (1970) speculate that Eurasia did not exist in the Devonian. "The sea covered about a half of the present area. The size of the biggest Devonian subcontinent did not exceed 10% of the present continental area" (Sinitsyn 1970, p. 5). Angaraland and Tobolia within East Siberia were the largest among such subcontinents (Atlas . . . 1969; Sinitsyn 1970), and Katasia and Malaysia in the south-eastern part of the Asiatic continent (Paleogeographic . . . 1962; The Geologic Evolution . . . 1968). Especially big marine basins in Eurasia occur in the Urals, Middle Asia, West Siberia, Taimyr, the Kolyma region, and Mongolia. There are grounds to suggest that the Tethys ran far west into the present Atlantic Ocean and was connected by seaways in southernmost North America.

If continental drift is not invoked, then all the seaways of the Tethys or Eurasia can be considered as ocean basins. In the Middle Devonian, according to similar reconstructions, the Arctic, Indian, and Atlantic Oceans apparently did not exist. Thus we have only one ocean basin within the Tethys.

Carbonate and terrigenous-carbonate sedimentation predominated in the Middle Devonian sea. Reef carbonate strata formed in many regions. Terrigenous and volcano-terrigenous red beds and evaporite sediments were wider developed than those of the Cambrian. However, gray-colored terrigenous strata became less in number.

In fact, terrigenous gray-colored sediments accumulated essentially in marine basins located in western and southern Gondwanaland, i.e., in South America, southernmost Africa, and in West Antarctica. Thus, in South America Middle Devonian formations are represented predominantly by dark green-gray subgraywacke, gray quartz sandstone, and gray-green mudstone. Sedimentary sequences of similar composition were found in Falkland Islands (the Monte Maria Group) and the Pre-Cordillera Basin (Villavicencio, Punta Negra and Chavello formations) in Argentina; in the Parana and Parnaiba Basins in Brazil, in Columbia, and Venezuela (Stibane 1967; Weisbord 1967); more fine-grained terrigenous rocks, usually dark gray mudstone and siltstone, formed in the Amazon Basin (Kroemmelbein 1967). In southern Africa dark gray sandstone, quartzite, and mudstone of the Bockweld Formation are Middle Devonian in age; deposits of similar composition with fish remains were reported from Antarctic (Boucot et al. 1967). They are the only known vast regions of Middle Devonian gray-colored beds on the Earth.

Carbonate accumulated in many marine basins of North America, Tethys, and Eurasia, however continuous carbonate sedimentation throughout the Middle Devonian took place only in a few regions. These were in North America: the western Rocky Mountains in Canada and USA (Danner 1967; Griffin 1967; Poole et al. 1967; Sandberg and Mapel 1967), North Kansas in southern Midcontinent (Amsden et al. 1967), the southern parts of the Franklinian Miogeosyncline, Canadian Islands (Kerr 1967b); in Eurasia: the Austrian Central Alps and Carnian Alps (Flügel 1967); south-eastern and eastern Russian platform, Central Asian Basin, Taimyr Basin, locally the Verkhoyansk-Chuchotka fold area (the Devonian System . . . 1973); in north-western Africa: locally Anti-Atlas and High Atlas (Hollard 1967). In general, areas of carbonate sedimentation occupied either outer parts of marginal seas or central parts of inland seas.

In Middle Devonian basins carbonates are usually intercalated with terrigenous and volcaniterrigenous rocks forming units facially nonpersistent. Four types of such units are to be mentioned: (1) gray-colored terrigenous-carbonate, (2) gray-colored volcaniterrigenous-carbonate, (3) variegated terrigenous-carbonate, and (4) variegated volcaniterrigenous-carbonate.

Middle Devonian variegated terrigenous-carbonate deposits have been found in the following regions: in the Tindouf, Reggan, Ahnet, Timimoun, Mouydir, Illisie and other basins in north-western Africa (Aliev et al. 1971), in Spanish-Portugese and Pyrenean Basins (Llado et al. 1967); in the Bohemian Basin in Czechoslovakia (Chlupac 1967); within the Massif Armorican, in France; in the Sudeten and Carpathian Mountains (Chlupac 1967; Pajchlowa 1967); in North and South Dobruja and the Rumanian Lowland (Patrulius et al. 1967); in central and western Afghanistan (Durkoop et al. 1967); in northern India, in the Spiti River Basin; in some basins of Birma, Laos, Central and North Vietnam, as well as south-western China (Hamada 1967; Mueller 1967); in the Carnarvon Basin, Australia (Johnstone et al. 1967); in South Transcaucasus; in the western miogeosynclinal zone of the Ural Basin and on Novaya Zemlya; locally within the Mongol-Okhotsk belt (The Devonian . . . 1973); in a miogeosynclinial zone of the Rocky Mountains in California and Nevada (Poole et al. 1967); in the Appalachian Basin (Oliver et al. 1967).

A distribution of gray-colored terrigenous-carbonate beds shows a regular pattern. They are common either in marine basins of the Tethys of marginal miogeosynclinal zones of major geosynclinal basins such as the Caucasian, Ural, and Mongol-Okhotsk Basins in Eurasia, the Rocky Mountains, and the Appalachians in North America.

According to a different proportion or rocks among gray-colored terrigenous-carbonate formations we recognize more terrigenous accumulating near a source area, and more carbonate associated with the center of the basins mentioned. To a certain degree they correspond to two "facies" proposed by Erben (1962), Rheinisch and Hercynian (or Bohemian?). In Erben's definition (Erben and Zagora 1967, p. 57) the Rheinish facies is a "dirty" facies consisting essentially of conglomerate, graywacke, sandstone, quartzite, and mudstone and subordinate carbonate. The Hercynian facies is considered as "pure" because of the almost complete absence of coarse-grained terrigenous rocks; however, limestone and/or mudstone predominate. Each facies is marked by a peculiar fossil assemblage. The Hercynian and Rheinish facies are traced onshore and offshore respectively over almost the entire Tethys from West Europe and north-western Africa to Malaysia. The facies are also known from gray-colored terrigenous-carbonate sedimentary basins farther north in Eurasia. Thus, in East Asia (The Geologic Evolution . . . 1968) a facies similar to the Hercynian facies is referred to as a biofacies of Japanese type, which was reported from the Japan Islands where carbonate rocks dominate the Middle Devonian. An equivalent of the Rheinish facies in mineral and faunal composition is the Manchurian biofacies developed mainly in near-shore areas of the Mongol-Okhotsk Basin where terrigenous rocks predominate. Sedimentary beds of the Appalachian Basin are also similar to the Rheinish facies. In other basins of North America and Eurasia, where gray-colored terrigenous-carbonate sequences accumulated, carbonate sedimentation was most important and so these sediments may be assigned to the Hercynian facies.

The Rheinish and Hercynian facies with their peculiar fossil assemblages are believed to indicate a warm humid climate. Such a conclusion is valid because their occurrence in the humid zone during the Devonian is confirmed by other lines of paleoclimatic evidence. However, the distribution of the facies was caused by sedimentary environments in near-shore and offshore areas.

Gray-colored volcaniterrigenous-carbonate beds accumulated as a rule in eugeosynclinal troughs exemplified by the Moravia Basin in Czechoslovakia (Chlupac 1967), the Greater Caucasus, the East Ural Basin and Mugodzhara, the inner zone of the Dzhungar-Balkhash geosyncline, the Mountain Altai (The Devonian System . . . 1973), the Rocky Mountains of California and Nevada (Poole et al. 1967), south-eastern Alaska (Gryc et al. 1967), the Sverdrup Basin in the Canadian Islands (Trettin 1967), and East Australia (Hill 1967). In general gray-colored volcaniterrigenous-carbonate sequences occur in the same areas as gray-colored terrigenous-carbonate series, but the former are shifted to inner zones of geosynclinal basins with intense volcanic activity.

Variegated terrigenous-carbonate beds occur mainly in regions of carbonate sedimentation or in near-shore marine basins adjacent to the areas of red bed sedimentation. They are known from basins adjoining Laurasia (Namur, Dinant, Lvov, Morsovo, Timan-Pechora in Europe; Michigan, Pre-Appalachian and West Canadian in North America) or in basins around big subcontinents of Eurasia, namely, Angaraland (North Siberian) and Katasia (South Chinese).

Variegated volcaniterrigenous-carbonate formations were found in three far-spaced areas of Asia: (1) Central Asia encompassing Turgai Trough, middle zone of Dzhungar-Balkhash Geosyncline, Kuznetsk Basin, south-western Altai and Major Altai Anticlinorium; (2) North East where variegated strata occur in the Omolon and Okhotsk Massifs, Penzha Range and elsewhere (3) Uda-Shuntara.

Middle Devonian red beds and variegated deposits are known in the same regions, but the former occur along continental margins and on major land masses. Red beds are widespread in the Old Red Sandstone area which comprises Devonian red beds of England, Fennoscandia, the Russian Platform, Greenland, Spitsbergen, and the Maritime Provinces. This area covers most of Laurasia from West Canada to Novaya Zemlaya where both continental and near-shore red beds formed. The red beds show also wide distribution in Central Kazakhstan, the intermontane basins of the Sayan-Altai fold area, and in Angaraland. Another area of red bed sedimentation is known in Central Australia, namely in the Amadeus, Dulcie, Toko, and other basins.

In the Middle Devonian evaporites occupied vast areas. They filled the West Canada, Hudson, Central Iowa, Illinois, Michigan Basins in North America and in the Canadian Arctic Archipelago as well. In Eurasia evaporites were reported from the North Siberian, Tuva, Minusinsk, Morsovo, and Moesian-Wallachian Basins. In Africa they were found in the Tindouf Basin. In the southern hemisphere only one evaporite basin is known, the Adavale Basin in Australia.

In addition to the above sediments we should note some other climate indicators, namely bauxite, iron, and manganese ores and coal, suggesting humid sedimentation environments. These rocks occur in the following areas. The first major region runs from north-west to south-east through the Urals, Central Kazakhstan, Altai, and China. Bauxites are known on the eastern slope of the North and Middle Urals, the western slope of the Middle Urals and in Salair; iron oolite ores in the Volga-Uralian area, Central Kazakhstan, and China; manganese ores and shows in the Urals and in Central Kazakhstan; liptobiolith coal in Kuznetsk. The second region occupies the western and central Tethys where numerous iron and manganese deposits are known in England, Belgium, Western Germany, Turkey, Tunis, and Algeria. Moreover, some iron and manganese deposits were discovered in the north-east USSR, in eastern Kolyma Massif, eastern Mongol-Okhotsk Basin and in the Khabarovsk area, central Mongolia and in western North America in the Rocky Mountains of Arizona.

Different types of deposit formed in the Middle Devonian marine and continental basins occupy the following position with respect to the present equator and poles on the scheme which does not invoke continental drift.

There is no regular trend in distribution of gray-colored deposits. They stretch as a continuous belt along the western margin of Gondwanaland. On its route in most places it is subnormal to the present equator and crosses different climatic zones, i.e., equatorial, sub-equatorial, tropical, sub-tropical, and southern temperate zones. Such a position is difficult to explain using the recent climatic zonation and, moreover, sediments indicative of arid warm climate were found at the same latitudes along the eastern margin of Gondwanaland. It would be possible to expect that many gray-colored terrigenous sequences of western South America accumulated either in a temperate or even in the Arctic belt, being very similar to those of the Boreal realm of Mesozoic and Cenozoic time. One possible explanation is fossils found in these sediments. For example, Shirley (1968) determined Lower Devonian cold-water fossils of South Africa and South America as cold-enduring fauna of Australian type. H. Harrington places them in the Malvinokafric Province and agrees that they are cold-enduring fauna. However, to place these gray-colored sequences into the Arctic climate zone it would be necessary to shift continents of South America and Africa to near-polar areas of the southern hemisphere; this fact is also confirmed by paleomagnetic data and cold-loving fauna. Thus, the distribution of gray-colored terrigenous deposits on the globe in the Middle Devonian is inconsistent with a recent climatic zonation and position of the present equator.

All the areas of Middle Devonian carbonate sedimentation are situated in the northern hemisphere between 30° and 75° N on the litho-paleogeographic map which does not invoke continental drift. Gray-colored terrigenous-carbonate sequences essentially occupy the same position. Both carbonate and gray-colored terrigenous deposits occur within a zone between 30° and 50° N. The data available (widespread carbonate rocks, fossil assemblages, presence of argillaceous and terrigenous sediments, coaly shale, bauxites, hematite, and hydrogoethite-chamosite ores) on these formations imply the presence of a humid, possibly tropical, climate zone in the sense of Sinitsyn (1970) which comprises also equatorial and sub-equatorial belts. However, the recent extent of the zone is at variance with the present climatic zonation, it lies farther north not only of the present equatorial and sub-equatorial belts, but even north of the modern tropics.

The inconsistency between a position of the Middle Devonian and recent humid tropical zones is emphasized by the distribution of Middle Devonian biogenic carbonate rocks, namely reefs, bioherms, and other carbonate buildups which are very abundant in the Middle Devonian marine basins. They occur in all the seas of the Tethys, North America, and Eurasia from 20° to 75° N and even farther north both in areas of essentially carbonate sedimentation and in those of terrigenous-carbonate deposits of the Hercynian facies type. Reef carbonate rocks are known also in zones of gray-colored and variegated volcaniterrigenous-carbonate sediments and in regions of evaporite sedimentation. Two types of Middle Devonian biogenic buildups are distinguished, namely rather extensive barrier reefs and relatively small isolated reefs, bioherms, and carbonate banks. The former are associated with marginal parts of evaporite basins within transitional zones to seas of normal salinity. They were recorded in the northern West Canadian Basin along the southern boundary of the Michigan Basin, along the eastern and south-eastern margin of the Morsovo Basin. The latter are widespread in inner zones of platform and miogeosynclinal marine basins which once were shallow basins; isolated reefs and bioherms occur also in inner parts of evaporite basins. Such a common occurrence of biogenic carbonate rocks in the seas of the northern hemisphere confirms the location of all these seas in the tropical zone (Sinitsyn 1970). Thus, the belt of Middle Devonian biogenic carbonate was very wide; it roughly parallels the present equator. However, its position in the northern hemisphere is absolutely inconsistent with the recent climatic zonation.

A Middle Devonian arid zone will be situated in the northern hemisphere near the present Arctic area. Both terrigenous red beds and evaporites imply such a location of the arid zone. These sediments are found between 40° and 80° N. All the occurrences of the Old Red Sandstone in North America, Fennoscandia, and Angaraland fall into this zone. It encompasses also the West Canadian, Hudson, Michigan, Morsovo, Tuva, and North Siberian salt basins as well as sulfate basins of Central Iowa, Illinois, Moesian-Wallachian, Minusinsk, and Canadian Islands. Many workers determine this zone as a northern evaporite belt. Lotze (1968) also recognizes there a Devonian evaporite belt. Similarly, Sinitsyn (1970) establishes an arid zone in northern Eurasia, suggesting that areas of humid lithogenesis in the Middle Devonian were situated in the European Alps, Asia Minor, India, South and East China. Only the Tindouf sulfate basin is outside the belt, between 20° and 30° N.

The above data show that on the scheme which does not invoke continental drift the evaporite belts (without the Tindouf Basin) lie north of the humid belt in place of the present temperate and near-polar areas. It is considerably shifted northward as compared to the present northern zone of arid and semi-arid climate. There are no grounds to suggest that the displacement was due to an overall warming of the climate on the Earth as Meyerhoff (1970b) believes, because we can see a shift and widening of a tropical zone in a northern direction, which should not take place with a persistent position of the equator.

A statement put forward by Lotze (1968), Strakhov (1960), and Sinitsyn (1970) about a different position for the equator in the Middle Devonian seems to be more correct. Sinitsyn (1970, p. 114) notes that " . . . geography of climates in the Devonian in Eurasia differs greatly from that of the Mesozoic and Cenozoic. The difference lay in the position of an arid zone in Siberia and in the Arctic . . . and in another orientation of thermal boundaries crossing northern latitudes at an angle of 50 to 60 degrees. Such a pattern and geography of Devonian climates in Eurasia imply that the then North pole was situated within the north-western Pacific".

In the southern hemisphere it is impossible to show even an approximate latitudinal position for the evaporite belt with respect to the present equator. A single Middle Devonian salt basin, the Adavale in Australia, was found. Its belonging to an arid zone is beyond doubt, because terrigenous red beds are widespread within the same zone. In Australia evaporites and red beds occur between 15° and 35° S. On other continents in the southern hemisphere sediments indicative of arid climate have not been found at the same latitudes. It is impossible to suggest that sediments of temperate and Arctic climate belts accumulated in the Middle Devonian.

Thus, the distribution of different types of deposit (carbonate and biogenic carbonate, terrigenous-carbonate, terrigenous red beds and evaporites) does not help in establishing for the Middle Devonian tropical, sub-tropical, and temperate belts or even arid and humid zones which without regard for continental drift would occupy a latitudinal position with respect to the present equator or climatic zones. In this respect conclusions made by Meyerhoff (1970a,b; Meyerhoff and Meyerhoff 1972) about an invariable position of the equator and continents and oceans on the earth are not confirmed by observational facts concerning Middle Devonian formations. To locate all these deposits to fit climatic zonation of rocks indicative of climate, some speculations should be made either about pole displacements and a different position for the equator or different arrangement of continents during the Middle Devonian epoch.

A somewhat better fit is obtained for an areal distribution of arid and humid zone sediments for the Middle Devonian in paleogeography which does not invoke continental drift, but with a different position of the equator. One of the versions is shown on the litho-paleogeographic scheme (Fig. 28) by a heavy line. All the humid sedimentation areas in Africa, Europe, and southern Asia fall into a near-equatorial zone. A humid, maybe equatorial zone is a little wider than the present equatorial and sub-equatorial belts. Arid tropical and sub-tropical zones associated with the areas of evaporite and red bed sedimentation will lie farther north and south, the northern and southern evaporite belts will extend from about 15°–20° to 50°–60° N and 20° to 35°–40° S, respectively. The Tindouf sulfate basin will lie within the southern belt. On a reconstruction proposed by Lotze (1968) and Sinitsyn (1970) a climatic zonation of evaporite sedimentation for the Middle Devonian is essentially similar to that of the Recent. However there are still some inconsistencies because many indicators of humid climate in Eurasia (bauxites, iron and manganese ores, coal of the Urals, Turgai, and Kazakhstan) fall into the northern arid zone. It makes Sinitsyn (1970) suggest that the arid area of northern Eurasia was marked by " . . . diverse geographic arrangement of deposits of different climatic types . . . which results in common occurrence of gypsum and dolomite with coaly argillite and ferrous sediments. Apparently the temperate arid climate of that time in North Eurasia was favorable for intense weathering, resulting in local accumulation of autigenic formations, and did not preclude sulfate precipitation in shallow sea" (p. 30). In this reconstruction the greatest inconsistency concerns the southern hemisphere. To obtain a suitable distribution for Middle Devonian deposits of all types it is necessary to suggest the displacement of South America and Africa to the Arctic regions to form a single continent, Gondwanaland.

Strakhov (1960), in an attempt to eliminate the above discrepancies, proposed a different position for the equator. He believes that it is possible "to consider a band running through the Urals and West Siberian Lowland, using all the data available, as a tropical humid zone" (p. 189) or, in other words, as an equatorial zone. The position proposed by N.M. Strakhov for the equator is shown by a dashed line in Fig. 28. On his reconstruction a northern arid zone comprises evaporites and red beds of the USA and Canada, and north-eastern Asia. An equatorial zone forms a wide arcuate band running through China, Mongolia, Altai, Central Kazakhstan, and the Urals into the eastern coast of North America. It is with this band that bauxites, iron, and manganese ore and coal deposits are associated. The occurrences of gypsinate, salt deposits and red beds in some regions (Tuva Basin, Sayan-Altai area, Central Kazakhstan) of the equatorial tropical zone are apparently due to local arid areas in basins between mountain chains formed in Caledonian time (Strakhov 1960, p. 188). A southern arid zone on Strakhov's reconstruction contains deposits of the Old Red Sandstone of England, evaporites of the Russian platform and Dobruja area, and red beds of the Chu-Sarysu Basin and Talas Alatau. The entire Tethys area with peculiar features of warm humid sedimentation (sediments indicative of warm humid conditions) is assigned to a southern temperate humid zone. A northern temperate humid zone is determined in the north-easternmost part of Asia. Southern Africa and South America are placed in a near-polar zone. Thus, according to N.M. Strakhov, all the Middle Devonian climatic zones were arranged in bands convex northward. A.A. Ivanov (Ivanov and Voronova 1972) proposes a similar distribution pattern for arid zones in the Devonian.

A picture drawn by N.M. Strakhov is not without drawbacks, the main being the following. It seems unreasonable to place the Ural-Kazakhstan area of humid sedimentation and that of eastern North America in a single equatorial tropical belt crossing a continent of the Old Red Sandstone. The latter being part of Laurasia made a single whole in the Middle Devonian and was a zone of arid sedimentation evidenced by terrigenous red beds in Greenland and Spitsbergen very similar to those of Great Britain. Generally speaking no evidence exists for extending the Ural-Kazakhstan equatorial tropical zone to the north and west. Rather questionable seems a trend of a southern arid zone because it sets against a humid zone of the Tethys and becomes untraceable in the southern hemisphere. In general it is not clear where to place an arid zone in Australia, because to locate it in either a southern or a northern arid belt one should displace this continent eastward or westward. An idea about the difference between an equatorial humid belt and a temperate humid zone within the Tethys seems inconclusive, because they are similar both in type of

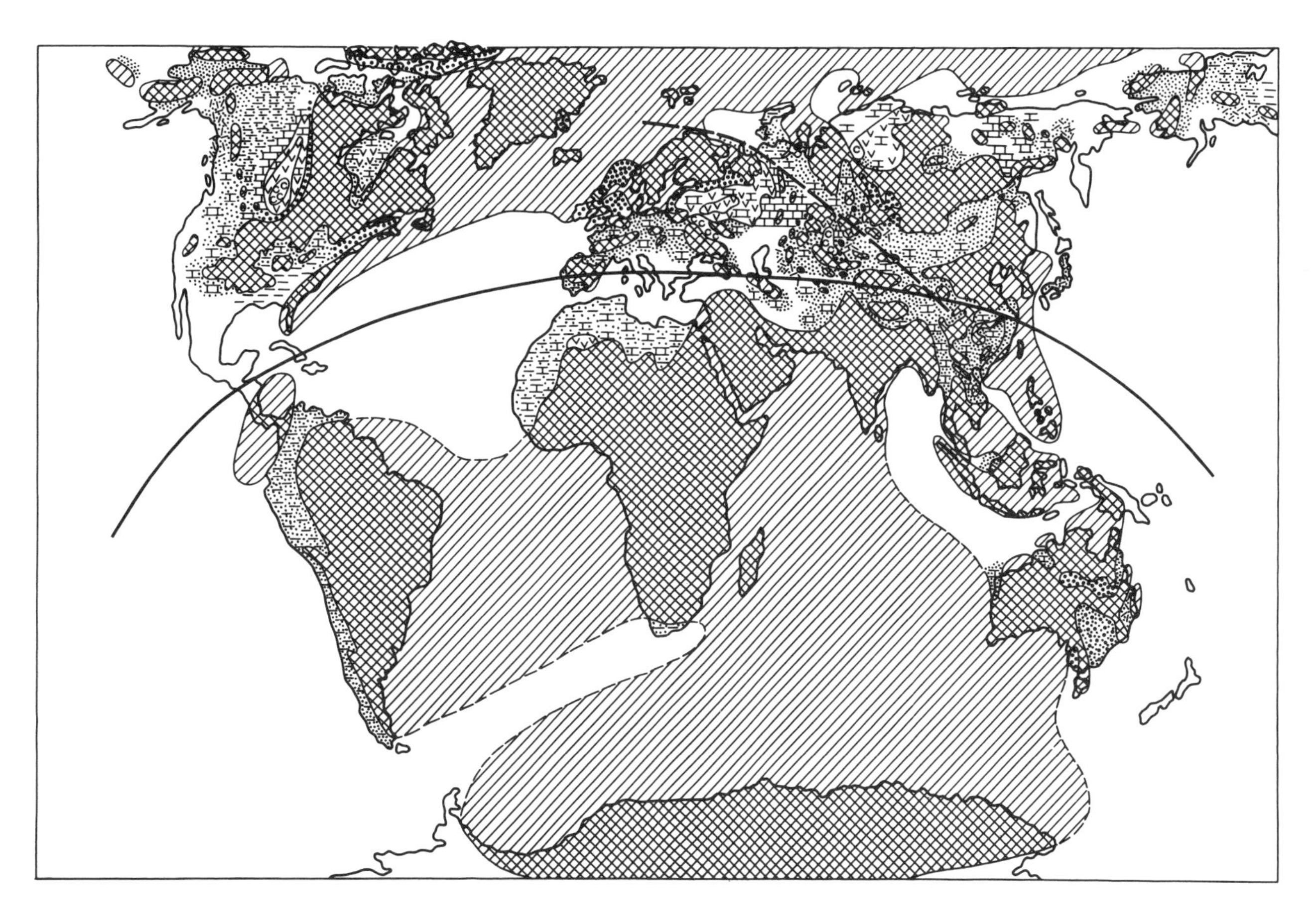

Fig. 29. Litho-paleogeographic map. Late Devonian epoch. For explanation see Figs. 26 and 28

rocks accumulated and in climate-indicator rocks, and they may be considered as warm and humid zones. Nevertheless, a climatic zonation proposed by N.M. Strakhov remains one of the most valid.

There was no major reassembly of land areas, oceans, and marine basins in the Late Devonian. As is evident from the litho-paleogeography which does not invoke continental drift (Fig. 29) a big supercontinent, Gondwanaland, continued to exist in the southern hemisphere, and in the northern hemisphere Laurasia became more pronounced, encompassing Angaraland in addition to North America, Greenland, and the Baltic Shield. The subcontinents Katasia and Malaysia, as well as archipelagoes within central and north-eastern Asia, remained in eastern Eurasia, say, with slightly changed outlines.

The extent of the Tethys remains almost the same, it consists of a number of epi-continental seas between Laurasia and Gondwanaland stretching from the Atlantic Ocean through North West Africa and Eurasia to Katasia. The ocean basin persisted only within the Pacific. The Late Devonian Arctic and Indian Oceans appeared to be occupied by land areas; there were no oceans there because several continents or their fragments could form supercontinents.

We can reconstruct a general distribution pattern for different types of sediment in the Late Devonian seas. Gray-colored terrigenous, terrigenous-carbonate, and volcaniterrigenous-carbonate sequences continued to accumulate essentially in marginal seas in western Gondwanaland, in miogeosynclinal and eugeosynclinal marine basins of the Rocky Mountains, Alaska, and the Appalachians in North America; in epi-continental seas in north-western Africa, in Central and South Europe, in Central South Eastern Asia, in the Ural Geosyncline and in the Mongol-Okhotsk Basin. The accumulation of carbonate deposits took place along margins of evaporite sedimentation regions in zones transitional to the seas of normal salinity. They were widespread around the north-western periphery of the West Canadian Basin, on the eastern Russian Platform, Taimyr, and in some regions of north-eastern Asia. The carbonate sedimentation dominated many areas of the inner marine basins of the Tethys and in the Central Asia Basin as well. Variegated terrigenous-carbonate and volcaniterrigenous-carbonate formations occurred mainly in areas of red bed sedimentation. Evaporite and terrigenous red beds continued to accumulate essentially in marginal seas of Laurasia and Angaraland, and in areas of continental sedimentation within Central Kazakhstan, Sayan-Altai, and Australia.

However, a composition of sediments in the Late Devonian basins has slightly changed. Firstly, gray-colored terrigenous sequences began to form in a great number of seas. If Middle Devonian gray-colored terrigenous strata consist chiefly of more coarse-grained rocks (subgraywacke and sandstone) and occur only in South America, southern Africa and Antarctica, then Late Devonian gray-colored deposits are made up usually of mudstone, shale, and siltstone, minor amounts of coarse-grained material, and in addition to the above regions they occur in many other basins. Thus, essentially argillaceous sequences are common in North America. They were recorded in Alaska (Nation Formation), in the Mackenzie Basin, and in the Rocky Mountains of British Columbia, Canada (Imperial, Fort Simpson, Redknife Formations), in southern Midcontinent (Chattanooga, Woodfort, Sheffield, Lime Creek Formations and others), in the Appalachian Basin (Amsden et al. 1967; Bassett and Stout 1967; Churkin and Brabb 1967; Collinson et al. 1967; Griffin 1967; Norris 1967; Oliver et al. 1967). Argillaceous gray-colored beds accumulated in seas of the Tethys, in the Caucasus, Urals, and in the Mongol-Okhotsk Basin.

Secondly, the Late Devonian gray-colored terrigenous-carbonate rocks became more argillaceous in composition. They may be called argillaceous (or marly)-carbonate rocks. They are most common in the Tindouf, Reggan, Ahnet, Timimoun, Mouydir, Bechar, and other basins of north-western Africa (Hollard 1967; Aliev et al. 1971), in West Europe within the Dimant and Namur Basins (Lecompte 1967), in Harz, Thüringia, and Lausitze (Erben and Zagora 1967), in the Pyrenees and Carnian Alps (Flügel 1967), in miogeosynclinal zones of the Caucasus, Urals, Central and South Eastern Asia marine basins. Great thicknesses of argillaceous rocks, accumulated in the Late Devonian basins, are often considered as an evidence for wide marine transgression. However, as has been shown above, no important changes were found in reassembly of land areas and seas in that time as compared to the Middle Devonian. Apparently, the formation of argillaceous gray- and dark-colored sediments was due to precipitation in humid zones where we found most of the above basins. A speculation that both inland and marginal seas in the Late Devonian were much deeper and land areas more peneplaned seems not improbable. Great thicknesses of siliceous and argillaceous siliceous rocks found, for example, in southern Midcontinent in Arkansas, USA, locally in the Ural Geosyncline and in Novaya Zemlya show that some seas were rather deep.

Thirdly, sulfate deposits, gypsum, and anhydrite became more widespread during the Late Devonian. In regions of evaporite sedimentation sulfate rocks occupied extensive areas which were peculiar transitional basins, whereas salt accumulation took place near the land masses in intensely subsiding zones far from the open sea.

An areal distribution of different types of deposit was in the Late Devonian very similar to that of the Middle Devonian discussed earlier. Gray-colored terrigenous and terrigenous-carbonate deposits were concentrated in five major zones, i.e., (1) North America, (2) South America, (3) the Tethys, (4) Urals-Central Asia, and (5) Mongol-Okhotsk. The North American zone encompasses marine basins around the western periphery of Laurasia, while the South American zone comprises seas situated along the western margin of Gondwanaland. A zone associated with the Tethys stretches west-east through southern West Europe and north-western Africa into Central Asia and farther into the south-eastern Asiatic continent in Malaysia. The Ural-Central Asiatic zone is situated in the center of Eurasia and runs north-south from Novaya Zemlya to the Tien Shan. The Mongol-Okhotsk zone occupies a geosynclinal basin between Angaraland and Katasia and stretches sublatitudinally. A sedimentation in most of these zones took place under the conditions of warm humid climate evidenced not only by reef carbonate widespread in the Tethys, Ural-Central Asiatic, and North American zones but by iron, bauxite, manganese, and coal deposits and ore shows in the Timan, Urals, West Europe (Harz, Sudetes and others); kaoline-bearing rocks are known there as well.

The Late Devonian evaporites and terrigenous red beds occur mainly in the following four regions: (1) northern North America (West Canada and Hudson Basins, Appalachians and in the Maritime Basin), (2) northern Europe (Upper Devonian Basin of the Russian Platform and red beds of Great Britain), (3) Angaraland, Sayan-

Altai fold area and Central Kazakhstan (North Siberian and Chu-Sarysu salt basins, Kuznetsk, Teniz, Turgai sulfate basins, intermontane basins of the Sayan-Altai area, Karatau and other regions), (4) Australia (red beds and variegated strata of the Dulcie, Amadeus, Grampians, Avon Rivers, Western Province Basins, etc.). These regions were situated in arid zones.

Thus, areas of arid and humid sedimentation can be discerned for the Late Devonian as well. However, they show a rather complex pattern of areal distribution on continents.

First let us see if it is possible to integrate arid and humid zones of Late Devonian sediments to form belts striking parallel to the present equator and which could coincide with present climatic zones. The North American zone of warm humid climate containing gray-colored terrigenous and terrigenous-carbonate, mainly argillaceous, deposits lies now between 30° and 70° N. It is impossible to stretch it parallel to the present equator because evaporites and red beds in North America occur at the same latitudes. Though the areas of humid sedimentation in the Tethys form a latitudinal belt, it lies farther north of the present tropical zone and entirely within the northern hemisphere. The Ural-Central Asiatic Late Devonian humid zone is impossible to make coincident with modern climatic zonation, it strikes subnormal to the present humid and temperate humid zones. A position of the Mongol-Okhotsk area of gray-colored apparently humid sedimentation does not allow combination with other humid areas into a single zone parallel to the present climatic belts.

Similar difficulties arise as to the arid regions of Late Devonian sedimentation. If we link arid zones in North America and Europe to form a single belt parallel to the present equator, then both in the east and in the west the belt will be crossed by North American and Ural-Central Asiatic humid zones, respectively. It is impossible to obtain a single zone including the arid area of northern Europe and those of Angaraland, the Sayan-Altai area, and Central Kazakhstan, and not crossing the Ural-Central Asiatic humid zone.

Generally, in an attempt to recognize the Late Devonian humid and arid belt according to the present climatic zone trends we do not obtain a clear-cut pattern of zonal deposit distribution. In doing so we have a single wide belt of warm climate situated in the northern hemisphere between 20° and 80° N; within this belt areas of arid sedimentation and humid sedimentation were located near continents and in marine basins, respectively. A warm, essentially arid climate is marked all over Australia (Brown et al. 1970). Thus, we can postulate that in the Late Devonian the climate of the Earth was warm. The above data and common occurrence of warm-loving fauna (primarily, crossopterygians and earliest amphibians) enable some workers to suggest a warm tropical climate on the globe. Thus, Sinitsyn (1970) writes that "a monotony of the Devonian fish fauna on all the continents and at all latitudes suggests a global presence of tropical dry climate for this spell of the geological history" (p. 85).

Now we shall try to outline arid and humid belts for the Late Devonian using a different position for the equator and poles without continental drift. Similar reconstructions were made by Strakhov (1960), Lotze (1968), Meyerhoff (1970b), Sinitsyn (1970), Meyerhoff and Teichert (1971), Meyerhoff and Meyerhoff (1972), Ivanov and Voronova (1972), and other workers.

We have already mentioned the viewpoint of Lotze (1968). He established an evaporite belt for the entire Devonian. It embraces areas of evaporite and red bed sedimentation in Eurasia and North America. In Eurasia the evaporite belt extends from south-east to north-west, encompassing both areas of red bed and salt deposits in England and the Russian Platform and those of evaporite sedimentation in Kazakhstan, East Siberia, and the Far East. A south-western boundary of the belt runs through England into the Caucasus, and a north-eastern boundary follows the eastern margin of the North Siberian salt basin. The Devonian evaporites of the north-eastern USSR not being involved, F. Lotze placed the north-eastern boundary of the belt farther west of its real position. The Ural-Central Asiatic humid zone was included into the evaporite belt in Eurasia and the belt went along the strike southward into humid zones of the Mongol-Okhotsk area and in the Tethys. In northern direction it ran through the Arctic Ocean and then southward, encompassing Greenland and almost the entire central part of North America. Such a placement of all the evaporite deposits of Eurasia and North America into a single evaporite belt with displacement of continents seems improbable. In this connection N.M. Strakhov emphasized that such reconstructions did not include all the data available, primarily those which indicate humid climate.

Sinitsyn (1970) proposed a different reconstruction for the Late Devonian arid and humid zones of Eurasia. In his opinion there was a tropical climate with a mean annual temperature of 27°–29° to 32°–33° C. He outlined more arid and more humid zones. The arid zone comprises such northern subcontinents as Fenno-Sarmatia, Tobolia, and Angaraland, i.e., almost the entire northern Eurasia (probably except for its north-eastern part) where evaporites and red beds were common. A humid area encompasses southern subcontinents, namely Katasia, India, and the Tethys including its European and African parts. A transitional area covering Central Europe, Central Asia, South Mongolia, and the Mongol-Okhotsk Basin was situated between the above two zones. The southern humid zone could be an equatorial one. Thus, in the Late Devonian the equator could occupy a position shown by a heavy line in Fig. 29. This reconstruction places northern and southern regions of North America with all the areas of evaporite and red bed sedimentation in an arid and humid zone, respectively. Another arid area known in Australia will be in the southern hemisphere. Thus, extending V.M. Sinitsyn's reconstruction for Eurasia to the entire area of the globe we recognize two arid belts (belts of arid climate): northern and southern belts with a wide equatorial zone in between. A zone of evaporite sedimentation will be confined to the northern arid belt. Note that reconstructions using a permanent position of continents result in the conclusion of a very hot global climate in Late Devonian time and the distribution of arid belts both in the northern and southern near-polar areas. Additionally, arid zones in Eurasia are very often replaced by humid areas at the same latitudes (for example, by the Ural-Central Asian). V.M. Sinitsyn attributes this, as in the Middle Devonian, to an overall high humidity and high content of CO_2 in the atmosphere which gave rise to an isothermal climate with a weak atmospheric circulation, and promoted the accumulation of chemical weathering products even in arid areas. Despite the fact that these assumptions are quite probable, it is difficult to give an unambiguous climatic zonation for the Late Devonian using the above-mentioned position of the equator. A discrepancy cannot be eliminated unless we unite all the southern continents into a single supercontinent, Gondwanaland.

Strakhov's reconstructions (1960) do not provide either a harmonious picture for the distribution of humid and arid climatic zones on the globe. Though N.M. Strakhov has not compiled a separate climatic zonation for the Late Devonian of the Earth, his ideas are well represented on schemes of Eurasia (Strakhov 1960, Figs. 53, 54). An equatorial tropical humid zone like in the Middle Devonian is in the Urals and within West Siberia. In this case an approximate position for the equator is shown by a heavy dashed line in Fig. 29. For the lower Late Devonian an equatorial humid zone is extended into the Russian Platform where it encompasses almost the entire Volga-Ural area, Moscovian Syneclise, and Baltic Schield. It was based on some occurrences of iron ores indicative of humid environment. In the Fammenian the equatorial zone is shifted to the east and north-east. A southern arid zone is outlined within the Russian Platform including the Upper Devonian salt basin and it stretches in Central Kazakhstan where the Chu-Sarysu salt basin is situated. Its farther south-eastern trend is not shown for lack of evidence. There is no northern arid zone in Strakhov's reconstruction, however it could run through the Siberian Platform. Some suggestions may be put forward because Late Devonian evaporite deposits are known through most of the North Siberian salt basin. However, again we cannot trace a trend of the northern arid zone to the south-east because we have there a humid zone of the Mongol-Okhotsk Basin. A southern temperate (humid) zone is drawn by N.M. Strakhov within the European part of the Soviet Union, Caucasus, and Central Asia. Apparently, more southern regions situated in the Tethys are to be assigned to this zone. It is very difficult to stretch all the above climatic zones into the western hemisphere and into North America, and it is doubly difficult to correlate them with arid zones of the southern hemisphere, in Australia. N.M. Strakhov attributes inconsistencies and uncertainties in the Late Devonian climatic zonation to the fact of a critical reassembly of climatic zones commenced in the Late Devonian, when a Middle Devonian zonation pattern was disappearing and a new Upper Paleozoic pattern characterized by an overall humidization of the Earth's climate was taking shape (Strakhov 1960, p. 193). However we should state that Strakhov's reconstruction does not provide an unambiguous climatic zonation and belts of evaporite sedimentation for the Late Devonian.

Thus, the proposed positions for humid and arid zones of the Late Devonian based on reconstructions which do not invoke continental drift are either not confirmed or in poor agreement with evidence on arid distribution of different sediment types, and thus do not provide a clear-cut picture of the Late Devonian climatic zonation.

Reconstruction Which Invokes Continental Drift

Paleogeography and distribution of evaporite and other Middle Devonian deposits which invoke continental drift are as follows (Fig. 30).

Creer (1973), on the basis of paleomagnetic data, places the North and South Poles in the north-western Pacific east of Kamchatka and in the Indian Ocean off

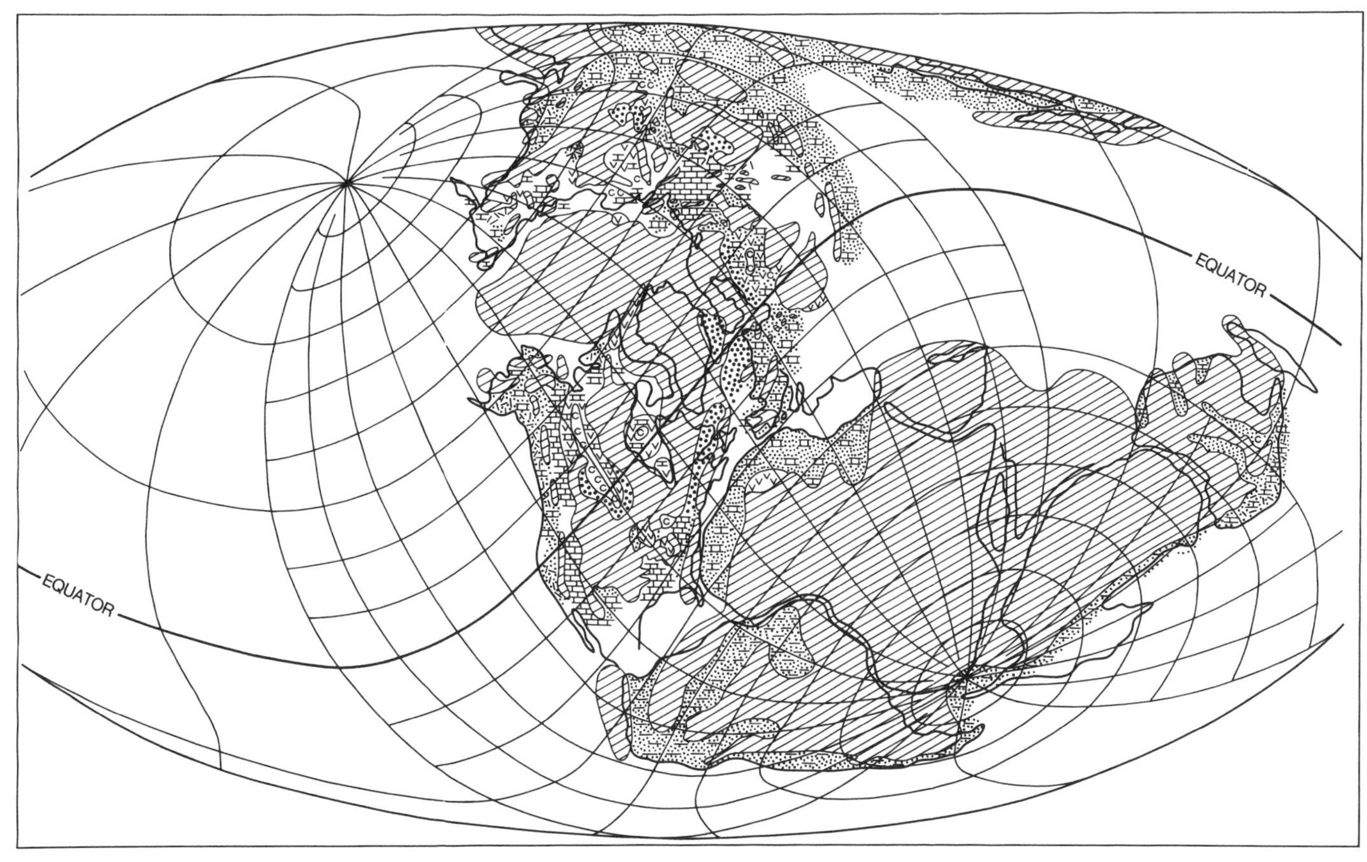

Fig. 30. Litho-paleogeographic map which invokes continental drift. Middle Devonian epoch. Mollweide projection. For explanation see Figs. 26–28

southern Africa, respectively. In the whole, land areas on the globe moved to the south relative to these poles according to paleogeography which invokes continental drift. The southern hemisphere is largely occupied by a single large continent, Gondwanaland; it is smaller in size than in paleogeography which does not invoke continental drift, owing to the connection of South America with Africa and Antarctica with Africa, Australia, and the Hindustan Peninsula. Thus, southern Gondwanaland is situated in the polar area. Laurasia also shifted to the south. North America and Europe are largely in the southern hemisphere. Land areas in Eurasia (Tobolia, Angaraland, Katasia, Malaysia, and others) are situated in the northern hemisphere south of 60° N. An ocean basin, the Tethys, situated in the near-equatorial zone, lies now between Gondwanaland, on the one side, and Malaysia, Katasia, and Laurasia, on the other. Owing to the fact that many former inland basins around the Tethys periphery turned into marginal basins, the western Tethys, which was situated in the Atlantic, became much narrower and marine basins became inland ones because of North America's approach to Africa. Laurasia was near Gondwanaland, giving rise to a future supercontinent, Pangaea.

All these reassemblies result in a considerably new distribution pattern of Middle Devonian deposits of various types and hence in a different position with respect to the ancient equator.

It should be noted that there is no room for the climatic zones proposed by Strakhov (1960) and Sinitsyn (1970) in such a reconstruction. A northern arid zone of Sinitsyn embracing northern regions of Eurasia occupies a longitudinal position with respect to a Middle Devonian equator. Equatorial tropical, arid, and temperate humid zones of Strakhov are also arranged independently of any relation to poles and equatorial belts. Despite this fact some earlier uncertain problems of climatic zonation acquire a more logical explanation.

A position of the above-mentioned belt of gray-colored terrigenous deposits confined to western margins of Gondwanaland becomes more evident. It is situated partly on the South Pole or in the Arctic area and lies south of 40° S. It is well in accord with ideas about the formation of gray-colored terrigenous beds containing fossils occurring in a cold climatic belt of the Earth. Such an assumption could not be confirmed without reference to continental drift.

All the areas of Middle Devonian carbonate sedimentation are assigned to equatorial and tropical zones. They lie between 50° N and 30° S. Most biogenic carbonate buildups are also placed within this belt. An extent of the above discussed Hercynian facies became more obvious: it is entirely related to a humid belt along the Tethys margin and lies close to the equator.

Thus, continental drift and paleomagnetic data enabled recognition for the Middle Devonian of three major global zones: two cold Arctic zones and one warm zone between them, the near-equatorial or tropical zone.

Evaporite basins and areas of the Middle Devonian terrigenous red beds fall in a zone between 60° N and 35° S. It is possible to identify an equatorial belt of evaporite sedimentation including basins of North America and Europe. Some basins of the belt lie at the equator or in its vicinity. They are: the West Canadian, Hudson, Michigan, Illinois, Central Iowa, Canadian, Morsovo, East Ural, and Moesian-Wallachian Basins. Evaporite basins of Asia (East Siberian, Minusinsk, Tuva Basins) shifted farther north of the equator up to 40°–45° N and are separated from the equatorial belt by a zone of humid sedimentation within Turgai, the Urals, and Kazakhstan. An area of arid sedimentation in Australia lies between 25° and 35° S. The Tindouf sulfate basin of north-western Africa is associated with the equatorial evaporite belt.

By reconstructions involving continental drift an unusual arrangement of the Middle Devonian arid and humid climatic zones is observed. An arid zone of Laurasia easily recognized on evaporites and red beds lies essentially at the equator in the place which, judging from the present climatic zonation, should be occupied by a humid zone. In turn, a humid zone of the Tethys within southern Europe and Africa, which could be considered as an equatorial one away from the equator, lies in the place of an arid tropical zone on the basis of recent evidence. Areas (probably equatorial) of humid sedimentation in Central and South-East Asia also lie far from the equator between 10° and 30° N, which is at variance with the recent data. The Ural-Kazakhstan humid zone lies outside the near-equatorial area, as before it had an unusual trend. Only areas of evaporites and red beds in Australia and in north-western Africa are consistent with the present location of arid zones. Such an arrangement of arid and humid sedimentation belts in the Middle Devonian when continental drift is involved may be due to a peculiar position of Laurasia and the Tethys at the equator. However, the inconsistency of a climatic zone distribution in the Middle Devonian with that of the present time is an important fact. It may suggest that recent climate indicators cannot be simply used for unambiguous establishment of climatic zones of the old geological epochs, because their arrangement on the globe may be different depending on the position of continents, oceans, and marine basins with respect to each other in the past, as well as on the current orientation and zonal atmospheric circulation.

Now let us try to imagine an arrangement of land areas and seaways in the Late Devonian using continental drift and paleomagnetic data (Fig. 31). Since the position of the equator and arrangement of continents remains the same as those of the Middle Devonian and we know an areal distribution pattern of sediments in the Late Devonian marine basins, we can trace how an areal distribution of different types of sediments changes depending on the postulated position of the equator.

Firstly, it is to be noted that in reconstructions involving continental drift all the Late Devonian marine basins of North America, Eurasia, north-western Africa, and Australia fall within a belt with a southern limit at 50° S and a northern limit at 60° N. Essentially, they lie outside the Late Devonian polar zones. Hence we put forward an idea inconsistent with the conclusions made by Meyerhoff and Meyerhoff (1972) and Sinitsyn (1970) that in the Devonian climate was not hot-tropical all over the globe. The reconstruction discussed implies that a number of basins situated in South America and southern Africa lie in a southern polar area which is consistent with their sediment lithology and faunal assemblages. In the northern Arctic area there was an ocean and all the seas of Eurasia and North America lay farther south, which explains the absence of the Boreal Devonian sediments on the continents mentioned.

Thus, we can propose three major climatic zones for the Late Devonian, as was the case for the Middle Devonian, namely southern Arctic, warm near-equatorial, and apparently warm northern Arctic zones. Of interest is a considerable width of the warm near-equatorial zone which according to Sinitsyn (1970) was always tropical. However it is impossible to postulate arid and humid belts within this zone which could be situated relative to the equator as at the present time. The map clearly shows that areas of arid climate around and inside Laurasia (West Canadian and Hudson salt basins, Moose River sulfate basin, red beds of the Appalachians and England, Maritime, Canadian Arctic Basins, Upper Devonian salt basin of the Russian Platform) are confined to the equator or lie not far from it. This arid belt is situated between 20° N and 20° S. Again we see that it is in the place which at present belongs to the equatorial humid belt. It should be noted that on reconstructions with regard to continental drift the arid zone should occupy a position proposed by F. Lotze when he suggested his Devonian evaporite belt. And again, Sinitsyn's and Strakhov's reconstructions are inconsistent with a position for the Late Devonian equator obtained using paleomagnetic data. A northern arid zone proposed by V.M. Sinitsyn for Eurasia is longitudinal relative to this equator, and his tropical zone confined to the Tethys crosses an equatorial area from south-west to north-east. Climatic zones suggested by N.M. Strakhov are in general at variance with reconstructions which invoke continental drift. All these zones (equatorial and arid) are situated in the northern hemisphere and are not latitudinal relative to the Devonian paleoequator.

In a reconstruction involving continental drift a Late Devonian humid zone of the Tethys cannot be considered as near-equatorial and corresponding to the present equatorial and sub-equatorial belts. Areas of the Tethys situated in southern West Europe and north-western Africa lie south of the equator between 10° and 40° S, as if in the place of an arid zone of the present tropical and sub-tropical belts. The areas of the Tethys, South, Central, and South-East Asia would be situated in the northern hemisphere between 20° and 40° N, i.e., it is again in the place which is at present partly occupied by the sub-tropical arid climatic zone.

An arid zone within Angaraland, Sayan-Altai area and Central Kazakhstan by a Late Devonian paleogeography which invokes continental drift is shifted northward and lies between 25° and 50° N. This zone relative to the equator occupies the same position as the present day zones of arid climate within the tropical and sub-tropical belts. The same holds for the arid zone of Australia situated between 25° and 35° S.

A Late Devonian Ural-Central Asian humid zone which N.M. Strakhov considers as an equatorial tropical zone is also inconsistent with a position of the equator. It lies between 20° and 30° N separating two arid zones; southern European, and northern Angarida-Kazakhstan.

Thus, it is impossible by the reconstruction involving continental drift to locate all the arid and humid Late Devonian zones so that they would form single belts of arid or humid climate. Apparently, one should assume, as for the Middle Devonian, that their position was controlled mainly by the arrangement of oceans and seas, coastal lines of continents, and land areas, as well as current direction and zonal atmospheric circulation. These zones probably being different, the position of humid and arid zones could have differed from that of the present zones.

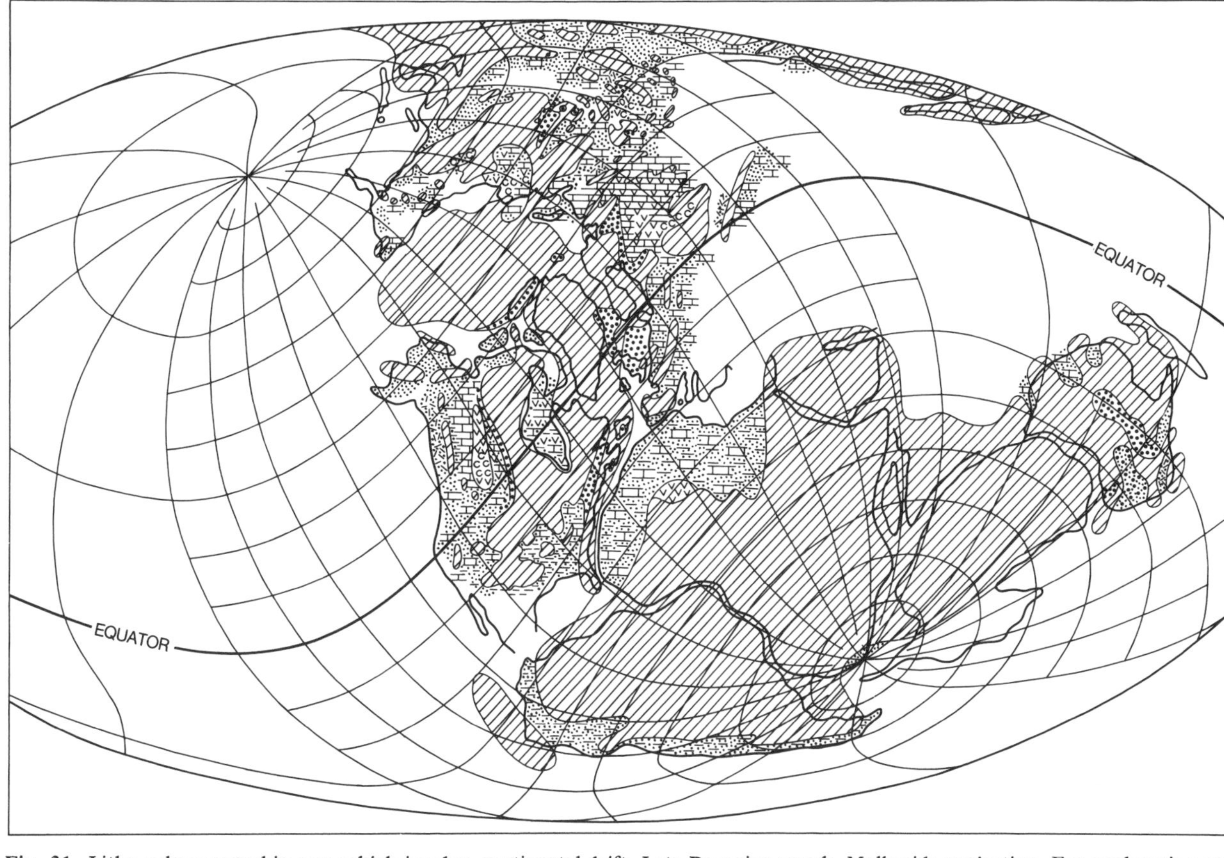

Fig. 31. Litho-paleogeographic map which invokes continental drift. Late Devonian epoch. Mollweide projection. For explanation see Figs. 26 and 28

Conclusions Concerning Paleoclimate and Paleogeography of Evaporite Basins During the Middle and Late Devonian

The above analysis implies that it is impossible to establish an unambiguous belt of equatorial, tropical, sub-tropical, temperate, and Arctic climate for the Middle and Late Devonian either by a reconstruction which does or does nor take into account continental drift. Thus, one should rule out reconstructions suggesting an unchangeable position for continents and oceans as well as for the poles and the equator. Reconstructions which invoke continental drift should be favored because they are supported by paleomagnetic data and enable us to comtemplate, though roughly, areas of cold climate which gravitate toward polar zones, and a warm climate zone within a belt probably embracing sub-tropical, tropical, and equatorial zones. Even in this case the arrangement of Late and Middle Devonian zones of arid and humid climate could differ from that of the present-day zones. The arid zones would be associated with the equator, while humid zones, though having some features characteristic of the equatorial belt, are remote from the equator. One should assume that a climatic zonation in Middle and Late Devonian time differed from the present one. Thus, one should extrapolate the present climatic zonation for the Devonian with great care.

It is impossible to propose unambiguously evaporite belts either for the Middle Devonian or for the Early Cambrian using the data available. Reconstructions based on a different position for the equator or those involving continental drift imply that evaporites accumulated in arid zones situated mainly between 60° N and 35° S. This conclusion is consistent with paleomagnetic data obtained by Blackett (1961), Irving and Briden (1962), Opdyke (1966), and Briden and Irving (1968), which suggest such a position for a warm tropical climate zone in the Devonian.

The paleogeography of Middle Devonian evaporite basins suggests that they were typical epi-continental seas. However, all these seas, unlike those of the Early Cambrian, though occupying a marginal position, were situated within continents. Such an arrangement is confirmed by reconstruction both with and without regard to continental drift. Marginal salt and sulfate basins may have been absent during the Middle Devonian and thus major paleogeographic changes in the history of Paleozoic evaporite sedimentation occurred since the Middle Devonian. Despite great distances separating Middle Devonian salt basins from oceans, they could be connected by broad seaways in the form of medium basins of barrier and island type (Zharkov 1971b). This idea is supported by numerous paleo-biogeographical data (Dubatolov and Spassky 1964, 1970, 1973; Boucot et al. 1967; House 1968; Sinitsyn 1970; Dubatolov 1972; The Devonian . . . 1973). Thus, on the basis of coral distribution Dubatolov and Spassky (1970) suggest that, starting from the Eifelian, connections between biogeographic provinces became wider and during the upper Middle Devonian the pre-existing Ural-Tien Shan, Sayan-Altai, Dzhungar-Balkhash, and Indigirka-Kolyma Provinces formed a single Ural-North Asian Province. Connections with the Mediterranean Province also became easier. However, the connection of the above provinces with the Appalachian in North America did not take place. The latter remained as a separate province throughout the entire Middle Devonian (Dubatolov 1972, p. 191). This fact may imply that sea water could come into salt basins of Eurasia from the Tethys and to those of North America from the Pacific Ocean.

House (1968) believes that data on ammonoid distribution imply the existence of a permanent connection between European seas and those of North America through the Arctic and not through the Atlantic Ocean. Moreover, it may suggest the absence of the present Atlantic in Middle Devonian time. Nevertheless, broad connections between basins of Europe and North America through the Arctic seem highly improbable. Paleogeographic reconstructions and data imply that similar connections were possible through marine basins on the Urals, Taimyr, north-eastern Asia, and Alaska rather than through the Arctic Ocean which apparently did not exist at that time. Similar conclusions can be arrived at on the basis of brachiopod distribution in the Devonian marine basins (Boucot et al. 1967). All this points to the fact that a supply of sea water into salt zones of evaporite basins in Middle Devonian time took place via seas connected with the Tethys or Pacific Ocean, which were situated either in the Ural-North Asian biogeographic province or in the North American Province. However, this connection is inconsistent with the ideas of Meyerhoff (1970b), who postulates the supply of warm salt water into evaporite basins of Asia from the Atlantic through the Arctic Ocean.

Neither can we obtain a clear-cut picture of evaporite belts for the Late Devonian. Reconstructions of climatic zonation with regard to continental drift seem more probable for that time. However, they make us suggest that the arrangement of arid and humid climate zones both in Late and Middle Devonian differed greatly from the present one due to a different arrangement of the then-existing continents, oceans, and seas.

In the paleogeographic position of Late Devonian evaporite basins and means of supply of sea water of normal salinity no great changes occurred as compared to Middle Devonian time. All salt basins were still epi-continental seas situated mainly along continental margins. They were separated from oceans by broad medium basins of archipelago or barrier type with predominantly carbonate, sulfate-carbonate, and sulfate sedimentation (Zharkov 1971b; Zharkov 1976). The connections between salt basins of North America and those of Eurasia and the oceans separating them were provided by seas of the Tethys or Pacific Ocean rather than by the Atlantic Ocean, as is confirmed by paleo-biogeographic data (Dubatolov and Spassky 1964, 1970, 1973; Sinitsyn 1970; Dubatolov 1972; The Devonian . . . 1973).

Paleogeography of Continents and Paleoclimatic Zonation of Evaporite Sedimentation in the Permian

Reconstruction Which Does Not Invoke Continental Drift

The litho-paleogeographic maps not invoking continental drift and worked out for the Artinskian and Kungurian (Early Permian) (Fig. 32) and for the Permian in general (Fig. 33) are almost similar. Thus, we shall discuss them together, emphasizing changes characteristic of these time intervals in the Permian period.

Primarily, attention is drawn to considerable changes in the distribution of land masses, oceans, and seas on the Earth as compared to those of the Devonian.

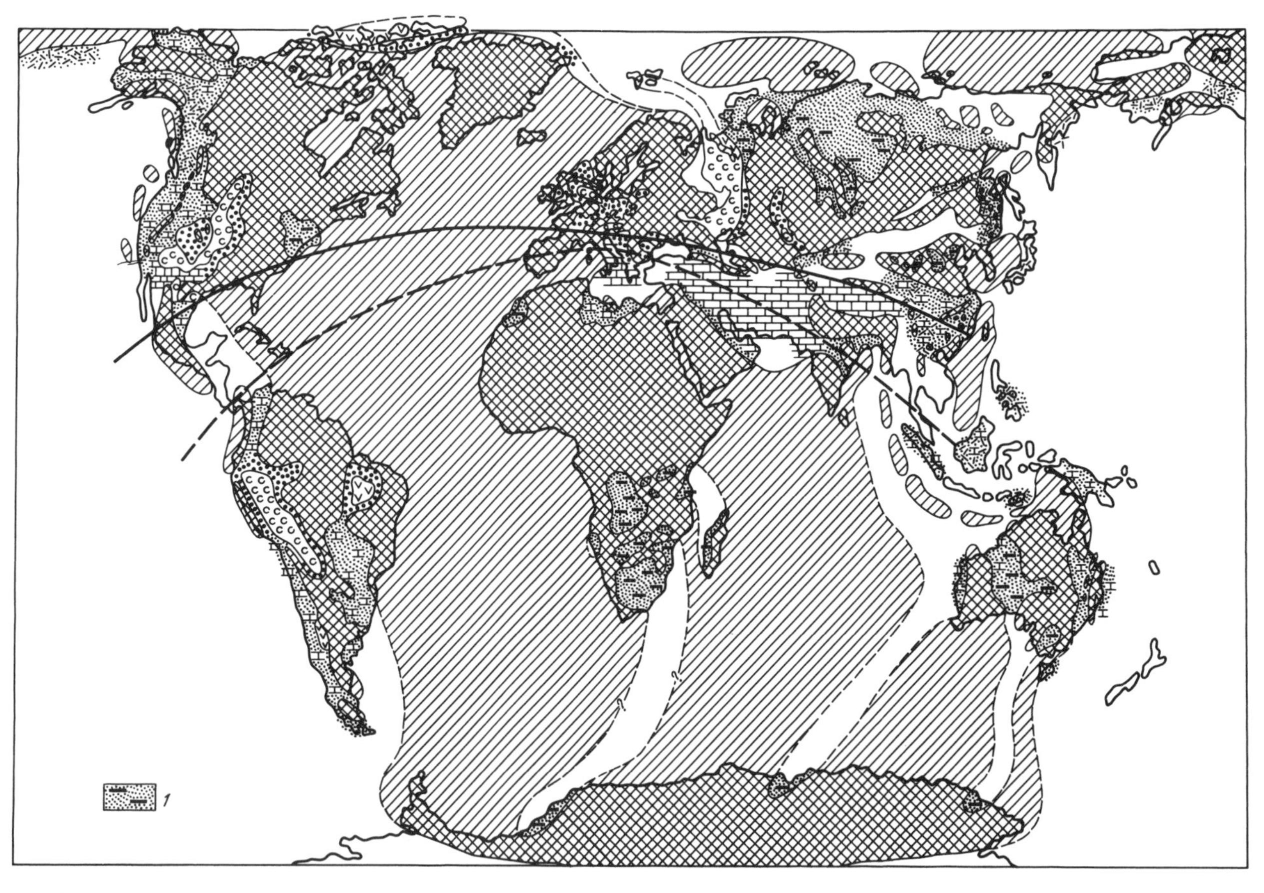

Fig. 32. Litho-paleogeographic map. Early Permian epoch (Artinskian and Kungurian ages). *1* areas of occurrence of coal-bearing terrigenous gray-colored beds. For other explanations see Figs. 26 and 28

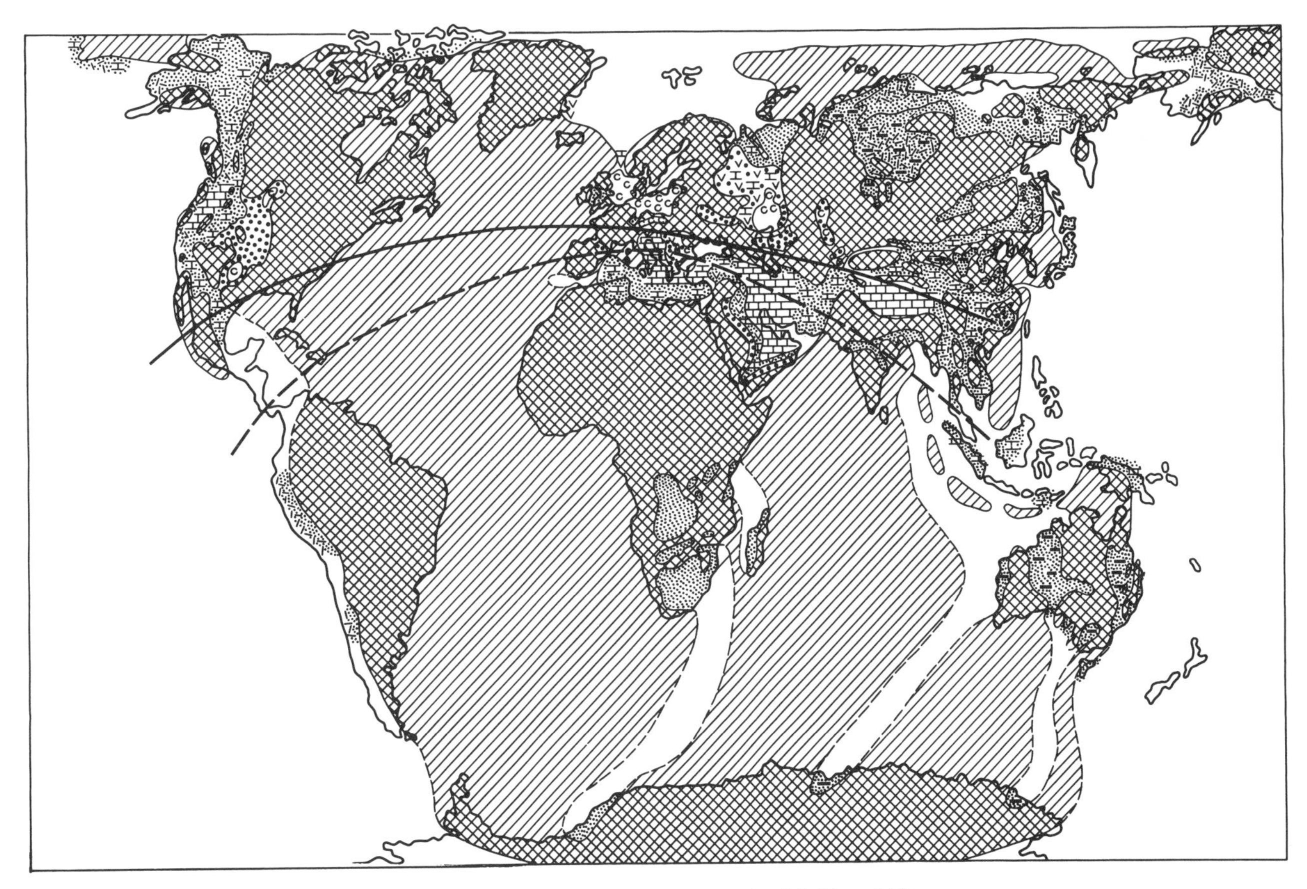

Fig. 33. Litho-paleogeographic map. Late Permian epoch. For explanation see Figs. 26, 28, and 32

Both in the Early and Late Permian the existence of the supercontinent Pangaea is clearly established; it probably comprised Africa with the Arabian and Hindustan Peninsulas, Antarctica, Australia, South and North America, Greenland, and almost the whole of Europe. In the north-east this continent joined a large land area, Angaraland, separated at the beginning of the Early Permian from Pangaea by a relatively narrow strait, and at the end of the Early Permian becoming connected by a narrow isthmus. In eastern and south-eastern Asia there were also large land masses, such as Katasia, the Chinese-Korean mainland, and Chukotka. In the Early Permian a land mass took shape in the Kara Sea north of Angaraland; in the Late Permian it merged with Angaraland and Chukotka. Smaller land masses and archipelagoes are known in Central and North-East Asia, in Alaska, in western North America, and Malaysia.

The increase of the entire land mass resulted in abrupt decrease in the number of marine basins. They were preserved only in the west of North and South America, in northern and north-eastern Asia and in some regions of Australia. The maps which do not invoke continental drift show the Tethys as a single large epi-continental marine Basin, extending from southern Europe eastward through Turkey, Iran, Afghanistan, and Pakistan into Central Asia. Many epi-continental seas appeared within continents rather remote from the ocean. They were either zones of evaporite and red bed accumulation, or areas where coal of the paralic type formed. Freshwater and brackish intracontinental basins were widespread where, under different environments, either red or gray alluvial-lacustrine sediments or limnic coal-bearing deposits accumulated.

It is only within the Pacific that the presence of an oceanic basin in the Early and Late Permian can be tentatively assumed. In place of the Arctic Ocean was a marine basin separated from the Pacific by the island "chains" and larger land masses (Chukotka and Kara mainlands during the Early Permian or a joint Kara-Chukotka mainland in the Late Permian). The Atlantic and Indian Oceans cannot be distinguished for these time intervals in maps which do not invoke continental drift.

Variegated sediments accumulated in Early and Late Permian marine and continental basins: marine gray carbonate and terrigenous-carbonate, marine gray terrigenous and coal-bearing carbonate-terrigenous, red terrigenous and volcaniterrigenous, marine and continental evaporitic, alluvial-lacustrine gray terrigenous and coal-bearing terrigenous, glacial-terrigenous sediments, etc.

During the Permian, carbonate deposits mostly formed within the Tethys. In the second half of the Permian they were especially widespread. Leven (1975) states that "the middle of the Permian period was marked by an extensive marine transgression resulting in the ubiquitous formation of carbonate strata. On the vast territory extending from the Mediterranean in the west to China, Indochina, and Japan in the east, carbonate sediments formed, in fact, a single formation. It should be noted that up to that time sedimentation usually took place in separate troughs, and then became very extensive, so we find carbonate strata resting on older formations, including Proterozoic" (p. 52). In the Late Permian carbonate accumulation in the Tethys was slightly reduced, and terrigenous-carbonate deposits became predominant there. Another region of carbonate formation is in western North America, in the Rocky Mountains. It should be emphasized that carbonate accumulation there was most intense during the Late Permian, whereas the Early Permian is marked by deposition of terrigenous-carbonate sediments (McKee et al. 1967).

Gray-colored terrigenous-carbonate marine strata usually formed in marginal and inland seas confined to the Pacific margin of Pangaea, as well as to the Tethys. Gray-colored terrigenous beds developed in sea basins around Angaraland, Katasia, and the Chinese-Korean mainland. In the off-shore zones of these seas coal-bearing terrigenous and sometimes terrigenous-carbonate formations accumulated.

The areas of evaporite and red bed accumulation of the Early and Late Permian were often the same. In the Early Permian these sediments accumulated in the following evaporite basins: Midcontinent, Supai, North Mexican, Peru-Bolivian, Rio Blanco, Parnaiba, Central European, North Italian, Moesian, East European, Karasu Ishsay, Darvaza, and Sverdrup Basins. The Early Permian terrigenous red beds were found in the Central Appalachians and the Michigan Basin in the United States, north-eastern Greenland and north-western Africa, and in some regions of southern Europe. The Early Permian continental salt and red beds are known from the Chu-Sarysu Basin. The Late Permian evaporite and red beds occur in the Central European, East European, Chu-Sarysu, Karasu Ishsay, Darvaza, Moesian, Dobruja, Rakhov, Mecsek, Alpine, North Italian, Dinarids, Arabian, Midcontinent, and East Greenland Basins. Additionally terrigenous red beds without evaporites occur in Peru and Chile in western South America, and in the westernmost Tethys, southern Spain, and north-western Africa. The Early Permian volcaniterrigenous red beds are widespread in the Central European Basin and in more southern regions of Europe.

During the Early and Late Permian continental alluvial-lacustrine gray-colored beds and coal-bearing terrigenous deposits covered vast areas of Angaraland, the Chinese-Korean mainland, Africa, Antarctica, Australia, and India. The Permian glacial deposits are at present found in many regions, within both the southern and northern hemispheres. In the northern hemisphere they were found in India and in northern Pakistan, as well as in north-eastern Asia, and in the southern hemisphere in Antarctica, South America (in the Andes within Chile, Bolivia, and Argentina, in southern Brazil and northern Uruguay), central and southern Africa, Australia, New South Wales, and Tasmania.

From the brief account of the distribution of various types of deposits it follows that either evaporites and terrigenous red beds, or terrigenous gray beds and coal-bearing formations of continental and marine origin, often associated with glacial terrigenous sediments, mostly accumulated in the Early and Late Permian sedimentary basins. Carbonate deposits, extremely widespread during the Early Cambrian, Middle and Late Ordovician, recede into the background. This, undoubtedly, was caused by a considerable increase of land masses, and thus the Permian period is generally known as geocratic.

Let us now discuss the distribution of climatic zones of the second half of the Early Permian and the Late Permian in the reconstructions not invoking continental drift. Many investigators gave great attention to such reconstructions (Strakhov 1960; Stehli 1963; Opdyke 1966; Helsley and Stehli 1968; Lotze 1968; Meyerhoff 1970a,b; Sinitsyn 1970; Ivanov and Voronova 1972; Ustritsky 1972; etc). Let us first try to find out if the Permian climatic zonation can be established in such a way that the zones have a latitudinal position relative to the recent equator, as A. Meyerhoff assumes, suggesting that in the Phanerozoic no displacements of continents and no polar wandering and displacement of the equator took place.

The Permian evaporites in North America occur between 25° and 85° N. i.e., occupy almost the entire continent. Terrigenous red beds occur in the same latitudes. Thus, the presence of a hot arid climate in the Early and Late Permian seems to be proven for this region. This belt of arid climate can be traced eastward on the basis of the presence of red beds and evaporites in Greenland, Europe, north-western Africa, Turkey, Siberia, and on the Arabian Peninsula. The Central European and East European salt basins at present lie between 45° and 70° N. Such Late Permian salt and sulfate basins as the Alpine, Moesian, Dinarides, Arabian, etc. lie farther south between 15° and 40° N. Such a distribution of evaporites and red beds in Europe and in the Middle East shows that the arid climate belt might be extended from North America parallel to the present equator to the Urals. However, this belt cannot be traced farther. The arid zone might also be distinguished in south-western Angaraland, because in Central Kazakhstan Early and Late Permian continental evaporites and red beds are known in the Chu-Sarysu, Karasu Ishsay, Darvaza Basins, and in north-western China, where variegated and red beds occur within the Tarim Massif and Ordos. However, in more eastern regions of the northern hemisphere coal-bearing gray formations are ubiquitously developed, marking a humid zone of sedimentation all over Angaraland and South East Asia. These data do not allow us to distinguish an arid climatic belt as extending through the entire northern hemisphere. It is not so simple when the southern hemisphere is considered. There the Permian arid zone of red beds and evaporite sedimentation can be distinguished only within South America where it is traced between the equator and 40° S. On other continents of the southern hemisphere coal-bearing and glacial deposits occur at the same latitudes, delineating temperate damp or cold climatic areas. Such an areal distribution of humid Permian deposits in the southern hemisphere, naturally, does not allow us to outline, even approximately, a southern arid zone. However, it might be proposed for those regions of Africa where Permian sediments were not recorded.

In the southern hemisphere, against the background of the recent position of continents, a very wide belt of temperate and cold climate is clearly established, comprising southern South America, Africa with Madagascar, Australia, and Antarctica. However, the data available do not enable us to establish a similar belt of temperate and cold climate in the northern hemisphere. A similar belt can be proposed only for North and North-East Asia (Ustritsky 1972) on the assumption that the North Pole was displaced farther south-east than its present position. All the discrepancies do not allow us to draw Permian climatic belts so as to correspond to the recent climatic zonation.

Numerous reconstructions using recent distribution of continents, but presupposing a different position of the Permian equator and poles in general give a similar pattern. Thus, Strakhov (1960), assuming that the old climatic zonation was essentially similar to the recent one, distinguished the northern moderate humid zone, northern arid belt, tropical humid, southern moderate humid, and southern circumpolar zones for the close of the Lower Permian and beginning of Late Permian. They were as follows: northern moderate humid zone comprising north-eastern Asia and north-western North America where coal-bearing and gray-colored sediments are developed; a northern arid belt encompassing all the areas of evaporite and red bed sedimentation in North America and Eurasia; and a tropical humid zone traceable through the Tethys and comprising the areas of coal-bearing and glacial deposits in India and Pakistan. A southern arid belt is established tentatively on the basis of two widely spaced regions of halogenic rocks in South America and Australia; a southern, moderate humid zone comprising areas of terrigenous gray-colored beds and coal-bearing deposits of South America and southern Africa; a southern circumpolar zone distinguished on the basis of glacial deposits occurring in southern regions of South America, southern Africa, and Antarctica.

N.M. Strakhov pointed out that the weakest point of these reconstructions is a recognition of an equatorial tropical zone (in Figs. 32 and 33 of Strakhov the equator is shown by a solid dashed line). He emphasized that the equatorial tropical zone is "a hypothetical zone for the lack of facial control" (Strakhov 1960, p. 183). Nevertheless, the equatorial tropical zone is of particular significance in such reconstructions, for the strike of all the other climatic zones is shown in accordance with that of the equatorial zone. The latter zone being hypothetical, the strike of all the other zones cannot be regarded as reliable, either. This is especially obvious from the southern arid belt. The existence of the western part of this belt confined to the South American Permian arid region is beyond any doubt, whereas the eastern Australia part resulted from erroneous data by Lotze concerning the occurrence of the Permian gypsum in the Canning Basin of Australia. Later Lotze (1968) amended this error. It is well known at present (Brown et al. 1970) that in the western basin of Australia Permian sediments are mostly represented by gray-colored beds among which glacial formations are widespread. It is also quite obvious that it is impossible to extend the southern arid belt eastward from South America, as was done by N.M. Strakhov. A position of the southern moderately humid zone seems quite improbable for it is distinguished in regions where sediments of glacial origin are widespread.

A position of the northern arid and northern moderately humid zones determined by Strakhov (1960) was confirmed by many workers. For example, Lotze (1968) delineates a Permian northern evaporitic belt in a similar way. A.A. Ivanov (Ivanov and Voronova 1972) places the northern arid zone within the same boundaries. Sinitsyn (1970) gave similar reconstructions for Eurasia. He distinguishes four climatic regions there: (1) Fenno-Sarmatian-South Tobol, of arid sedimentation, comprising all the areas of red beds and evaporite development in Europe, Kazakhstan, and Central Asia; (2) Siberian, of boreal-humid sedimentation; (3) an area of humid-tropical sedimentation of the East Sino-Gobi and Katasia; (4) Gondwana with boreal-humid sedimentation in the Early Permian and tropical-humid sedimentation in the Late Permian. The Fenno-Sarmatian-South Tobol arid region coincides completely with a northern arid zone proposed by N.M. Strakhov, and the Siberian boreal-humid with his northern moderately humid zone. Sinitsyn draws an important conclusion concerning a spatial position of the Permian equatorial tropical belt. In his opinion, the Fenno-Sarmatian-South Tobol arid region together with the humid-tropical region of Sino-Gobi and Katasia formed a single tropical European-Chinese belt. Two aspects are noteworthy. Firstly, it shows that in Permian time an equatorial tropical zone could not be outlined in the place proposed by Strakhov, but rather north of it (in Figs. 32 and 33 this position of the equator is approximately shown by a solid line); and, secondly, in the Permian epoch the distribution of climatic zones could be slightly different from the present one.

As was indicated by Strakhov and Sinitsyn, it is rather difficult to distinguish a Permian tropical belt in regions of occurrence of glacial and coal-bearing deposits in India and Pakistan. Most workers assume that these rocks accumulated in the cold climatic zone, and on the basis of Gondwana type fauna and flora found in glacial conglomerate breccias unite these regions with the other southern continents forming a single supercontinent, Gondwanaland, separating the Hindustan Peninsula from the Tethys. Opposed to this, N.M. Strakhov (1960) has to assume that glacial deposits of India formed during "mountain type glaciations resulting from rather high uplifts in . . . the equatorial zone" (p. 181). These glacial deposits make Sinitsyn (1970) assign the Tethys to the Gondwana boreal-humid realm and suppose that "the severe climate of Hindustan in Early Permian time was caused by a short-term existence of an ice cover in the highest parts of the mountains" (p. 107), and the origin of glacial conglomerate breccia çan be explained "in terms of mudflow genesis" (p. 55).

It should be noted that a climatic zone confined to northern and north-eastern Eurasia is determined differently by N.M. Strakhov and V.M. Sinitsyn. The former regards it as a moderately humid zone, whereas the latter as a boreal humid zone. At first sight these differences do not seem very significant. But they will look otherwise if we refer to Ustritsky (1972) who considers that the same area was boreal throughout the Permian; he outlines it as a northern Arctic zone and places the North Pole in the lower Lena area. According to Ustritsky, southward the boral realm just grades into a tropical area. This assumption may also imply that distribution of climatic zones in the Permian differed greatly from the present one.

The above ideas show that it is impossible to obtain an unambiguous Permian climatic zonation even for the northern hemisphere if reconstructions do not invoke continental drift. The equatorial and tropical zones can be distinguished in different ways. The problem of existence of separate moderately humid and Arctic belts has not been solved. Many problems arise as to the areal distribution of arid zones and the northern evaporite belt. It becomes even more complicated if data on the southern hemisphere are taken into account, since it is impossible to correlate climatic zones distinguished in the north with a distribution of sediments serving as climate indicators in the south without their displacement. One of the major drawbacks of the above assumptions lies in the fact that they, like previous assumptions, do not take into account outlines of old continents, distribution of sea and oceans on the Earth, direction of sea currents, and atmospheric circulation, i.e., those factors which strongly affect climate and can lead to a quite different distribution of climatic zones through geological epochs.

Reconstruction Which Invokes Continental Drift

Let us discuss how the problem of a Permian climatic zonation is solved in reconstructions invoking continental drift. The litho-paleogeographic maps constructed for the second half of the Early Permian (Fig. 34) and for the Late Permian (Fig. 35) show a considerable change in the distribution of land masses and oceans over the Earth as compared to earlier reconstructions not invoking continental drift. When the North and South America, Africa, Antarctica, Australia, the Hindustan and Arabian

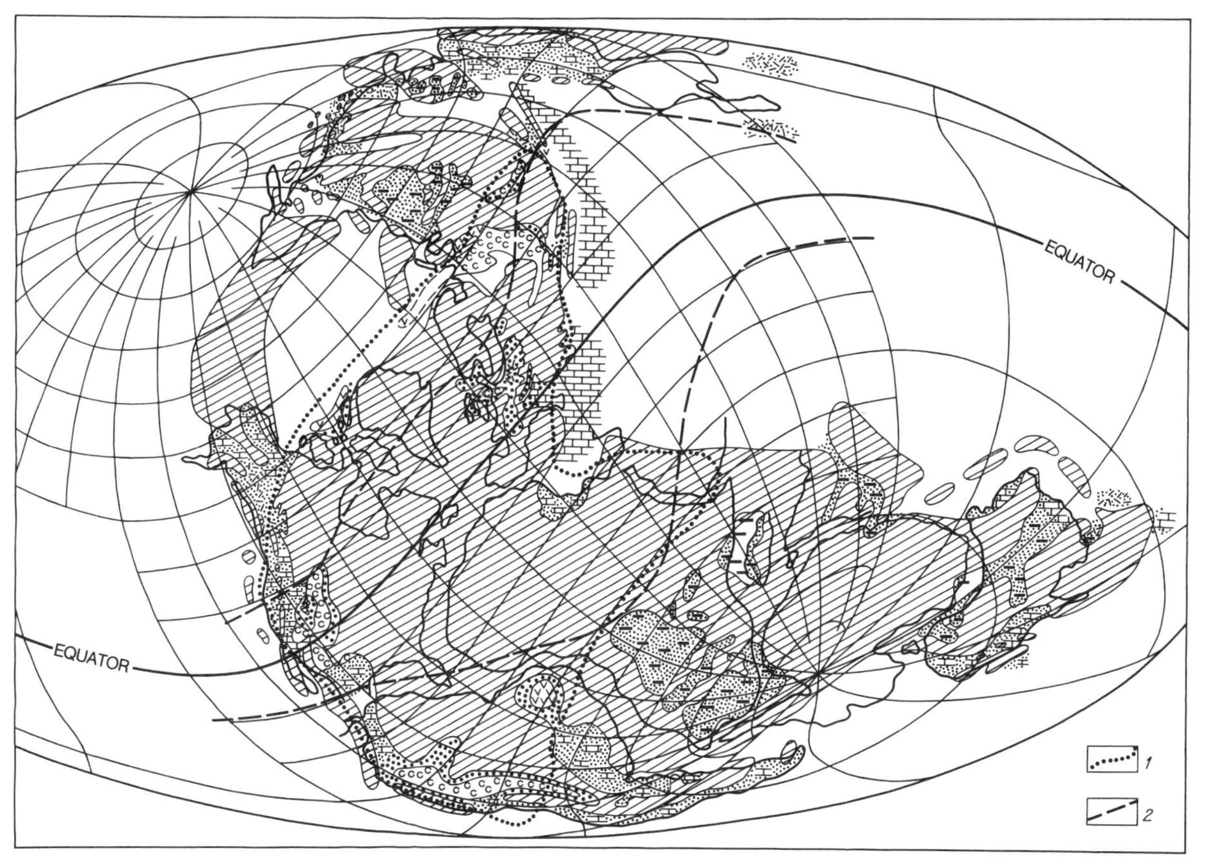

Fig. 34. Litho-paleogeographic map which invokes continental drift. Early Permian epoch (Artinskian and Kungurian ages). Mollweide projection.

1 boundaries of the arid belt, *2* boundaries of the Intertropical Convergence Zone (ITCZ) from Robinson (1973). For other explanations see Figs. 26, 29, and 32

Peninsulas formed a single supercontinent, Pangaea, it displaced southward and lay mostly in the southern hemisphere according to reconstructions proposed by Bullard et al. (1965), Smith and Hallam (1970), and Creer (1973). A vast oceanic basin, Tethys, occurs east of Pangaea within the pre-equatorial zone and together with the Pacific. As before, an epi-continental sea is shown within the Arctic Basin; this sea is separated from the Pacific by a relatively narrow land mass, probably a group of islands. The South Pole lies in Antarctica, and the North Pole in the north-western Pacific, not far from Kamchatka.

By such an arrangement of continents and oceans the climatic zonation correlates well with a spatial distribution of rocks, the indicators of climate. Areas of Arctic climate confined to the polar zones are clearly distinguished. The southern Arctic area comprises all the coal-bearing and glacial deposits of South America, southern Africa, Australia, and India. It extends from the pole to about 45°–50° S. Dimensions of the northern Arctic region prove to be the same; judging by the presence of coal-bearing and partly glacial deposits in northern and north-eastern Asia this region also extends from the North Pole to 40°–45° N. Both the northern and southern Arctic regions seem to be not entirely boreal, but also occupied temperate humid climatic zones and, thus, it would be more correct to call them not Arctic, but boreal-temperate, as was proposed by V.M. Sinitsyn for the Siberian area.

A broad warm climatic belt is postulated between 40° and 45° S between northern and southern boreal-temperate regions. To this belt all the Early and Late Permian evaporite basins are confined, as well as areas of terrigenous red beds and areas of carbonate accumulation within marginal seas of the Tethys and in the Rocky Mountains in western North America. Salt and sulfate deposits are not confined to any particular zone within a warm belt and do not form separate evaporite zones. They occur both near the equator (Midcontinent, Supai, North Mexican, Alpine, Central European, Moesian Basins, etc.) and at middle latitudes, between 20° and 35° N (East European, Chu-Sarysu, Karasu Ishsai, Darvaza, East Greenland Basins) and 25° and 53° S (Peru-Bolivian, Parnaiba, Rio Blanco Basins, etc.). The same holds for red beds. In general, if all the areas of evaporite and red bed sedimentation are united, we shall find a single belt of arid climate with complex boundaries extending through Pangaea from south-west to north-east across the equator. This arid belt does not correspond to the location of the present arid climatic zones relative to the recent equator, which undoubtedly should be accounted for by different outlines of continents and oceans and their different distribution over the Earth.

This conclusion was confirmed by Robinson (1973). Let us discuss in more detail her arguments, since they are of great importance for paleoclimatic reconstructions.

P. Robinson emphasized that climatic zonation on the Earth is caused not only by general regularities in atmospheric circulation and directions of currents in the hydrosphere, but also by the location and dimensions of continents. At present it is known that climatic belts are established using the existing extensive zones of an overall atmospheric circulation. Two sub-tropical high pressure belts lie at about 30° north and south of the equator. Masses of dry air come into these belts from the upper troposphere, forming two major systems of wind movement. One wind system (trade winds) is directed toward the equator, and the other toward the poles. The trade winds merge near equatorial areas forming the Intertropical Convergence Zone

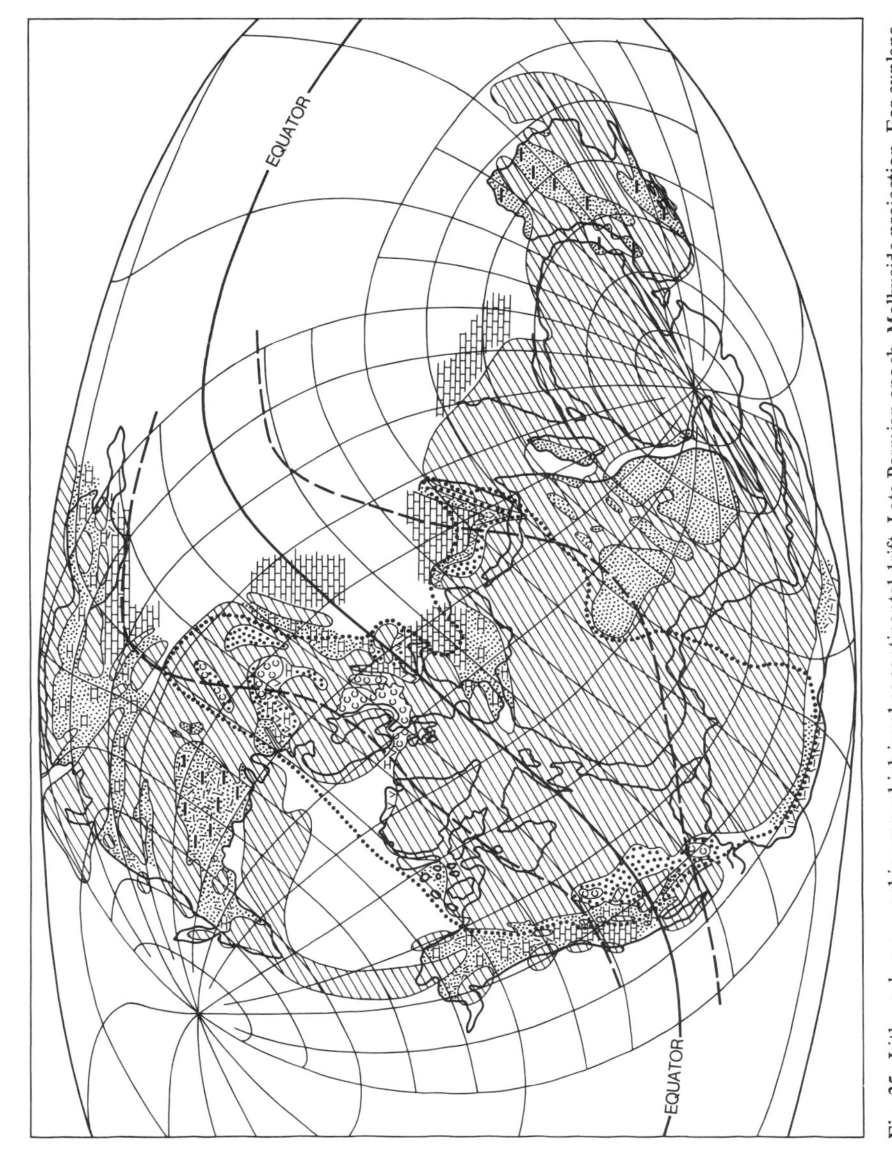

Fig. 35. Litho-paleogeographic map which invokes continental drift. Late Permian epoch. Mollweide projection. For explanation see Figs. 26, 28, 32, and 34

(ITCZ) or the meteorological equator (Lukjanov 1973; Robinson 1973; Samoilenko 1973). These three belts (northern high-pressure, Intertropical Convergence Zone, and southern high-pressure) exert particular influence on the formation of tropical and sub-tropical zones of arid climate and the equatorial humid zone. Shape and position of the belts change with seasons and are influenced by some other factors. The high-pressure belts are widened off western coasts of recent continents. The boundaries of the Intertropical Convergence Zone are displaced northward during the northern summer (reaching maximum in July), and southward during the winter (with maximum displacement in January). Thus, the ITCZ is a wide zone encompassing equatorial and sub-equatorial climatic belts. Additionally, boundaries of the ITCZ are influenced by the dimensions and position of continents. The effect of large continents may be exemplified by Eurasia. There, the winter in Siberia is characterized by another high-pressure maximum resulting in a dry winter in South East and East Asia. In summer a low-pressure belt forms throughout the whole of Eurasia, and affects the position of the adjacent boundary of the ITCZ. It appears to be displaced farther north toward the pole, relative to its usual position. The same is observed on other continents. In general, summer conditions on large continents result in displacement of their ITCZ poleward. According to Robinson this phenomenon should be taken into account determining paleoclimates in the geological past, since the development of the high-pressure maxima similar to those of Siberia might take place in winter seasons on large continents.

Robinson (1973) pointed to the influence of the large continents on climatic zonation for Late Permian time. Primarily, she emphasized the existence of the supercontinent Pangaea in the Late Permian, formed as a result of junction of Gondwanaland and Laurasia. The equator crossed the northern part of this supercontinent. Major land masses were displaced into the southern hemisphere. This continent is inferred to have had winter high-pressure maxima and summer deviations of the Intertropical Convergence Zone comparable to present-day maxima of Eurasia: the first in the inner part of Gondwanaland and the second, a smaller one, in Laurasia. During the summer period in Laurasia and the winter period in Gondwana trade winds from a high-pressure center of Gondwanaland spread far beyond the paleoequator due to the northern displacement of the ITCZ. Winds were directed toward the eastern part of southern Laurasia, passed the Tethys, and brought summer monsoon rains to the coastal areas. Winds directed toward the western part of southern Laurasia, in their turn, were always overland, which resulted in a dry summer in these regions. The opposite is observed during the winter period in Laurasia and the summer period in Gondwana. Trade winds in January blew over the Tethys toward the eastern part of northern Gondwanaland bringing summer monsoon rains to South Arabia, and western and northern India, whereas western and central Gondwana had an arid summer due to the influence of trade winds blowing overland. Thus, the western and central Pangaea, lying between two tropics, showed a tendency toward arid climate both in winter and in summer.

Additionally, strong southward ITCZ deviation recorded for Gondwana could give rise to arid climate setting up to 50° S, i.e., in western Argentina, North Brazil, and in Central Africa. The arid zone at high-pressure northern latitudes comprised part of Alaska, North Greenland, and the Canadian Arctic Archipelago, whereas the arid zone of high-pressure southern latitudes covered northern South America within Ecuador, Peru, northern Chile, Bolivia, and northern Argentina.

A marked asymmetry in the arrangement of continents relative to the paleoequator could result in a strong influence of oceanic currents on climate in the two regions. One of these regions lay along the eastern boundary of Gondwanaland. There southern equatorial current, of east-west direction, could diverge first southward at low latitudes along the coast off India and then eastward, into the Australian part of the coast. Although the direction of the southern equatorial current could change with the seasons, it always passed at low latitudes, bringing warm water along the eastern coast of Gondwana. Robinson believes that this current might be called the Gulf Stream of the southern hemisphere of Late Permian time. Thus, the Indian part of Gondwanaland could have had a temperature rather high for these latitudes. To a lesser extent, this holds for North Australia. India and Australia, situated not far from the Tethys, in general had a more rainy climate than that of the eastern coasts and middle latitudes. The eastern coasts of Asian Laurasia probably also had a mild winter owing to the warm current coming from adjacent high-pressure centers. The Antarctic-Parana Sea could also have had higher winter temperatures than those of land masses lying in the southern Arctic region.

The reconstructions by Robinson allowed us to distinguish the following major climatic areas for Late Permian time: (1) an arid area within the continent near the paleoequator extending to 30° N and about 50° S in central and western Gondwana, (2) monsoon areas with humid summers and arid winters lying mostly in southern Laurasia and Arabia, (3) humid belts of middle and high latitudes confined to northern and north-eastern Asia, Antarctica, Australia, southern South America, southern Africa, and India.

It is very important to emphasize that, according to Robinson, an arid area was situated not only within the tropical belt, but also covered the intracontinental zones of the equatorial belt due to a different arrangement and dimensions of Permian continents. It means that in the Permian a spatial position of climatic zones differed from the present one. Without due regard for this fact we might make serious mistakes in determining climatic zonation using rocks, the indicators of climate. It is evident, for instance, that an arid zone of Eurasia and North America, regarded by N.M. Strakhov as a northern zone, might, in fact, lie near the equator, whereas the equatorial tropical zones distinguished by Sinitsyn and Strakhov (confined to the Tethys) might, in fact, mark temperate climatic humid belts.

The reconstructions of P. Robinson correlate well with the distribution of evaporites, red beds, coal-bearing and other deposits shown on maps which invoke continental drift. According to P. Robinson, the boundaries of the Intertropical Convergence Zone, marking approximately areas of arid climate of Pangaea, are shown by solid dotted lines on maps of Figs. 34 and 35.

Conclusions Concerning Paleoclimate and Paleogeography of Evaporite Basins During the Permian

The analysis of various reconstructions concerning paleogeography of continents and paleoclimatic zonation shows quite definitely that climatic belts for the second half of the Early Permian cannot be distinguished if continental drift is not invoked, even on the assumption of a different position for the poles and the equator. It is the climatic zonation reconstructed with regard to paleomagnetic data and continental drift that correlates best of all with the evidence available; thus, it might be at present regarded as the most valid zonation. These reconstructions suggest that Early and Late Permian evaporite belts may be considered as single latitudinal zones encompassing all the areas of arid sedimentation in Gondwanaland and Laurasia between 50° S and 30° N.

The paleogeography of the Permian salt basins has not changed essentially since the Devonian stage of salt accumulation. They were mostly epi-continental seas. However, some major differences can be mentioned. Firstly, Permian salt seas often lie in the inner parts of continents and were surrounded by land on all sides. They were linked with the open sea by narrow straits. These are, for instance, the Central and East European Basins. Even the marginal salt basins, similar to that of Midcontinent, cut deep inland and were separated from the sea of normal salinity by intermediate basins of archipelago type with a series of extensive islands dissected by one or several narrow straits through which sea water came into salt-producing zones. The appearance of deeply embayed salt basins and intermediate basins of strait type is a peculiar feature of paleogeography of the Permian salt accumulation stage. Secondly, in the Permian period there were typical intracontinental basins not linked with the sea where salt accumulation was caused by evaporation of continental surface water. The Chu-Sarysu Basin provides an example of such basins. Such a paleogeography is accounted for by a general increase of land mass resulting in the formation of a supercontinent, Pangaea. It caused considerable changes of climate and led to an abrupt increase in both arid and humid precipitation, influencing also the composition of salt deposits.

The relationship between epi-continental salt basins and seas in the Permian period was not so broad as in the Devonian and Cambrian. Most evident is the supply of water of normal salinity into the Midcontinent Basin linked with the Pacific. The North Mexican, Peru-Bolivian, Rio Blanco, and Parnaiba Basins were also recharged from the Pacific. The relations of the Central and East European Basins with the ocean are not so clear. The data available demonstrate that during the Early and Late Permian the Central European salt basin could have been linked with the North Sea, and by this sea with the Pacific Ocean (Richter-Bernburg 1955, 1957, 1959, 1972; Ustritsky 1967; Dutro and Saldukas 1973; Watson and Swanson 1975; Ziegler 1975). However, we should not deny the possibility of sea water supply into the Central European Basin from the Tethys through narrow straits. During the Permian period the East European salt basin was also linked with the North Sea (Strakhov 1962; Ustritsky 1967; Fiveg and Banera 1968; Grunt 1970; Tikhvinsky 1971; Fiveg 1972; Ivanov and Voronova 1972). At the same time quite reliable data are available showing that this basin was linked with the Tethys in the first half of the Permian, up to the Kungurian. Thus, Grunt (1970) emphasizes that at the close of the Artinskian and in the Kungurian brachiopod faunas of the Boreal and Tropical realms were mixed. The migration of brachiopods took place in north-east Asia, where broad relationships between the Arctic and Tethys were renewed. Dutro and Saldukas (1973) suggested relationships through the East European Basin. Sea water from the Tethys was supplied into the Alpine and Moesian salt basins.

The above evidence allows the conclusion that sea water of normal salinity was supplied into salt basins of North America and Eurasia either from the Pacific or from the Tethys.

Main Paleogeographic and Paleoclimatic Features of Paleozoic Evaporite Sedimentation

The data presented enable the following conclusions concerning the paleogeographic setting of salt accumulation during the Paleozoic.

Most Peleozoic salt basins were epi-continental seas. This conclusion is valid for Early Cambrian, Middle and Late Devonian, Early and Late Permian salt basins discussed in this chapter and for Middle and Late Cambrian, Ordovician, Silurian, and Carboniferous salt basins as well.

Paleogeography, dimensions, and outlines of Paleozoic epi-continental salt basins changed with time and depended on their distribution on continents and oceans. Vast Cambrian salt basins were predominantly marginal seas. They were separated from the ocean by a relatively narrow but extensive group of islands. These basins are similar to a certain degree in size and position to the recent marginal seas of the Pacific Ocean, such as the Bering and South China Seas and the Sea of Okhotsk, but they were situated on continents. Cambrian salt basins can be regarded as seas because for a long time they were connected with the ocean, being separated only during short periods of salt accumulation.

Starting from the Silurian and especially since the Eifelian age, paleogeography of salt basins changed. They occupied marginal parts but within continents and became similar to inland marine basins. There were no marginal salt seas in the Middle and Late Paleozoic. Ordovician, Late Silurian, and Devonian salt basins were rather widely connected through intermediate basins of archipelago and barrier type.

Permian salt basins which became mainly typical inland seas were situated already in interior parts of continents. They were connected with the ocean either by narrow seaways or by intermediate basins of archipelago type. Salt accumulation in the Permian was greatly influenced by fresh water supplied from continents.

Continental basins appeared along with marine salt basins during the second half of the Paleozoic. The Chu-Sarysu Basin, situated entirely in a continent, provides a typical example. Salt deposition took place there owing to water of continental origin.

The changes mentioned in the paleogeography and type of salt basins were caused by redistribution of continents, oceans, and seas. Throughout the Paleozoic era in any reconstructions invoking or not continental drift, one can see an overall increase in land masses and subsequent reassembly of separate continents and land masses into a single supercontinent, Pangaea. During the Cambrian period three relatively large continents existed: Gondwanaland, America, and Fennoscandia. The other land masses (Katasia, Beringia, and others) were much smaller. Vast areas of the Earth were occupied by oceans (Pacific, North, Pre-Atlantic). Most of Eurasia, North America, Australia, South America, and Antarctica were covered by seas. There were broad relationships between oceans and marine basins.

Salt basins under such paleogeography were during the Cambrian predominantly very large marginal seas.

Their size increased greatly during the Devonian. A large continent, Laurasia, came into being in the northern hemisphere. An even larger continent, Gondwanaland, continued to exist in the south. New land masses similar to Angaraland and Malaysia took shape. The area under seas of the Tethys decreased. In reconstructions which invoke continental drift Laurasia drew closer to Gondwanaland. The oceanic basin remained only within the Pacific. At that time the Atlantic and Arctic Oceans apparently did not exist. The overall land area was much larger than that of the Cambrian. It is probably due to all these transformations that salt accumulation in the Devonian took place mainly in inland seas situated along continental margins. However, connections between open seas and salt basins still remained wide. Conditions favorable for carbonate sedimentation and formation of biogenic buildups have set in in intermediate basins.

A supercontinent, Pangaea, came into existence in the Carboniferous and mainly in the Permian. Most of Asia was occupied by land. Oceanic basins continued to exist in the Pacific and probably in the Tethys. At that time areas of salt accumulation were displaced farther inland and were almost completely situated in interior parts of continents. Their connection with oceans was not so broad. Large amounts of fresh water was supplied into salt basins from continents. The formation of continental salt basins began at that time.

It was mainly due to influx and evaporation of marine water that salt was deposited in almost all of the epi-continental Paleozoic salt basins. Continental water was of importance only in some Carboniferous and Permian basins. Water of normal salinity came into salt basins mainly either from the Pacific or from the Tethys. Even during the Early and Middle Paleozoic when an ocean basin could still exist in the northern part of the present Atlantic it did not supply warm salt water into evaporite basins of the northern hemisphere.

The reconstructions discussed above of paleoclimatic zones of Early Cambrian, Middle and Late Devonian, Early and Late Permian when continental drift is not involved show that in most cases one cannot restore climatic zones in such a way as to make them strike parallel to the present equator and latitudinally relative to the present climatic belts. Such a reconstruction can be done only for the Early Cambrian. Since it is impossible to prove the validity of such reconstructions for the remaining stages of the Paleozoic, one can conclude that if they were made it was by mere chance. To obtain a more or less clear picture of an ancient climatic zonation a different arrangement of the equator and poles or continental drift should be inferred.

Reconstructions not invoking continental drift but suggesting a different arrangement of the equator and poles cannot provide an unambiguous pattern of paleoclimatic zonation. One of the major restraints of these reconstructions is that the arrangement of ancient climatic zones coincides with the present one, and no assumptions are made for another possible arrangement of arid and humid areas which could be controlled by dimensions of continents and oceans in the geological past and their distribution relative to the ancient equator. This enables us to group in different ways arid and humid areas into belts depending on the interpretation of the material available, which is quite obvious from reconstructions for Devonian and Permian time.

Paleoclimatic zonation seems best verified in reconstructions which invoke continental drift. They allow the conclusion that arid and humid zones in the Paleozoic could be arranged in quite a different way relative to the equatior than at present due to different outlines, dimensions, and distribution of ancient continents, oceans, and seas, direction of currents and zonal atmospheric circulation. All this should be taken into account when the present arrangement of climatic zones is projected into the geological past. In this respect reconstructions for the Late Permian are the most representative. They imply that an arid climatic belt was situated in near-equatorial regions of Pangaea between 50° S and 30° N, as if areas of hot dry climate using nomenclature of the present climatic zones had occupied not only tropical but equatorial and sub-equatorial belts. This implies a probable absence in the Permian period of two arid belts that we have now, which is why their establishment seems unlikely. Apparently only one belt of arid climate existed in the Permian which at the same time was a belt of evaporite sedimentation.

Using the data available for the Permian it is possible to specify the distribution of arid climatic zones and belts of evaporite sedimentation for the Devonian period. In reconstructions which invoke continental drift both Middle and Late Devonian climatic zonations differ from the present one. Arid areas lie near the equator while humid areas occur in place of the present tropical zones. Such a zonation must have existed by a certain arrangement of continental position relative to the paleoequator. The supercontinent Gondwana, being in the southern hemisphere, lay mainly south of 30° S. The southern belt of high pressure ran, except for Australia, through oceans and seas. Laurasia was partly situated at the equator close to Gondwana. The northern belt of high pressure was above central Laurasia where evaporites and red beds are widespread. In Australia conditions were also favorable for an arid climate. This evidence suggests for the Devonian the existence of a single belt of arid sedimentation stretching from south-west to north-east and embracing marginal and central parts of Laurasia. On the basis of the present climatic zonation this belt lay within equatorial and tropical zones.

It is impossible to determine the effect of size and position of a continent relative to the paleoequator on the climatic zonation of the Early Cambrian. Outlines of ancient continents can be established only hypothetically. The same holds for ancient oceans and seas. However, a climatic zonation restored by means of paleomagnetic data seems most probable. It suggests that the Cambrian areas of arid climate and evaporite sedimentation were situated either in an equatorial zone or within tropical belts but not farther than 30° to 35° on either side of the equator. There is no evidence at present for the recognition of Cambrian evaporite belts. Reconstructions which invoke continental drift imply the entire absence of such belts in the Cambrian.

Throughout the entire Paleozoic, evaporite deposits were associated climatically with certain areas. In reconstructions which invoke continental drift, and based on paleomagnetic data, they are always situated in a near-equatorial zone. Briden and Irving (1968) paid attention to this feature when they studied paleolatitudes of evaporite deposits. They stated that in the Paleozoic paleolatitudes were concentrated at the paleoequator, in the Mesozoic northern paleolatitudes predominated, while in the Cenozoic all the paleolatitudes were temperate northern. According to J. Briden and E. Irving, recent continental evaporite deposits show peaks at about 25° S and 40° N.

Such a distribution of evaporite deposits also implies a different arrangement of areas of halogenic sedimentation relative to the paleoequator in the geological past as compared to the present-day arrangement. However, they always remained within a warm climatic belt. Temperate and cold climatic zones lay north and south of the marine belt. This evidence suggests an existence of an overall latitudinal climatic zonation throughout the entire geological history, which is confirmed by laws of physical paleoclimatology (Bernar 1968). Therefore, the width of a Paleozoic near-equatorial warm belt may be roughly used as an indicator of an overall warming or cooling of the Earth's climate with regard to dimensions and distribution of land areas on the Earth's surface.

The main conclusions are as follows:

1. The most valid climatic zonation may be restored when continental drift is invoked and paleomagnetic data are used.

2. The Paleozoic paleoclimatic zonation and distribution of arid and humid areas could only differ greatly from the present one due to different dimensions of continents, oceans, and seas, their position relative to the paleoequator, directions of currents and effect of zonal atmospheric circulation; therefore in reconstructions all these changes are to be taken into account.

3. A latitudinal climatic zonation relative to the paleoequator persisted throughout the Paleozoic which allows us by a comprehensive analysis to single out a wide tropical zone and belts of temperate and cold climate.

4. During the Paleozoic evaporites always accumulated in areas of arid climate, however the distribution of belts of evaporite sedimentation, if they existed, was different from that of the present. In particular, in the Permian and Devonian there was only a single belt of evaporite sedimentation; apparently it did not exist at all in the Cambrian and halogenic deposits accumulated in major isolated areas of arid climate.

Epilog

Evolution of Evaporite Sedimentation in the Paleozoic

There are two major problems to be solved with respect to the evolution of evaporite sedimentation in the Earth's history. Firstly, the evolution of geological environment of halogenic sedimentation (paleogeographic, paleoclimatic, and paleotectonic changes in environments of sulfate, halite, and potassium sedimentation, changes in intensity of halogenic sedimentation with time, etc.) and, secondly, changes in mineral composition of evaporite deposits caused by the various amounts of different rocks and minerals are to be established. Let us treat these problems separately.

All the material discussed in this book unambiguously points to changes in geological environments and intensity of evaporite sedimentation in Paleozoic history. However, these changes take place in a peculiar way and their impact is different on sulfate, halite, and potassium sedimentation.

These differences become obvious even by the recognition of stages of evaporite sedimentation. As to sulfate sedimentation in the Paleozoic, it was continuous from the Cambrian to the Permian, therefore the Paleozoic era is considered as a single salt epoch. Minor stages of sulfate deposition can be distinguished only on the basis of the area of distribution and extent of sulfate rocks during different periods, epochs, and ages of the Paleozoic history. As a result three major stages can be proposed: two stages marked by an important sulfate sedimentation (Early Cambrian and Middle Devonian/Permian) and one stage characterized by a restricted sulfate deposition (Middle Cambrian/Early Devonian). In the Paleozoic, halite accumulation, unlike sulfate deposition, is characterized by easily recognizable stages. The salt stages (Cambrian, Middle Devonian/Early Carboniferous and Permian) stand out sharply against the Paleozoic history of the Earth when deposition of salt beds took place in many basins. They alternated either with episodical salt stages or salt-free stages. Salt stages in turn differed in size of the salt basins then existing and in the overall continental area of salt accumulation, as well as in net mass of rock salt deposited. This evidence enabled determination of the intensity of halite accumulation and demonstration that in the Paleozoic it was absolutely nonuniform, therefore most rock salt accumulated during two epochs of immense salt accumulation, i.e., Early Cambrian and Kungurian. However, only the latter was an epoch of intense halite formation. Potassium sedimentation differed greatly from that of sulfate and halite. It is characterized by clear-cut stages when deposition of potash salt took place in a salt basin of a certain age.

The changes in paleogeography of sedimentary environments strongly affected sulfate, halite, and potassium sedimentation. Mainly it concerns salt accumulation environment, thus in the time interval from the Cambrian to Permian both types of salt basins and their links with oceans and seas of normal salinity changed. If in the Cambrian salt basins were mainly marginal seas separated from the ocean by intermediate basins of archipelago type then in the Middle and Late Paleozoic these basins ceased to exist. In the Middle Paleozoic there came into existence salt seas situated along continental margins; they were separated from the ocean by intermediate basins either of archipelago or barrier type. The latter is a paleogeographic feature known only since the Middle Paleozoic. In the Permian salt basins became predominantly inland seas situated in interior parts of continents. They were either separated from the ocean by archipelago intermediate basins or linked with it through intermediate basins of a new type, i.e., by straits. All these changes were related primarily to the evolution of paleotectonic conditions of salt accumulation.

Ancient platforms remained the most favorable sites for the formation of salt basins throughout the entire Paleozoic. In the first half of the Paleozoic (from Cambrian to Silurian inclusive) they were single tectonic elements favorable for halogenic sedimentation. Apparently there were no favorable conditions in Early Paleozoic geosynclinal or miogeosynclinal basins for evaporite sedimentation. Paleotectonically the distribution of salt basins changed greatly in the Devonian. After the emplacement of folded structures of Caledonian systems some of the pre-existing land masses became new areas of salt accumulation. However, peculiar features of Caledonian systems gave rise to the appearance of tectonically very peculiar salt basins situated within the inherited troughs and superimposed basins. In the Devonian great changes took place in paleotectonic conditions of salt accumulation on ancient platforms. Salt zones were associated mainly with graben-like depressions. The evolution of tectonic processes during the Paleozoic resulted in the formation of true intermontane basins and foredeeps only in the Carboniferous when Hercynian fold systems came into existence. Apparently there were no structures within Caledonian systems. The appearance in the Carboniferous of intermontane basins in zones of epigeosynclinal orogeny favored the formation of a tectonically new type of salt basin, intermontane basin type, which was not known in earlier periods of the Paleozoic. Salt basins in epiplatform orogenic areas take shape for the first time in the Carboniferous. The formation of a sedimentary cover of a number of younger platforms began within Hercynian and Caledonian systems at the close of the Paleozoic. Some of them, especially those in West Europe, became favorable for the initiation of salt basins. During the Permian paleotectonic conditions of salt accumulation changed greatly on ancient platforms as well. Salt zones were situated then in foredeeps and marginal syneclises or egsogonal depressions.

The changes in paleogeography and paleotectonic processes strongly affected potassium sedimentation. Most chloride potash pools of Devonian age formed in interior marine salt basins separated from seas of normal salinity by intermediate basins of barrier type. Salt accumulation took place there in zones most remote from source areas situated as a rule in intensely subsided structures: graben-like depressions (Pripyat Depression) or in depressions of syneclises (West Canadian Basin). These zones were separated from the open sea by areas of sulfate accumulation and reefogenic barriers. During the Carboniferous and Permian major accumulations of potash salt of chloride type predominantly formed either in intermontane basins (Paradox Basin) or foredeeps (Solikamsk and Pechora Depressions of the East European Basin) or within major syneclises of young platforms (Central European Basin). However, all these structures were, firstly, intensely subsided and, secondly, most remote from the sea.

The changes in paleoclimate strongly affected the entire process of evaporite sedimentation. However, it was shown that the paleoclimatic zonation in the Paleozoic could differ greatly from that of the present-day due to different outlines, dimensions, and arrangement of ancient continents, oceans and seas relative to the paleoequator. Reconstructions which invoke continental drift suggest that belts of halogenic sedimentation could have been completed during the early stages of the Paleozoic and evaporite basins could have occupied isolated areas on different continents. In all probability there was only one arid belt to which all salt and sulfate basins were related in the Middle and Late Paleozoic.

The second problem is: has or has not a mineral composition of evaporite deposits changed during the Paleozoic? This problem has been repeatedly discussed (Zharkov 1971a, 1974a,b, 1976, 1977).

Discussing the peculiar features of distribution with age or potash salt of chloride and sulfate type attention was paid to the fact that sulfate potash salts predominantly accumulated at the close of the Paleozoic, from the Kungurian to Tatarian age. During the earlier period in the Paleozoic potash salt of such a composition has not been found in these amounts in salt basins. This enables us to subdivide the Paleozoic history of potassium sedimentation into two unequal stages: (1) from the Cambrian to the first half of the Permian (Artinskian age inclusive) and (2) from the Kungurian to Tatarian age of the Permian. It was only potash salt of chloride type (sylvine and carnallite) which was deposited episodically throughout the first epoch. During the second epoch potash salt of sulfate type began to be deposited along with potash salt of chloride type. In other words regular changes in the mineral composition of potash salt can be easily distinguished in the history of Paleozoic potassium sedimentation. Additionally, starting with the Carboniferous, continental salt basins also appear where sodium-sulfate salt deposits accumulated (Zharkov 1971a).

References

Aaloe A, Kaljo D, Klaamann E, Nestor H, Einasto R (1976) Stratigraphical classification of the Estonian Silurian (in Russian). Izv Akad Nauk Est SSR 25: 38–45

Ahmad F (1969) The position of Australia in Gondwanaland reconstruction. In: Gondwana stratigraphy. IUGS Symposium Buenos Aires, 1–15 October. Unesco, Paris, pp 487–520

Ahmad F (1973) Have there been major changes in the Earth's axis of rotation through time? In: Implications of continental drift in the Earth sciences, vol I. Academic Press, London New York, pp 487–501

Aitken JD, Cook DG (1974) Careajon Canyon map-area, District of Mackenzie, Northwest Territories. Geol Surv Can Pap 74-13: 28 pp

Aitken JD, Macqueen RW, Usher JL (1973) Reconnaissance studies of Proterozoic and Cambrian stratigraphy, Lower Mackenzie River Area (Operation Norman), District of Mackenzie. Geol Surv Can Pap 73-9: 178 pp

Aizenshtadt GEA, Slepakova GI (1977) Some new data on the conditions of salt structures formation. In: Problems of salt accumulation (in Russian), vol II. Nauka Press, Novosibirsk, pp 26–29

Ala MA (1974) Salt diapirism in Southern Iran. Am Assoc Petrol Geol Bull 58: 1758–1770

Aleksandrova MA (1971) The Carboniferous System. The Eastern Bet-Pak-Dala (in Russian). In: Geology of the USSR, vol 40 (South Kazakhstan), Geological descriptions, B 1. Nedra Press, Moscow, pp 282–292

Aleksandrova MA (1973) Sarysu-Chu-Balkhash watershed (in Russian). In: Stratigraphy of USSR, Devonian System, B 1. Nedra Press, Moscow, pp 504–510

Aleksandrova MI, Borsuk BI (1955) Geological structure of the Paleozoic basement of Eastern Bet-Pak-Dala (in Russian). Tr VSEGEI Nov Ser 7: 304 pp

Aliev MM, Lausin NA, Korzh MV, Mkrtchyan OM, Orudzheva DS, Said A, Yakovlev BM (1971) Geology and oil and gas potential of the Algiers Sahara (in Russian). Nedra Press, Moscow, 329 pp

Allen JRZ, Dineley DL, Friend PF (1967) Old red sandstone basins of North America and North-West Europe. In: International Symposium on the Devonian System, vol I. Calgary, Alberta, Canada, pp 69–98

Alling HL (1928) The geology and origin of the Silurian salt of New York State. NY State Mus Bull 275: 139 pp

Alling HL, Briggs LI (1961) Stratigraphy of Upper Silurian Cayugan evaporites. Am Assoc Petrol Geol Bull 45: 515–547

Amsden TW, Caplan WM, Hilpman PL, McGlasson EH, Rowiand TL, Wise OAH (1967) Devonian of the Southern Midcontinent Area, United States. In: International Symposium on the Devonian System, vol I. Calgary, Alberta, Canada, pp 913–965

Anderson JJ (1965) Bedrock geology of Antarctica: A summary of exploration, 1831–1962. In: Geology and paleontology of the Antarctic, Antarctic Research Series, vol VI. American Geophysical Union, pp 1–70

Anderson RY (1968) A "type" stratigraphic time series for the Permian Castile Formation (abs.). Geol Soc Am Progr Annu Meet, Mexico City, p 6

Anderson SB, Eastwood WP (1968) Natural gas in North Dakota. In: Natural gases of North America, vol II. Am Assoc Petrol Geol Mem 9. Tulsa, Oklahoma, pp 1304–1326

Anderson SB, Hansen DE (1957) Halite deposits in North Dakota. North Dakota Geol Surv Rep Invest 28: 3 pp

Andrichuk JM (1958) Stratigraphy and facies analysis of Upper Devonian reefs in Leduc, Stettler and Redwater Areas, Alberta. Am Assoc Petrol Geol Bull 42: 1–93

Andrichuk JM (1960) Facies analysis of the Upper Devonian Wabamun Group in West Central Alberta, Canada. Am Assoc Petrol Geol Bull 44: 1651–1681

Asrarullah (1963) Rock salt resources of Pakistan. In: Symposium on industrial rocks and minerals, Lahore, Pakistan, December 1962. Central Treaty Organization, pp 303–313

Atlas of the lithological-paleogeographical maps of the Russian Platform and its geosyncline margins, Part 1: Late Precambrian and Paleozoic. Kasanian age (1960) (in Russian). Forsh NN (ed). Moscow, Leningrad

Atlas of the lithological-paleogeografical maps of the USSR, vols II, III (1969) (in Russian). Moscow

Atlas of lithological-paleogeographical maps of the Paleozoic and Mesozoic of the Northern Cis-Uralian Region (1972) (in Russian). Nauka Press, Leningrad

Atlas of paleogeographical maps of Ukrainian SSR and Moldavian SSR (1960) (in Russian). Izd Acad Nauk Ukr USSR, Kiev

Auboin J, Blauchet R, Cadet JP, Celet P, Charvet J, Chorowicz J, Cousin M (1970) Essai sur geologie des Dinarides. Bull Soc Geol Fr Ser 7 XII: 1060–1095

Baars DL, Parker IW, Chronic I (1967) Revised stratigraphic nomenclature of Pennsylvanian System, Paradox Basin. Am Assoc Petrol Geol Bull 51: 393–403

Bachman GO (1975) New Mexico. In: Paleotectonic investigations of the Pennsylvanian System in the United States, Part I. Geol Surv Prof Pap 853. US Government Printing Office, Washington, pp 233–243

Bailey EB (1948) Geology of the Salt Range of the Punjab. Nature (London) 265–266

Bain GW (1969) Climatic zone patterns through the ages. In: Gondwana stratigraphy. IUGS Symposium Buenos Aires, 1–15 October. Unesco, Paris, pp 651–671

Bakirov SB (1965a) Stratigraphy of the Upper Paleozoic deposits of the Maly Karatau Ridge and the Chuya Depression (in Russian). In: Tr Kazakh. PTI, B 25. Nedra Press, Leningrad, pp 148–154

Bakirov SB (1965b) On climatic peculiarities of the Maly Karatau in the Carboniferous (in Russian). Vestn Acad Nauk Kaz SSR 7: 47–51

Bakirov SB (1977) Paleogeographic conditions of salt accumulations in the Chuy Depression in the Permian (in Russian). In: Problems of Salt Accumulation, vol II. Nauka Press, Novosibirsk, pp 146–152

Bakirov SB, Vlasov VI, Li AB (1971) The Carboniferous System, Chuy Depression, Minor Karatau, Kirghiz Range (in Russian). In: Geology of USSR, vol 40 (Southern Kazakhstan), Geological descriptions, B 1. Nedra Press, Moscow, pp 267–289

Bancroft M (1957) Salt deposits at Malagash and Pugwash, Nova Scotia: In: Can Inst Miner Met Ind Miner Div. Geology of Canadian Industrial Mineral Deposits, pp 215–218

Barkhatova VP (1963) The Carboniferous system. Northern Russian Platform between the Baltic shield and Timan (in Russian). In: Geology of USSR, vol 2, (Arkhangelsk, Vologda Regions and Komi ASSR), part 1. Moscow, pp 347–420

Barth W (1972) Das Permokarbon bei Zudanez (Bolivien) und eine Übersicht des Jung-Paläozoikums im zentralen Teil der Anden. Geol Rundsch 61: 249–270

Bassett HG, Stout IG (1967) Devonian of Western Canada. In: International Symposium on the Devonian System, vol I. Calgary, Alberta, Canada, pp 717–752

Bayazitov SA (1963) Characteristics of the potassium salt deposits of Byelorussia, recording to exploratory work of recent years (in Russian). In: Potassium salt and methods of proceeding. Izd Acad Nauk USSR, Minsk, pp 3–14

Bazhenov ML (1976) Paleomagnetism and geotectonics (in Russian). In: Continental drift. Nauka Press, Moscow, pp 72–85

Belgovsky GL (1972) The Carboniferous System. Upper series (in Russian). In: Geology of USSR, vol 25 (The Kirghiz SSR), Geological description, B 1. Nedra Press, Moscow, pp 183–190

Bell WA (1960) Mississippian Horton Group of type Windsor-Horton District, Nova Scotia. Geol Surv Can Mem 314: 112 pp

Belt ES (1965) Stratigraphy and paleogeography of Mabou Group and related Middle Carboniferous facies, Nova Scotia, Canada. Geol Soc Am Bull 76: 777–802

Belyaevsky NA (1974) The earth's crust within USSR territory (in Russian). Nedra Press, Moscow, 280 pp

Belyea HR (1960) Distribution of some reefs and banks of the Upper Devonian Woodbend and Fairholme Groups in Alberta and eastern British Columbia. Geol Surv Can Pap 59-15: 7 pp

Belyea HR (1964) Upper Devonian, Woodbend, Winterburn and Wabamun Groups. In: Geological history of Western Canada. Alberta Soc Petrol Geol, Calgary, Alberta, pp 66–88

Benavides V (1968) Saline deposits of South America. In: Saline deposits. Geol Soc Am Spec Pap 88: 249–290

Benson DG (1967) Geology of Hopewell map-area, Nova Scotia. Geol Surv Can Mem 343: 58 pp

Benson DG (1970a) Notes to accompany geological maps of Antigonish and Cape George map-areas, Nova Scotia. Geol Surv Can Pap 70-8: 4 pp

Benson DG (1970b) Notes to accompany maps of the geology of Merigomish and Maeignant Cove map-areas, Nova Scotia. Geol Surv Can Pap 70-9: 4 pp

Bernar EA (1968) Laws of physical paleoclimatology and logical meaning of paleoclimatic data (in Russian). In: Problems of palaeoclimatology. Mir Press, Moscow, pp 189–200

Berry WBN, Boucot AJ (1970) Correlation of the North American Silurian rocks. Geol Soc Am Spec Pap 102: 289 pp

Berthlsen A, Noe-Nygaard A (1968) Precambrian of Greenland (in Russian). In: The geologic Systems. The Precambrian, vol II. Mir Press, Moscow, pp 107–235

Berzin LE, Ozolin NK (1967) Correlation of the Devonian deposits of the Latvian Republic on the basis of standard logging (in Russian). In: Problems of geochemistry of the Middle and Upper Paleozoic of Baltic. Zinatne Press, Riga, pp 10–15

Betekhtina OA (1973) Importance of non-marine Bivalves for paleogeographic reconstructions of the Late Paleozoic (in Russian). In: Environment and life in the geological past. Late Precambrian and Paleozoic of Siberia. Nauka Press, Novosibirsk, pp 18–32

Beurlen K (1969) Die Problematik paläogeographischer Rekonstruktionen. Geol Rundsch 58: 713–739

Bezborodova IV (1975) Biogerms of the anhydrite-carbonate Oka sequence (Visean) of the South-Eastern Russian Platform (in Russian). In: Lithology and paleogeography of biogerm massives. Nauka Press, Moscow, pp 124–138

Bgatov VI, Kazarinov VP, Matukhin RG, Nesterovski VS (1967) Perspectives of potassium-content of Devonian deposits in the North of the Siberian Platform (in Russian). Geol Geofiz 4: 44–47

Bgatov VI, Matukhin RG, Menner VV, Fradkin GS (1968) Salt potential of the Devonian in the Siberian Platform (in Russian). In: Material on regional geology of the Siberian Platform. Nauka Press, Novosibirsk, pp 169–173

Biske YuS, Kushnar LV (1976) Stratigraphy of the Upper Paleozoic deposits of the North-Eastern Fergana (in Russian). Vestn Leningr Gos Univ 24: 72–84

Biterman IM, Leonov BN, Natanov LM (1970) The Carboniferous. The Olenek Uplift (in Russian). In: Geology of the USSR, vol 18 (Western Yakutia), part 1, Geological description, B 1. Nedra Press, Moscow, pp 264–265

Blackadar RG (1963) Dumbbells Dome, Ellef Ringnes Island. In: Geology of the North-Central part of the Arctic Archipelago, Northwest Territories (Operation Franklin). Geol Surv Can Mem 320: 558–562

Blackett PMS (1961) Comparison of ancient climates with the ancient latitudes deduced from rock magnetic measurements. Proc R Soc London Ser A 1: 263 pp

Blagovidov VV (1970) Some peculiarities of structure and age of the Middle Paleozoic salt-bearing sequence of the Chu-Sarysu Depression in Central Kazakhstan (in Russian). In: Problems of regional geology and petrography of Siberia and methods of geochemical and geophysical investigations. Material of Conference of Young Scientist and Post-Graduate Students, vol 2. Rotaprint IGiG SO Akad Nauk USSR, Novosibirsk, pp 28–30

Blagovidov VV, Marzlyakov GA (1970) On the problem of correlation of the Upper horizons of the Cambrian salt-bearing carbonate deposits of the Kansk-Taseevo Depression and the internal regions of the Irkutsk Amphitheatre (in Russian). In: Geology and potassium potential of the Siberian Platform and other regions of salt accumulation in the USSR. Nauka Press, Moscow, pp 202–209

Blagovidov VV, Zharkov MA, Merzlyakov GA (1972) Some peculiarities of structure and formational conditions of the Morsovo salt-bearing sequence of the Middle Devonian in the Russian Platform (in Russian). In: Geology and formational conditions of the potassium salt deposits. Tr VNII Galurgii, vol 60. Leningrad, pp 6–17

Blanchet R (1970) Sur un profil des Dinarides, de l'Adriatique (Split-Omiŝ, Dalmate) au Bassin pannonique (Banja Luka-Doboj Bisnie). Bull Soc Géol Fr Sér 7, XI: 1010–1027

Blanchet R, Delda MD, Mullade M, Sigal J (1970) Contribution à l'étude du Crétacé des Dinarides internes: la région de Maglaj, Bosnie (Yougoslavie). Bull Soc Géol Fr Sér 7, XII: 1003–1009

Bobrinskaya OG, Bobrinsky VM, Bukatchuk PD et al (1964) Stratigraphy of the sedimentary formations of Moldavia (in Russian). Map of Moldavia Press, Kishinev, 132 pp

Bobrinsky VM, Kaptsan VH, Safarov EI (1964) The Upper Paleozoic (in Russian). In: Stratigraphy of the sedimentary formations of Moldavia. Map of Moldavia Press, Kishinev, pp 31–40

Bobrinsky VM, Bukatchuk PD, Burgelya NK et al (1965) Paleogeography of Moldavia (in Russian). Map of Moldavia Press, Kishinev, 466 pp

Bogatsky VI, Ivanov AV, Agulov SN (1977) Conditions of salt accumulation in the Verkhnepethora salt-bearing basin of the Komi ASSR (in Russian). In: Problems of salt accumulation, vol II. Nauka Press, Novosibirsk, pp 138–141

Bogdanov AA (1960) The main regularities of the Paleozoic folding development within USSR territory (in Russian). In: The Caledonian orogeny. Int Geol Congr, XXI Session. Contrib Sov Geol Problem 19. Izd Acad Nauk USSR, Moscow, pp 5–15

Bogdanov NA, Chugaeva MN (1960) Paleozoic deposits of Omulevskie mountains (in Russian). Izv Akad Nauk USSR, Ser Geol 5: 24–40

Bond DS, Bell AH, Meents WF (1968) Gas in Illinois Basin. In: Natural gases of North America, vol II. Am Assoc Petrol Geol Mem 9. Tulsa, Oklahoma, pp 1746–1753

Bond DS, Atherton EA, Bristol GM et al (1974) Possible future petroleum potential of region 9 (in Russian). In: Perspective oil-and-gas provinces of the USA. Nedra Press, Moscow, pp 553–576

Bonneau M (1970) Les lambeaux allochtones du revers septentrional du massif des Psiloriti (Crête moyenne, Grèce). Bull Soc Géol Fr Sér 7 XII: 1124–1129

Borchert H, Muir RO (1964) Salt deposits – the origin, metamorphism, and deformation of evaporites. D. Van Nostrand Ltd, London, 338 pp

Bordon VE, Makarov VG, Uryev II (1977) Geochemical peculiarities of halogenic formation of the Upper Devonian in Byelorussia and some problems of salt accumulation (in Russian). In: Problems of salt accumulation, vol II. Nauka Press, Novosibirsk, pp 181–184

Bosellini A, Hardie LA (1973) Depositional theme of a marginal marine evaporite. Sedimentology 20: 5–27

Boucot AJ, Doumani GA, Webers GF (1967) Devonian of South America. In: International Symposium on the Devonian System, vol I. Calgary, Alberta, Canada, pp 651–671

Bouroz A, Wagner RG, Gordon M, Meyen SV, Einor OL (1977) Proposals for an international chronostratigraphic classification of the Carboniferous. Izv Akad Nauk USSR Ser Geol 2: 5–24

Boyle RW (1963) Geology of the barite, gypsum, manganese, and lead-zinc-copper-silver deposits of the Walton-Cheverie Area, Nova Scotia. Geol Surv Can Pap 65-25, 36 pp

Briden JC (1969) Intercontinental correlations based on paleomagnetic evidence for recurrent continental drift. In: Gondwana stratigraphy, IUGS Symposium Buenos Aires, 1–15 October 1967. Unesco, Paris, pp 421–440

Briden JC, Irving E (1968) Palaeolatitude spectra of sedimentary palaeoclimatic indicators (in Russian). In: Problems in Palaeoclimatology. Mir Press, Moscow, pp 104–129

Britan IV (1971) Potassium potential of the Lower Cambrian deposits of the South-Western Siberian Platform (in Russian). VINITI Dep 2688, 108 pp

Britan IV, Zharkov MA, Kavitsky ML, Kolosov AS, Mashovich YaG, Chechel EI (1977) Structure and formational conditions of the Cambrian salt-bearing deposits within the territory of the USSR (in Russian). In: Problems of salt accumulation, vol II. Nauka Press, Novosibirsk, pp 203–227

Britchenko AD, Feshenko NI, Pioshko VV (1968) Stratigraphy and correlation of the Devonian sections of the North-Western Dnieper-Donets Depression (in Russian). In: Material on geology and oil-and-gas potential of Ukraine. Tr Ukr NIGRI, Issue XVI. Nedra Press, Moscow, pp 135–141

Britchenko AD, Vakarchuk GI, Tkachishin SV, Khmel FF (1977) The Paleozoic salt-bearing formations of the North-Western Dnieper-Donets Depression (in Russian). In: Problems of salt accumulation, vol II. Nauka Press, Novosibirsk, pp 170–175

Brown D, Kembell N, Kruk K (1968) The geological evolution of Australia and New Zealand. Pergamon Press, Oxford, 409 pp

Brown JS, Engel ARJ (1956) Revision of Grenville stratigraphy and structure in the Balmat–Edwards District, North-West Adirondacks. NY Geol Soc Am Bull 67: 1599–1622

Bruns EP (1956) The history of development of the Pripyat Depression in the Paleozoic (in Russian). Material of VSEGEI, Nov Ser, Issue 14. Leningrad, pp 185–207

Buggisch W, Flügel E, Leitz F, Tietz GF (1976) Die fazielle und paleogeographische Entwicklung im Perm der Karnischen Alpen und in den Randgebieten. Geol Rundsch 65: 649–690

Bukha V, Malkovski Z, Petrova GN, Rodionov VP, Roter K, Khramov AN (1976) The problem of continental drift on the territory of Eurasia according to the data of paleomagnetic research of the Lower Paleozoic (in Russian). In: Continental drift. Nauka Press, Moscow, pp 86–202

Bulgakova MD, Korobitsin AV, Semyonov VP, Ivensen VYu (1976) Sedimentary and volcanogenic-sedimentary formations of Verkhoyanye (the Paleozoic and Lower Mesozoic) (in Russian). Nauka Press, Novosibirsk, 134 pp

Bullard E, Everett J, Smith A (1965) The fit of the continents around the Atlantic. In: A Symposium on continental drift. Philos Trans R Soc London Ser A 268: 41–51

Bulle J, Rollet M (1970) Essai de définition des zones internes des Dinarides en Macédoine (Yougoslavie). Bull Soc Géol Fr Sér 7 XII: 1048–1059

Burshtar MS, Chernobrov BS, Shvemberger YN (1972) Geotectonic regionalization of the Fore-Caucasus part of the Scythian plate basement (in Russian). Sov Geol 6: 79–87

Bush VA, Garetsky RG, Ivanov YuA, Kiriukhin LG (1973) Structural-formation zones in Lower Permian deposits of North-Western Europe (in Russian). Geotektonika 1: 18–28

Cadet JP (1966) Sur des niveaux permiens associés à des gypses en Bosnie méridionale (Yougoslavie). CR Soc Géol Fr Paris 10: 403–404

Cadet JP (1970a) Sur la géologie des confins méridionaux de la Bosnie et de la Serbie: mise en evidence de la nappe du Semec (région de Visegrad et Rogatica, Yougoslavie). Bull Soc Géol Fr Sér 7 XII: 967–972

Cadet JP (1970b) Esquisse géologique de la Bosnie-Herzégovine méridionale et du Monténégro occidental. Bull Soc Géol Fr Sér 7 XII: 973–985

Caldwell WG (1959) The Lower Carboniferous rocks of the Carrick-on-Shannon Synecline. Geol Soc London QJ 115: 163–187

Carlson CG, Anderson SB (1965) Sedimentary and tectonic history of North Dakota part of the Williston basin. Am Assoc Petrol Geol Bull 49: 1833–1846

Carlson CG, Anderson SB (1966) Potash in North Dakota. North Dakota Geol Surv, Miss Ser 26: 12

Catacosinos PA (1973) Cambrian lithostratigraphy of Michigan Basin. Am Assoc Petrol Geol Bull 57: 2404–2418

Chabdarov NM, Sevostianov VG (1971) Carboniferous System, Trans-Ili Region (in Russian). In: Geology of the USSR, vol 40 (South Kazakhstan), Geological description, B 1. Nedra Press, Moscow, pp 298–311

Chamot GA (1965) Permian section at Apillapampa, Bolivia, and its fossil content. J Paleontol 39: 1112–1124

Charysz W (1975) Über die Genese des Zechsteins 3 in der Kujawy-Region (Mittlere VR Polen). Z Geol Wiss 3: 156–166

Charvet J (1970) Aperçu géologique des Dinarides aux environs du méridien de Sazsjevo. Bull Soc Géol Fr Sér 7 XII: 986–1002

Chaterji GC et al (1967) Devonian of India. In: International Symposium on the Devonian System, vol I. Calgary, Alberta, Canada, pp 557–564

Chechel EI (1969) On the distribution of the Middle Cambrian deposits within the Irkutsk Amphitheatre (in Russian). In: Biostratigraphy and paleontology of the Lower Cambrian of Siberia and the Soviet Far East. Nauka Press, Moscow, pp 202–208

Chechel EI, Mashovich YaG (1977) On the oldest salt-bearing deposits on the Siberian Platform, confined to the Irkutsk Horizon (in Russian). In: Problems of salt accumulation, vol II. Nauka Press, Novosibirsk, pp 227–229

Cherkesova SV (1968) Devonian system. The Taimyr fold system (in Russian). In: Geological structure of the USSR, vol I. Nedra Press, Moscow, pp 351–352
Cherkesova SV (1973) Taimyr (in Russian). In: Stratigraphy of the USSR. Devonian System, B 2. Nedra Press, Moscow, pp 139–148
Chermnykh VA (1962) Correlation of the Carboniferous sections of the Timan-Uralian Region (in Russian). In: Collections of papers on the geology and oil-and-gas presence in the Arctic. Gostoptekhizdat, Leningrad, pp 12–26
Chermnykh VA (1966) To the paleogeography of the Early Carboniferous epoch of the Northern Urals (in Russian). In: Paleozoic sediments of the North Cis-Uralian Region. Nauka Press, Moscow Leningrad, pp 40–67
Chlupac J (1967) Devonian of Czechoslovakia. In: International Symposium on the Devonian System, vol I. Calgary, Alberta, Canada, pp 109–126
Choroviez J (1970) La transversale de Zrmanja (Yougoslavie). Bull Soc Géol Fr Sér 7 XII: 1028–1033
Chosh PK, Bandyopadhyay SK (1969) Paleogeography of India during the Lower Gondwana times. In: Gondwana stratigraphy IUGS Symposium Buenos Aires, 1–15 October. Unesco, Paris, pp 523–536
Christie RL (1964) Geological reconnaissance of north-eastern Ellesmere Island District of Franklin. Geol Surv Can Mem 331: 79 pp
Churkin M, Brabb EE (1967) Devonian rocks of the Yukon-Porcupine to other Devonian sequence in Alaska. In: International Symposium on the Devonian System, vol II. Calgary, Alberta, Canada, pp 227–258
Ciric B (1970) Corrélation des phases tectogénétiques et magmatiques alpines dans les Dinarides. Bull Soc Géol Fr Sér 7 XII: 945–947
Clar E (1972) Anmerkung über weitere Vorkommen von Prebichlschichten. Verh Geol Bundesanst Mitt Ges Geol Bergbaustud (Austria) 20: 123–125
Clément B (1970) A propos des séries allochtones à l'Est du Parnasse (Grèce continentale). Bull Soc Géol Fr Sér 7 XII: 1118–1123
Collins RK (1964) Salt and the Canadian salt industry. Can Dep Mines Geol Surv Mines Branch Inf Circ 157: 33 pp
Collinson C, Becker LE, Carlson MP, Dorheim FH, James GW, Koenig IW, Swann DH (1967) Devonian of the North-Central Region, United States. In: International Symposium on the Devonian System, vol I. Alberta Soc Petrol Geol, Calgary, Alberta, pp 933–972
Cook PJ (1969) Rodinga, Northern Territory, 1:250 000 Geol Ser. Explanatory notes. Bur Miner Resour Aust, Canberra
Cooper BN (1966) Geology of salt and gypsum deposits in the Saltville Area, Smyth and Washington Countries, Virginia. In: 2nd Symposium on salt. N Ohio Geol Soc Cleveland, Ohio 1: 11–34
Coulomb J (1969) L'expansion des fondes océnaiques et la dérive des continents. Presses Universitaires de France, Paris, 232 pp
Cousin M (1970) Esquisse géologique des confins italo-yougoslaves: leurs place des Dinarides et les Alpes méridionales. Bull Soc Géol Fr Sér 7 XII: 1034–1047
Creer KM (1973) A discussion of the arrangement of paleomagnetic poles on the map of Pangaea for epochs in the Phanerozoic. Implications of continental drift in the Earth Sciences, vol I. Academic Press, London New York, pp 47–76
Crosby DG (1962) Wolfville map-area, Nova Scotia (21 H 1). Geol Surv Can Mem 325: 67 pp
Crosby EJ, Mapel WJ (1975) Central and West Texas. In: Paleotectonic investigations of the Pennsylvanian System in the United States, part I. Geol Surv Prof Pap 853. US Government Printing Office, Washington, pp 197–232
Cumming LM (1975) Ordovician strata of Hudson Bay Lowlands in Northern Manitoba. Geol Ass Can Spec Pap 9: 189–197
Daily B (1956) The Cambrian in South Australia. El sistema cámbrico, su paleogeografía y el problema de su base, T 2. México. XX Congr Int Geol, pp 49–61
Dal Cin R (1972) I conglomerati tard-paleozoici post-ercinci delle Dolomiti. Verh Geol Bundesanst, Mitt Ges Geol Bergbaustud (Austria) 20: 47–74

Dalyan IB, Posadskaya AS (1972) Geology and oil-and-gas content of Pre-Caspian Depressions eastern margin (in Russian). Nauka Press, Alma-Ata, 197 pp

Danner WR (1967) Devonian of Washington, Oregon and Western British Columbia. In: International Symposium on the Devonian System, vol I. Calgary, Alberta, Canada, pp 827–842

Davies GR (1974a) Paleozoic evaporites of the Canadian Arctic Archipelago. In: Coogan AH (ed) 4th Symposium on salt, Houston, 1973. N Ohio Geol Soc 1: 119–125

Davies GR (1974b) Submarine cementation, fracturing and internal sedimentation in Pennsylvanian-Permian carbonate build-ups, Arctic Archipelago. Am Assoc Petrol Geol Annu Meet Abs 1: 25

Davies GR (1975a) Upper Paleozoic carbonates and evaporites in the Sverdrup Basin, Canadian Arctic Archipelago. Geol Surv Can Rep Act Pap 75-1B: 209–214

Davies GR (1975b) Hoodoo L-41: Diapiric halite facies of the Otto Fiord Formation in the Sverdrup Basin, Arctic Archipelago. Geol Surv Can Rep Act Pap 75-1C: 23–29

Davies GR, Ludlam SD (1973) Origin of laminated and graded sediments of Middle Devonian of Western Canada. Geol Soc Am Bull 84: 3527–3546

Davies GR, Nassichuk WW (1975) Subaqueous evaporites of the Carboniferous Otto Fiord Formation, Canadian Arctic Archipelago: a summary. Geology 3: 273–278

Dedeev VA, Raznitsyn VA (1969) Pre-Timan Trough (in Russian). In: Tectonic of the northern part of the Russian Platform. Tr VNIIGRI, Issue 275. Nedra Press, Leningrad, pp 48–51

Dellwig LR (1955) Origin of the Salina salt of Michigan. J Sed Petrol 25: 83–110

Dellwig LF, Evans R (1969) Depositional processes in Salina salt of Michigan, Ohio, and New York. Am Assoc Petrol Geol Bull 53: 949–956

Del-Negro W (1960) Geologie der Österreichischen Bundesländer in kurzgefaßten Einzeldarstellungen. Salzburg Wien, 56 pp

De Villiers J (1967) Devonian of South Africa. International Symposium on the Devonian System, vol I. Calgary, Alberta, Canada, pp 303–307

Devonian deposits of the central regions of the Russian Platform (1958) (in Russian). Gostoptekhizdat, Leningrad, 405 pp (Authors: Filippova MF, Aronova SM et al)

Dewey JF, Pitman WG, Ryan WBF, Bonnin J (1973) Plate tectonics and the evolution of the Alpine systems. Geol Soc Am Bull 84: 3137–3180

Diarov MD (1974) Potassium bearing of evaporite formations of the Pre-Caspian Depression (in Russian). Nedra Press, Moscow, 128 pp

Diarov MD (1977) Paleogeographic formation conditions of potassium salts of the Pre-Caspian depression (in Russian). In: Problems of salt accumulation, part II. Nauka Press, Novosibirsk, pp 76–79

Diarov MD, Dzhumagaliev TN (1971) Formation conditions of potassium salt deposits of the Pre-Caspian depression (in Russian). Tr Zap Kaz NIGRI vol II: 126–137

Dickins JM (1969) Correlation chart for the Permian System in Australia. In: Gondwana stratigraphy, IUGS Symposium Buenos Aires, 1–15 October. Unesco, Paris, pp 475–477

Dikenshtein GK, Glushko VV, Solovjev BA, Tchernyshev SM et al (1975) Petroleum fields of North-West European province (in Russian). Nedra Press, Moscow, 208 pp

Dikenshtein GK, Jeremenko NA, Zhabrev IP, Krylov NA, Maximov SP, Semenovich VV (1976) Stage character of development of some major structural elements of Europe-Asia in connection with distribution of oil-and-gas provinces (basins) and their resources (in Russian). Sov Geol 4: 3–11

Dietz RS (1973) Morphologic fits of North America/Africa and Gondwana. In: Implications of continental drift in the Earth Sciences, vol II. Academic Press, London New York, pp 865–872

Dineley DL (1971) Arches and basins of Southern Arctic Islands of Canada. Proc Geol Assoc 82: 411–443

Ditmar VI (1961) Salt plugs in Bet-Pak-Dala, Southern Kazakhstan (in Russian). Dokl Akad Nauk SSSR 140: 1144–1147

Ditmar VI (1962a) Some peculiarities of structure of the Chu-SarysuDepression and its petroleum potential (in Russian). In: Tectonic problems of petroleum provinces. Izd Acad Nauk USSR, Moscow, pp 211–217

Ditmar VI (1962b) Peculiarities of geologic structure and petroleum potential of the Chu-Sarysu Depression according to new data. Izd Acad Nauk USSR, Moscow, p 11

Ditmar VI (1963) Peculiarities of geologic development of Chu-Sarysu Depression in the Middle and Upper Paleozoic (in Russian). Dokl Akad Nauk USSR 148: 406–409

Ditmar VI (1965) Peculiarities of geologic development and petroleum potential of Chu-Sarysu Depression (in Russian). In: Epi-Paleozoic platforms, their tectonic and petroleum potential. Nauka Press, Moscow, pp 192–206

Ditmar VI (1966) Tectonics and petroleum potential of Chu-Sarysu and adjacent Depressions in Kazakhstan (in Russian). Nauka Press, Moscow, 176 pp

Ditmar VI, Tikhomirov VI (1964) Permian halogenic deposits of the south-eastern part of Central Kazakhstan (in Russian). Dokl Akad Nauk USSR 158: 1089–1092

Ditmar VI, Tikhomirov VI (1967) Permian halogenic formation of Southern Kazakhstan (in Russian). Litol Polezn Iskop 6: 67–76

Divina TA, Matukhin RG (1975) Lithological peculiarities and strontium-bearing deposits of Upper Devonian and Lower Carboniferous strata of the Pyasino and of the Keta-Irbina Area (in Russian). In: Lithological-geochemical investigations of Paleozoic and Precambrian deposits of Siberia. Rotaprint IGiG SO Acad Nauk USSR, Novosibirsk, pp 153–162

Dixon OA (1972) Lower Carboniferous rocks between the Curlew and Ox Mountains, North-Western Ireland. J Geol Soc 128: 71–101

Dobretsov GL, Kumpan AS (1973) Lithology and facies of Upper Paleozoic deposits of the Minor Karatau and Chu Depression (in Russian). In: Problems of stratigraphy and tectonics of the East Kazakhstan. Nauka Press, Leningrad, pp 118–130

Döhner C (1976) Die Bromverteilung im Übergangsbereich Staßfurtflöz/Grauer Salzton des Südharz-Kalireviers. Z Angew Geol 22: 201–209

Döhner C, Elert KH (1975) Genetische Prozesse im Staßfurt-Salinar. Z Geol Wiss 2: 121–142

Douglass AG, Moullade M, Nairn AEM (1973) Causes and consequences of drift in the South Atlantic. In: Implications of continental drift in the Earth Science, vol I. Academic Press, London New York, pp 517–537

Drugov GM, Isakova VS, Panaev VA (1972) The results of potassium salt prospecting in the Irkutsk region and some recommendations for orientation in future investigations (in Russian). In: Potassium potential in Siberia. Nauka Press, Moscow, pp 20–29

Dubatolov VN (1972) Tabulate corals and Siberian Middle and Upper Devonian biostratigraphy (in Russian). Nauka Press, Moscow, 184 pp

Dubatolov VN, Spassky NYa (1964) New Devonian corals from the Soviet Union (in Russian). In: Stratigraphic and geographic review of USSS Devonian corals. Nauka Press, Moscow, pp 112–140

Dubatolov VN, Spassky NYa (1970) The corals from main Devonian paleo-biogeographic provinces (in Russian). In: Laws governing distribution of Paleozoic corals from the USSR. Nauka Press, Moscow, pp 15–31

Dubatolov VN, Spassky NYa (1973) The principles used in paleo-biogeographic zonation of the seas (in Russian). In: Medium and life in geologic past. Late Cambrian and Paleozoic of Siberia. Nauka Press, Novosibirsk, pp 11–18

Durkoop A, Mensink H, Plodowski G (1967) Devonian of Central and Western Afghanistan and Southern Iran. In: International Symposium on the Devonian System, vol I. Calgary, Alberta, Canada, pp 529–544

Dutkevich GA (1937) Permian deposits of Central Asia (in Russian). Probl Sov Geol 7: 11–15

Dutro JT, Saldukas RB (1973) Perm paleogeography of the Arctic. J Res US Geol Surv 1: 501–507

Edelstein AYa (1962) To the problem of the range of the Silurian deposits in the Dniester-Prut interfluve (in Russian). Izv Acad Nauk Moldavian SSR 4: 143–147

Edelstein AYa (1969) Silurian System (in Russian). In: Geology of the USSR, vol 45 (Moldavian SSR), Geological description. Nedra Press, Moscow, pp 75–88

Edie RW (1958) Mississippian sedimentation and oil fields in South-Eastern Saskatchewan. Am Assoc Petrol Geol Bull 42: 94–126

Egiazarov BK (1959) Geologic structure of the Severnaya Zemlya archipelago (in Russian). Tr NIIGA, Issue 94: 139 pp

Egiazarov BK (1969) Geologic structure of Alaska and Aleutian Islands (in Russian). Nedra Press, Leningrad, 264 pp
Egiazarov BK (1970) Severnaya Zemlya (in Russian). In: Geology of the USSR, vol 26 (Islands of Soviet Arctic), Geological description. Nedra Press, Moscow, pp 237–323
Egiazarov BK (1973) Severnaya Zemlya (in Russian). In: Stratigraphy of the USSR, Devonian System, B 2. Nedra Press, Moscow, pp 148–152
Ektova ZA, Belgovsky GZ (1972) Carboniferous System (in Russian). In: Geology of the USSR, vol 25 (Kirghiz SSR), Geological description, B 1. Nedra Press, Moscow, pp 170–183
Ells GD (1974) Summary of future petroleum potential or Region 8, Michigan Basin. In: Future petroleum provinces of the United States. Their geology and potential (in Russian). Nedra Press, Moscow, pp 537–552
Elston DP, Shoemaker EM, Landis ER (1962) Uncompahgre front and anticline region of Paradox Basin, Colorado and Utah. Am Assoc Petrol Geol Bull 46: 1857–1878
Emelyanov, EM, Lisitsin AP, Koshelev BA (1971) Distribution and composition of carbonates in the upper bed of the Atlantic Ocean bottom sediments (in Russian). Dokl Akad Nauk USSR, 196: 95–97
Erben HK (1962) Zur Analyse und Interpretation der rheinischen und hercynischen Magmafacies des Devons. In: Arbeitstagung Bonn-Bruxelles, 1960, Symposium-Band, pp 42–61
Erben HK, Zagora K (1967) Devonian of Germany. In: International Symposium on the Devonian System, Calgary, vol I. Alberta Soc Petrol Geol, Calgary, Alberta, Canada, pp 53–68
Ermakov VA (1971) Peculiarities of the Permian hydrochemical sequence salt series on the territory of the Volgograd part of the Volga Basin (in Russian). Sov Geol 2: 145–150
Ermakov VA, Grebennikov NP (1977) Regularities of structure of bishofite deposits of salt-bearing sequence of the Lower Volga Region and paleogeographic conditions of their accumulation (in Russian). In: Problems of salt accumulation, vol II. Nauka Press, Novosibirsk, pp 40–45
Ermakov VA, Koval'sky FI, Grebennikov NP (1977) Postsedimental leaching of the salt and its peculiarities in manifestation in the sequence of Volga monocline (in Russian). In: Problems of salt accumulation, vol II. Nauka Press, Novosibirsk, pp 45–49
Eroshina DM (1968) On some peculiarities in structure of salt-bearing sequence in the north-western part of the Pripyat Depression (in Russian). In: Lithology, geochemistry, and useful minerals of Byelorussia and the Baltic Area. Nauka Tekhnika Press, Minsk, pp 229–237
Eroshina DM, Zelentsov II (1969) On new potassium-bearing horizons in the Western Shatilkov Depression (in Russian). In: Geology and petrography of potash salts of Byelorussia. Nauka Tekhnika Press, Minsk, pp 112–130
Eroshina DM, Vysotsky EA, Kislik VZ (1976) Potassium-bearing horizons of Upper Frasnian halogenic formation of the Pripyat Depression (in Russian). Dokl Akad Nauk BSSR 20: 353–355
Esipchuk KE, Ivanushko AS. Lul'ev YuB (1963) On the age of the Kingir formation (in Russian). In: Data on geology, geophysics, and geochemistry of Ukrainia, Kazakhstan, Transbaikal. Izd KGU, Kiev, pp 34–37
Evans R (1965) The structure of the salt deposits at Pugwash. N S Mar Sediment 1: 21–23
Evans R (1967) Structure of the Mississippian evaporite deposit at Pugwash, Cumberland Country. N S Econ Geol 62: 262–274
Evans R (1970) Evaporites in the Mississippian of the Maritime provinces of Canada. In: 6th Congr Strat Geol Carboniferous, Sheffield, 1967, Maastricht (Netherlands), vol II, pp 725–736
Falke H (1972) Vergleich zwischen der Ablagerung des Verrucano in den Westalpen und des Rotliegenden in Süddeutschland und Frankreich. Verh Geol Bundesanst. Mitt Ges Geol Bergbaustud (Austria) 20: 11–32
Falke H (1974) Die Unterschiede in den Sedimentationsvorgängen zwischen den Autunien und Saxonien von Mittel- und Westeuropa. Geol Rundsch 63: 819–849
Fairbridge RV (1968) The importance of limestone and its Ca/Mg content to paleoclimatology (in Russian). In: Problems in Palaeoclimatology. Mir Press, Moscow, pp 258–309
Fairbridge RV (1970) Carbonate rocks and palaeoclimatology in biogeochemical history of the planet (in Russian). In: Carbonate rocks, vol I. Mir Press, Moscow, pp 357–386
Fettke CR (1955) Preliminary reports, occurrence of rock salt in Pennsylvania. P Geol Surv Ser 4 Prog Rep 145

Filippova LI (1973) Facies zoning of Silurian carbonate rocks in Eastern Timan-Pechora Province (Russian Platform and Uralian miogeosyncline articulation zone) (in Russian). In: Facies and geochemistry of carbonate deposits. Leningrad Tallin, pp 36–38
Filippova MF, Krylova AK (1973) Central Regions of the Russian Platform (in Russian). In: Stratigraphy of the USSR, Devonian System, B 1. Nedra Press, Moscow, pp 117–128
Filippova MF, Aronova SM et al (1958) Devonian deposits in the Central Regions of the Russian Platform (in Russian). Gostoptekhizdat, Leningrad, 405 pp
Fischer AD (1968) Brackish oceans as the cause of the Permo-Triassic marine faunal crisis (in Russian). In: Problems in Palaeoclimatology. Mir Press, Moscow, pp 362–370
Fiveg MP (1962) Geologic surroundings of sedimentation of salt-bearing deposits and their potassium-bearing horizons (in Russian). Avtoref Dokt Diss, Moscow, 58 pp
Fiveg MP (1970) On some peculiarities of paleogeography of salt accumulation stages (in Russian). In: State and aims of Soviet lithology, vol III. Nauka Press, Moscow, pp 24–31
Fiveg MP (1972) On some problems of paleogeography of salt accumulation stages (in Russian). In: Geology and formation conditions of potassium salt deposits. Leningrad, pp 69–93
Fiveg MP (1973a) Potassium content of Cambrian salt-bearing deposits (in Russian). In: Potassium salt deposits of the USSR. Nedra Press, Leningrad, pp 5–15
Fiveg MP (1973b) Potassium salt in Devonian salt-bearing deposits of the Dnieper-Donets Depression and Donbass (in Russian). In: Potassium salt deposits of the USSR. Nedra Press, Leningrad, pp 59–63
Fiveg MP (1973c) Permian potassium salt basins, Potassium salt of the Dnieper-Donets Depression (in Russian). In: Potassium salt deposits of the USSR. Nedra Press, Leningrad, pp 65–69
Fiveg MP (1973d) Verkhnekama potassium salt basin (in Russian). In: Potassium salt deposits of the USSR. Nedra Press, Leningrad, pp 104–142
Fiveg MP (1973e) Verkhnepethora potassium salt basin (in Russian). In: Potassium salt deposits of the USSR. Nedra Press, Leningrad, pp 142–152
Fiveg MP, Banera NI (1968) Paleogeography of Kungursk salt accumulation in the eastern part of the Russian Platform and the Fore-Uralian Depression (in Russian). Lithol Polezn Iskop 1: 33–43
Flenry JJ (1970) Sur les modalités d'installation du flysch du Pinde, au passage Crétacé-Eocène (Grèce continentale et Péloponnese septentrional). Bull Géol Fr Sér 7 XII: 1110–1117
Flügel HW (1967) Devonian of Austria. In: International Symposium on the Devonian System, vol I. Calgary, Alberta, Canada, pp 99–109
Flügel HW (1975) Einige Probleme des Variszikums von Neo-Europa. Geol Rundsch 64: 1–62
Fomenko KZ (1971) Salt-bearing basins of the world according to structure crust (in Russian). Byull Mosk Ova Ispyt Prir Otd Geol 46: 152
Fontaine H (1967) Le Devonien du Laos, du Cambodge et du Viet-Nam. In: International Symposium on the Devonian System, vol I. Calgary, Alberta, Canada, pp 569–582
Forman DJ (1966) The geology of the south-western margin of the Amadeus Basin, Central Australia. Bur Miner Resour Aust Rep 87: 56 pp
Forman DJ, Milligan EN, McCarthy NR (1967) Regional geology and structure of the north-eastern margin of the Amadeus Basin, Northern territory. Bur Miner Resour Aust Rep 103: 79 pp
Forsh NN (1976) Upper Permian deposits of the Pre-Caspian Depression and its northern margin (in Russian). In: Actual questions of geology and oil-and-gas potential of the Pre-Caspian depression. Leningrad, pp 67–77
Fortier YO (1963) Economic geology, gypsum. In: Geology of the North-Central part of the Arctic Archipelago, North-West Terrotories (Operation Franklin). Geol Surv Can Mem 320: 660–663
Fortier YO et al (1963) Geology of the North-Central part of the Arctic Archipelago. Can Geol Surv Mem 320: 671
Fradkin GS (1964) On Devonian halogenous formation of the Siberian Platform (in Russian). Geol Geofiz 11– 3–15
Fradkin GS (1967) Geologic structure and oil-and-gas potential of Western Vilyui Syneclise. Nauka Press, Moscow, 204 pp
Fradkin GS, Matukhin RG, Menner VV, Sokolov PN, Kolodeznikov KE (1977) Lithological and facial peculiarities of Devonian salt-bearing formations in Tungusska and Vilyui Syneclise (in Russian). In: Problems of salt accumulation, vol II. Nauka Press, Novosibirsk, pp 197–202

Francheteau J (1973) Plate tectonic model of the opening of the Atlantic Ocean South of the Azores. In: Implications of continental drift in the Earth Sciences, vol I. Academic Press, London New York, pp 197–202

Frank W (1972) Permoskyth im Pennin der Hohen Tauern (Seidlwinkldecke östlich der Großglockner-Hochalpenstraße). Verh Geol Bundesanst, Mitt Ges Geol Bergbaustud (Austria) 20: pp 151–153

Fuller JGCM, Porter IW (1969) Evaporite formations with petroleum reservoirs in Devonian and Mississippian of Alberta, Saskatchewan, and North Dakota. Am Assoc Petrol Geol Bull 53: 909–926

Fursenko AV (1957) On the stratigraphy of Devonian deposits of the Pripyat Depression (in Russian). Tr Leningr Ova Estestvoispyt 69: 5–24

Gabai NL (1969) To the possible occurrence of potassium salts in the Dzhezkazgan saliferous basin (in Russian). Tezis Dokl 3 Nautchno-Tekhn Konf Karaganda, pp 5–8

Gabai NL (1977) Potassium potential in the Dzhezkazgan Saliferous Basin (in Russian). In: Problems of salt accumulation, vol II. Nauka Press, Novosibirsk, pp 153–155

Gaetani M (1967) Devonian of Northern and Eastern Iran, Northern Afghanistan and Northern Pakistan. In: International Symposium on the Devonian System, vol I. Calgary, Alberta, Canada, pp 519–528

Gailite LK, Rybnikova MV, Ul'st RG (1967) Stratigraphy, fauna, and formation conditions of Silurian rocks of the Middle Baltic Area (in Russian). Zinatne Press, Riga, 304 pp

Gailite LK, Rybnikova MV, Ulst RG (1967) Stratigraphy, fauna and formation conditions of Silurian rocks of the Middle Baltic area (in Russian). Zinatne Press, Riga, 304 pp

Gansser A (1967) Geology of the Himalayas (in Russian). Mir Press, Moscow, 352 pp

Garetsky RG (1972) Tectonics of young platforms of Eurasia (in Russian). Nauka Press, Moscow, 300 pp

Garetsky RG, Konishchev VS (1977) Conditions of formation and internal structure of saliferous formation in the Pripyat Depression (in Russian). In: Problems of salt accumulation, vol II. Nauka Press, Novosibirsk, pp 175–178

Gee ER (1934) Recent observations on the Cambrian sequence of the Punjab Salt Range. India Geol Surv Rec 68: 115–120

Geological history of Western Canada, Calgary, Alberta (1964) Alberta Soc Petrol Geol

Geological atlas of Poland (1968) Warsaw, 10 tables

Geology and potassium bearing of Cambrian deposits of the South-Western Siberian Platform (1974) (in Russian). Nauka Press, Novosibirsk, 416 pp

Geology and useful resources of Africa (1973) (in Russian). Nedra Press, Moscow, 544 pp

Geology of Poland (1970) Vol 1, part 1. Publ House Wydawnictwa Geol, Warsaw, 651 pp

George TN (1958) Lower Carboniferous paleogeography of the British Isles. Proc Yorks Geol Soc 31: 227–318

George TN (1962) Tectonics and paleogeography in Southern England. Sci Prog (London) 50:192

George TN (1963) Tectonics and paleogeography in Northern England. Sci Prog (London) 51: 32

Gerrard TA (1966) Environmental studies of Fort Apache Member, Supai Formation, East-Central Arizona. Am Assoc Petrol Geol Bull 50: 2434–2463

Glennie KW (1972) Permian Rotliegendes of North-West Europe interpreted in light of modern desert sedimentation studies. Am Assoc Petrol Geol Bull 56: 1048–1071

Glushko VV, Dickenshtein GK, Shmidt K, Goldebekher K (1974) Regioning of the Northern part of the GDR according to the age of basement folding (in Russian). Sov Geol 5: 37–43

Glushnitsky OT (1977) Low governing structure of Silurian and Devonian boundary rocks in the North-Eastern part of the Siberian Platform (in Russian). Izv Vuzov Geol Razved 2: 12–20

Glushnitsky OT (1967) Certain peculiar features of Upper Devonian sedimentation in the Norilsk region (exemplified by data on the south-eastern coast of lake Piassino) (in Russian). Dokl Akad Nauk USSR 174: 423–424

Glushnitsky OT, Menner VV (1970) To detailed correlation of Middle and Upper Devonian sequences of the Norilsk region (in Russian). Byull Mosk Ova Ispyt Prir Otd Geol XLV (1): 71–83

Goffenshefer SJ (1971) Carboniferous System, Upper series (in Russian). In: Geology of the USSR, vol 4 (Centre of the European part of the USSR), Geological description. Nedra Press, Moscow, pp 291–314

Goldring R, Hause MR, Selwood EB, Simpson S, Lambert RSJ (1967) Devonian of Southern Britain. In: International Symposium on the Devonian System, vol I. Calgary, Alberta, Canada, pp 61–66

Golubev BM (1977) Peculiarities of internal structure of the salt-bearing sequence of Verkhnekama potassium salt deposit (in Russian). In: Problems of salt accumulation, vol II. Nauka Press, Novosibirsk, pp 115–118

Golubtsov VK (1973) Pripyat Depression (in Russian). In: Stratigraphy of the USSR, Devonian System, B 1. Nedra Press, Moscow, pp 181–190

Golubtsov VK, Makhnach AS (1961) Paleozoic and Early Mesozoic facies of Byelorussia (in Russian). Izd Akad Nauk BSSR, Minsk, pp 177–182

Gorbatkina TE, Strok NI (1971) Permian System (in Russian). In: Geology of the USSR, vol 4 (Centre of the European part of the USSR), Geological description. Nedra Press, Moscow, pp 314–348

Gorbatkina TE, Strok NI (1971) Permian System (in Russian). In: Geology of the USSR, vol 4, Central European part of the USSR. Nedra Press, Moscow, pp 314–348

Gorbov AF (1973) Pre-Caspian potassium salt basin (in Russian). In: Potassium salt deposits of the USSR. Nedra Press, Leningrad, pp 70–104

Gorkun OP (1964) Structure, periodicity, and long-term of sedimentation of salt sequence of the Starobino deposit (in Russian). In: Materials on the geology of saliferous regions. Tr VNII Galurgii, Issue 45. Nedra Press, Leningrad, pp 5–22

Gorodnitchev VI, Drobot DI (1977) Most ancient evaporites in Late Precambrian deposits of the Southern Siberian Platform (in Russian). In: Problems of salt accumulation, vol II. Nauka Press, Novosibirsk, pp 229–231

Grachevsky MM, Kalik NG (1976) Features of structure and oil-gas possibilities of the South-Western regions of the Russian Platform (in Russian). Izv Akad Nauk USSR Ser Geol 11: 123–129

Grachevsky MM, Berlin YuM, Chepelyugin AB, Sheremetieva GA (1971) New data concerning reef origin of edge scarp of the Near-Caspian Depression (in Russian). Dokl Akad Nauk USSR, Ser Geol 5: 1133–1136

Grachevsky MM, Berlin YuM, Dubovsky IT, Ul'mishek GF (1969) Correlation of different facies of deposits according to oil-and-gas prospecting (in Russian). Nedra Press, Moscow, 295 pp

Grayston LD, Sherwin DF, Allan JF (1964) Middle Devonian, Chapter S. In: Geological history of Western Canada. Alberta Soc Petrol Geol, Calgary, Alberta, pp 49–59

Greiner HR (1962) Facies and sedimentary environments of Albert Shale, New Brunswick. Am Assoc Petrol Geol Bull 46: 219–234

Greiner HR (1963) Malloch Dome and Vicinity, Ellef Ringnes Island. In: Geology of the North-Central part of the Arctic Archipelago, Northwest Territories (Operation Franklin). Geol Surv Can Mem 320: 563–570

Griffin DL (1967) Devonian of Northeastern British Columbia. In: International Symposium on the Devonian System, vol I. Calgary, Alberta, Canada, pp 803–826

Grunt TA (1970) Brachiopod biogeography of Permian Tethys (in Russian). Byull Mosk Ova Ispyt Prir Otd Geol 45: 90–101

Gryc G, Dutro JT, Brosge WG, Tailleur IL, Churkin M (1967) Devonian of Alaska. In: International Symposium on the Devonian System, vol I. Calgary, Alberta, Canada, pp 703–716

Gulyaeva LA, Itkina IS, Tikhomirov VI (1968) Geochemistry and facies of marine and continental saliferous basins of the Chu-Sarysu Depression (in Russian). Nauka Press, Moscow, 118 pp

Gurary FG (1975) On oil-and-gas content of Permian deposits of the Polish-German megasynclinorium (in Russian). Sov Geol 7: 48–61

Gurevich KYa, Zavyalova EA, Nikiforova OI, Pomyanovskaya GM, Khizhnyakov AV, Shulga PL (1973) Lvov Depression, Volyno-Podoliya (in Russian). In: Stratigraphy of the USSR, Devonian System, B 1. Nedra Press, Moscow, pp 166–180

Gussow WG (1953) Carboniferous stratigraphy and structural geology of New Brunswick, Canada. Am Assoc Petrol Geol Bull 37: 1713–1816

Haas W (1967) The Devonian of Bithynia, northwest Turkey. In: International Symposium on the Devonian System, vol II. Calgary, Alberta, Canada, pp 61–66

Hailwood EA, Tarling DH (1973) Palaeomagnetic evidence for a proto-Atlantic Ocean. In: Implication of continental drift in the Earth Sciences, vol I. Academic Press, London New York, pp 37–46
Halogenic formations from North-Western Donbass and Dnieper-Donets trough and their potassium content (in Russian). Nedra Press, Moscow, 239 pp
Hamada T (1967) Devonian of East Asia. In: International Symposium on the Devonian System, vol I. Calgary, Alberta, Canada, pp 583–596
Hamilton JB (1961) Salt in New Brunswick. In: New Brunswick Dep. of lands and mines. Miner Resour Rep 1: 77 pp
Hamilton W (1970) The Uralides and motion of the Russian and Siberian Platforms. Geol Soc Am Bull 81: 2553–2576
Harding SR, Gorrell HA (1967) Distribution of the Saskatchewan potash beds. Can Mines Met Bull, 682–687
Harland WB (1964) An outline structural history of Spitsbergen (in Russian). In: Geology of the Arctic. Mir Press, Moscow, pp 11–77
Harrington HJ (1962) Paleogeographic development of South America. Am Assoc Petrol Geol Bull 46: 1773–1814
Haughton SH (1963) Stratigraphic history of Africa south of the Sahara. Hafner, New York, 365 pp
Haun JD, Kont HC (1965) Geologic history of Rocky Mountain Region. Am Assoc Petrol Geol Bull 49: 1781–1800
Heissel W (1972a) Verrucano in Westösterreich. Verh Geol Bundesanst, Mitt Ges Geol Bergbaustud (Austria) 20: 79–81
Heissel W (1972b) Permoskythische Ablagerungen im Tiroler Raum. Verh Geol Bundesanst, Mitt Ges Geol Bergbaustud (Austria) 20: 163–165
Helsley CA, Stehli FG (1968) Comparison of geomagnetic and zoogeographic evidences on the position of the pole in Permian (in Russian). In: Problems in Palaeoclimatology. Mir Press, Moscow, pp 356–361
Helwig J (1972) Stratigraphy, sedimentation, paleogeography and paleoclimates of Carboniferous ("Gondwana") and Permian of Bolivia. Am Assoc Petrol Geol Bull 56: 1008–1033
Hemman M (1972) Zur feinstratigraphischen Gliederung des Leinesteinsalzes im Ostteil des Subherzynen Beckens. Jahrb Geol, vol IV. Academie-Verlag, Berlin, pp 291–301
Herman G, Barkell CA (1957) Pennsylvanian stratigraphy and productive zones, Paradox salt basin. In: Symposium on stratigraphic type oil accumulations in Rocky Mountains. Am Assoc Petrol Geol Bull 41: 861–881
Herman G, Sharps SL (1956) Pennsylvanian and Permian stratigraphy of the Paradox salt embayment. In: Geology and economic deposits of East-Central Utah. Intern Assoc Petrol Geol, 7th Annu Field Conf, Salt Lake City, Utah. Intern Assoc Petrol Geol, 77–84
Heywood WW (1955) Arctic Piercement Domes. Can Inst Mines Met Bull 48: 59–64
Hilgenberg OC (1966) Die Paläogeographie der expandierenden Erde vom Karbon bis zum Tertiär nach paläomagnetischen Messungen. Geol Rundsch 55: 878–904
Hill D (1964) Archaeocyatha from Antarctica and a review of the phylum. Sci Rep Trans-Antarct Exped London 10: 129–140
Hill D (1967) Devonian of Eastern Australia. In: International Symposium on the Devonian system, vol I. Calgary, Alberta, Canada, pp 613–630
History of geologic development of the Russian Platform and its margin (1964) (in Russian) Nedra Press, Moscow Leningrad, 327 pp
Hite RJ (1960) Stratigraphy of the saline facies of the Paradox Member of the Hermosa Formation of Southeastern Utah and Southwestern Colorado. In: Geology of the Paradox Basin fold and fault belt. Four Corners Geol Soc 3rd Annu Field Conf, pp 86–89
Hite RJ (1961) Potash-bearing evaporite cycles in the salt anticlines of the Paradox Basin, Colorado and Utah. US Geol Surv Prof Pap 424-D: 135–138
Hite RJ (1968) Salt deposits of the Paradox Basin, Southwestern Colorado and Southeastern Utah. In: Saline deposits. Geol Soc Am Spec Pap 88: 319–330
Hollard H (1967) Le Devonian du Maroc et du Sahara Nord-Occidental. In: International Symposium on the Devonian System, vol I. Calgary, Alberta, Canada, pp 203–244

Holter ME (1969) The Middle Devonian Prairie evaporite of Saskatchewan. Sask Dep Miner Resour Rep 123: 134 pp
House MR (1968) Devonian northern hemisphere ammonoid distribution and marine links (in Russian). In: Problems in Palaeoclimatology. Mir Press, Moscow, pp 162–170
Howie RD, Barss MS (1975) Upper Paleozoic rocks of the Atlantic Provinces, Gulf of St. Lawrence and adjacent continental shelf. In: Offshore geology of Eastern Canada. Geol Surv Can Pap 74-30: 35–50
Howie RD, Cumming LM (1963) Basement features of the Canadian Appalachians. Geol Surv Can Bull 89: 18 pp
Hoyningen-Huene E (1967) Stratigraphische Korrelationsschemata für das Siles and das Perm der Deutschen Demokratischen Republik. Abh Zentr Geol Inst Berlin, 7: 133 pp
Humphrey WE (1958) Diapirs and diapirism in Persia and North Africa. Am Assoc Petrol Geol Bull 42: 1736–1744
Illing LV (1959) Upper Paleozoic carbonate sediments in Western Canada. Oilweek 10: 34–45
International Symposium on the Devonian System, Calgary (1967). Calgary, Alberta, Canada, vol I: 1377 pp, vol II: 1055 pp
Irving E, Briden JC (1962) Paleolatitude of evaporite deposits. Nature (London) 196: 425–428
Ivanov AA (1953) Principles of geology and methods of prospecting and valuation of salt deposits (in Russian). Gosgeoltekhizdat, Moscow, 204 pp
Ivanov AA (1975) Natural sulfate and nonsulfate potassium salts (in Russian). Sov Geol 8:75–84
Ivanov AA, Levitsky YuF (1960) Geology of halogenic formation of the USSS (in Russian). Gosgeoltekhizdat, Moscow, 422 pp
Ivanov AA, Voronova ML (1972) Evaporite formations (in Russian). Nedra Press, Moscow, 327 pp
Ivanov AA, Voronova ML (1975) Verkhnekama potassium salt deposit (stratigraphy, mineralogy and petrography, tectonics, genesis) (in Russian). Nedra Press, Leningrad, 219 pp
Ivanov YuA, Komissarova IM, Eventov YaS (1977) Peculiarities of structure and origin of Lower Permian salt-bearing formation of North Pre-Caspian (in Russian). In: Problems of salt accumulation, vol II. Nauka Press, Novosibirsk, pp 16–26
Johnson GAL (1973) Closing of the Carboniferous Sea in Western Europe. In: Implications of continental drift in the Earth Sciences, vol II. Academic Press, London New York, pp 843–850
Johnson GD, Vondra CF (1969) Lithofacies of Pella Formation (Mississippian), Southeastern Iowa. Am Assoc Petrol Geol Bull 53: 1894–1908
Johnstone MN, Jones PJ, Koop WJ, Roberts J, Gilbert-Tomlinson J, Veevers JJ, Wells AT (1967) Devonian of Western and Central Australia. In: International Symposium on the Devonian System, vol I. Calgary, Alberta, Canada, pp 599–612
Jordan SP (1967) Sask-reef trend looks big. Oilweek 16: 10–12
Jordan SP (1968) Will Zama be duplicated at Quill Lake, Saskatchewan? Oilweek 26: 10–12
Jung W (1968a) Über Gesteinstypen, Faziesdifferenzierungen und zyklisch-rhythmische Sedimentation im deutsch-polnischen Zechstein. In: Report of the 23rd Session of International Geological Congress, vol VIII. Prag, pp 211–226
Jung W (1968b) Zechstein. Grundriß Geol DDR 1: 219–237
Kahler F (1972) Das Perm der Karnischen Alpen. Verh Geol Bundesanst (Austria) 1: 139–141
Kalberg AA (1948) New data on stratigraphy and tectonics of the Central Taimir (in Russian) Sov Geol 33: 31–43
Kalinko MK (1951) Geology of salt stock of the Nordvik dome (in Russian). Tr NIIGA 10: 83–91
Kalinko MK (1953) Geology and oil content of the Nordvik peninsula (in Russian). Tr NIIGA 75: 132 pp
Kalinko MK (1959) History of geologic development and oil-and-gas potential of the Khatanga Depression (in Russian). Tr NIIGA 104: 360 pp
Kalinko MK (1973a) Salt accumulation, origin of salt structures and their influence on oil and gas content (in Russian). Nedra Press, Moscow, 132 pp
Kalinko MK (1973b) Nature of relationship between the salt and oil-and-gas possibilities within the continental and water areas (in Russian) Izv Acad Nauk USSR Ser Geol 10: 95–106

Kamashev Kh, Diarov M, Doghalov A, Tukhfatov K, Tursungaliyev A (1974) Geologo-lithological features of potassium salts in the Satimola locality (in Russian). Izv Akad Nauk USSR Ser Geol 12: 158–160

Kameneva GI (1975) Structure of the central part of the Vrangel Islands (in Russian). In: Geology and useful resources of Novosibirsk Islands and Vrangel Island. Rotaprint NIIGA, Leningrad, pp 72–77

Kaptsan VKh, Safarov EI (1965a) To the stratigraphy of the Permian deposits in Moldavia (in Russian). Izv Acad Nauk Moldavian SSR 8: 192–196

Kaptsan VKh, Safarov EI (1965b) On the stripping of native deposits of the Carboniferous in Moldavia (in Russian). Dokl Akad Nauk USSR 161: 659–662

Kaptsan VKh, Safarov EI (1969a) Carboniferous System (in Russian). In: Geology of the USSR, vol 45 (Moldavian SSR), Geological description. Nedra Press, Moscow, pp 91–94

Kaptsan VKh, Safarov EI (1969b) Permian System (in Russian). In: Geology of the USSR, vol 45 (Moldavian SSR), Geological description. Nedra Press, Moscow, pp 94–96

Kaptsan VKh, Polev PV, Safarov EI (1963) Recent data on the Upper Paleozoic deposits of Moldavia (in Russian). Dokl Akad Nauk USSR 150: 882–884

Kapustin IN, Milnichuk VS, Shafranov AP (1977) On influence of subsalt bed's structure upon constitution of salt-bearing deposits in the northern part of the Pre-Caspian depression (in Russian). In: Problems of salt accumulation, vol II. Nauka Press, Novosibirsk, pp 57–62

Kargatjev VA (1970) Anhydrite in diopside rocks of the Central Aldan region (in Russian). In: Mineral raw material, issue 22. Nedra Press, Moscow, pp 65–75

Kashik DS, Alekseeva IA, Nelzina RE, Polozova AN, Rostovtsev VN (1969) The stratigraphy of Lower Permian deposits on the North of the Russian Platform (in Russian). Dokl Akad Nauk SSSR 187: 399–402

Katzung G (1968) Rotliegendes. Grundriß Geol DDR 1: 201–218

Katzung G (1975) Tektonik, Klimat und Sedimentation in der Mitteleuropäischen Saxon-Senke und in angrenzenden Gebieten. Z Geol Wiss 11: 1453–1472

Kelley DG (1967) Baddeck and Whycocomagh map-areas with emphasis on Mississippian stratigraphy of central Cape Breton Island, Nova Scotia (11 K/2 and 11 F/14). Geol Surv Can Mem 351: 65 pp

Kent DM (1967) Devonian of Manitoba and Saskatchewan. In: International Symposium on the Devonian system, vol I. Alberta Soc Petrol Geol, Calgary, Alberta, pp 781–802

Kent DM (1968) The geology of the Upper Devonian Saskatchewan Group and equivalent rocks in Western Saskatchewan and adjacent areas. Sask Dep Miner Resour Rep 99

Kent PE (1958) Recent studies of South Persian salt plugs. Am Assoc Petrol Geol Bull 42: 2951–2972

Kent PE (1967) Progress of exploration in North Sea. Am Assoc Petrol Geol Bull 51: 731–741

Kent PE (1970) The salt plugs of the Persian Gulf Region. Reprinted from: Trans Leicester Lit Philos Soc LXIV, 88 pp

Kent PE (1975) Review of North Sea Basin development. J Geol Soc 131: 435–468

Kent PE, Hedberg HD (1976) Salt diapirism in Southern Iran, Discussion. Am Assoc Petrol Geol Bull 60: 458

Kerr IW (1967a) New nonmenclature for Ordovician rock units of the eastern and southern Queen Elizabeth Islands, Arctic Canada. Bull Can Petrol Geol 15: 91–113

Kerr IW (1967b) Devonian of the Franklinian miogeosyncline and adjacent Central Stable Region, Arctic Canada. In: International Symposium on the Devonian System, vol I. Calgary, Alberta, Canada, pp 677–692

Kerr JW (1968) Stratigraphy of Central and Eastern Ellesmere Island, Arctic Canada. Geol Surv Can Pap 67-27: 92 pp

Kerr JW (1974) Geology of Bathurst Island Group and Byam Martin Island, Arctic Canada (Operation Bathurst Island). Geol Surv Can Mem 378: 152 pp

Kerr JW (1975) Grinell Peninsula, Devon Island, District of Franklin. Rep Act Part B Geol Surv Can Pap 75-1: 545

Kerr JW, Morrow DW, Savigny KW (1973) Grinnell Peninsula, Devon Island, District of Franklin. Rep Act Part A, April to October 1972. Geol Surv Can Pap 73-1: 262–263

Keyes DA, Wright IY (1966) Geology of the I.M.C. potash deposit, Esterhazy, Saskatchewan. In: 2nd Symposium on salt. N Ohio Geol Soc Inc 1: 95–101
Khain VE (1971) Regional geotectonic, North and South America, Antarctica and Africa (in Russian). Nedra Press, Moscow, 546 pp
Khaiznikova KB (1970) Biostratigaphy and the Devonian tabulates of the Sette-Daban Ridge (South Verkhoyanye) (in Russian). Avtoref Kand Diss, Novosibirsk, 24 pp
Khalturina II, Bocharov VM, Avrova NP (1977) Mineralogic-lithological peculiarities and specificity of formation of some potassium magnesium salts of Pre-Caspian halogenic formation (in Russian). In: Problems of salt accumulation, vol II. Nauka Press, Novosibirsk, pp 73–76
Khizhnyakov AV, Pomyanovskaya GM (1967) Devonian of Volyno-Podolia region, Russian Platform. In: International Symposium on the Devonian System, vol II. Calgary, Alberta, Canada, pp 359–367
Kholodov VN (1973) Associations of ore components in Vendian and Cambrian deposits of Asia and some problems of their genesis (in Russian). In: Conditions of origin of geosynclinal phosphorites. Nauka Press, Moscow, pp 50–61
Khomentovsky VV (1976) The Vendian (in Russian). Nauka Press, Novosibirsk, 272 pp
Khomentovsky VV, Repina LN (1965) Lower Cambrian in Siberian stratotype section (in Russian). Nauka Press, Moscow, 200 pp
Khramov AN (1967) Earth geomagnetic field in Late Paleozoic (analysis of world geomagnetic data) (in Russian). Izv Akad Nauk USSR 1: 86–108
Khramov AN, Goncharov GI, Komissarova RA et al (1974) Paleozoic paleomagnetism (in Russian). Nedra Press, Leningrad, 238 pp
King L (1973) An improved reconstruction of Gondwanaland. In: Implications of continental drift to the Earth Sciences, vol II. Academic Press, London New York, pp 851–863
Kirikov VP (1959) Rhythm in salt sequences structure in the Pripyat Depression (in Russian). Inf Zb VSEGEI 11: 81–90
Kirikov VP (1962) On the comparison of Devonian halogenic sequences of the Pripyat and Dnieper-Donets-Depressions (in Russian). In: Materials on geology of nonmetal useful resources occurrences. Tr VSEGEI Nov Ser 83, Gostoptekhizdat, Leningrad, pp 129–136
Kirikov VP (1963a) Main stages of formation of halogenic Devonian deposits in the Pripyat and Dnieper-Donets-Depressions (in Russian). Tr VSEGEI Nov Ser 91: 107–129
Kirikov VP (1963b) On the conditions of occurrence of the potassic horizons of the Starobino deposit (in Russian). In: Geology of the potassium salt deposits. Tr VSEGEI Nov Ser 99, Gostoptekhizdat, Leningrad, pp 233–245
Kirikov VP (1973) Byelorussia massif and western limb of the Moscovian syneclise (in Russian). In: Stratigraphy of the USSR, Devonian System, B 1. Nedra Press, Moscow, pp 113–117
Kislik VZ (1966) Zones of replacement of sylvinite with rock salt in the 2-d potassium-bearing horizon of the Starobino deposit and conditions of their formation (in Russian). Avtoref Kand Diss, Minsk, 36 pp
Kislik VZ, Lupinovich YuI (1964) Peculiarities of border zone structure of the 2-d potassium-bearing horizon of the Starobino deposit (in Russian). Dokl Akad Nauk BSSR 8: 740–744
Kislik VZ, Lupinovich YuI (1968) Year rhythm in sylvinite sedimentation of the Starobino deposit (in Russian). In: Lithology, geochemistry, and useful minerals of Byelorussia and Baltic region. Nauka Tekhnika Press, Minsk, pp 207–218
Kislik VZ, Vysotsky EA, Lupinovich YuI (1971) Structural and compositional peculiarities of the salt-bearing bed of the Pripyat Depression. Dokl Akad Nauk BSSR 15: 54–56
Kislik VZ, Vysotsky EA, Golubtsov VK, Krucheck SA (1976a) The Middle Devonian salt-bearing deposits of the Pripyat Depression (in Russian). Sov Geol 4: 139–143
Kislik VZ, Vysotsky EA, Protasevich BA, Zelentsov II (1976b) Regularity of distribution of potassium salts in the Pripyat Depression and the perspectives of their industrial exploitation (in Russian). In: Geology of nonmetalliferous useful minerals of BSSR. Rotaprint BelNIGRI, Minsk, pp 3–20
Kislik VZ, Britchenco AD, Ditmar VI, Menner VV, Fradkin GS (1977) Devonian salt accumulation on USSR territory (in Russian). In: Problems of salt accumulation, vol II. Nauka Press, Novosibirsk, pp 156–167

Kityk VI (1970) Salt tectonics of the Dnieper-Donets Depression (in Russian). Naukova Dumka Press, Kiev, 203 pp

Kityk VI, Galaburda NI (1977) On the conditions of origin of Devonian halogenic formations in the Dnieper-Donets Depression (in Russian). In: Problems of salt accumulation, vol II. Nauka Press, Novosibirsk, pp 167–170

Klaus W (1972) Möglichkeiten der Stratigraphie im „Permoskyth". Verh Geol Bundesanst, Mitt Ges Geol Bergbaustud (Austria) 20: 33–34

Klaus W (1972) Möglichkeiten der Stratigraphie im „Permoskyth". Verh Geol Bundesanst (Austria) 33–34

Klingspor AM (1966) Cyclic deposits of potash in Saskatchewan. Bull Can Petrol Geol 14: 193–207

Klingspor AM (1969) Middle Devonian Muskeg evaporites of Western Canada. Am Assoc Petrol Geol Bull 53: 927–948

Knauf VI, Rezvoi DP (1972) Principal features of geological structure (in Russian). In: Geology of the USSR, vol 25 (Kirghiz SSR), Geological description, B 1. Nedra Press, Moscow, pp 23–37

Kolchanov VP (1971) O.H. Hildenberg's paleogeographical reconstructions for an expanding Earth (in Russian). Geotectonics 4: 99–107

Kolosov AS, Pustylnikov AM (1967) On the discovery of compound iron haloids in Devonian saline deposits of the Tuva (in Russian). Dokl Akad Nauk USSR 172: 946–948

Kolosov AS, Matukhin RG, Pustylnikov AM, Sokolov PN (1974) Composition of the Middle Devonian salts in the basin of the Mikchanda River (North-West of the Siberian Platform) (in Russian). Geol Geofiz 2: 134–138

Kolosov AS, Pustylnikov AM, Zaikov VV, Zaikova EV (1977) Structure geochemical characteristic and formation conditions of salt-bearing deposits in Tuva Basin (in Russian). In: Problems of salt accumulation, vol II. Nauka Press, Novosibirsk, pp 193–197

Kolotukhin AT, Mavrin KA, Romanov VV (1977) The position of Kungurian evaporites in the complex of formations in the Southern Cis-Ural Trough (in Russian). In: Problems of salt accumulation, vol II. Nauka Press, Novosibirsk, pp 110–111

Kolotikhina SE, Klanovskaya LI, Rozhanets AV (1974) Geology and economics of rare elements occurrences of Australia. Nauka Press, Moscow, 270 pp

Komar VA (1966) Upper Precambrian stromatolites of the Northern Siberian Platform and their stratigraphic significance (in Russian). Nauka Press, Moscow, 122 pp

Konishchev VS (1976) Saline Alpine glaciers and secondary salt-bearing rocks (in Russian). Byull Mosk Ova Ispyt Prir Otd Geol 51: 129–134

Konstantinovich E (1972) The genesis of Permian copper deposits in Poland (in Russian). Sov Geol 8: 101–117

Koop WJ (1966) Sahara No. 1 Well, stratigraphic nomenclature, Appendix 1. Bur Miner Resour Aust Petrol Search Subsidy Acts 80: 19 pp

Kopnin VI (1977) Study of salt accumulation conditions using methods of modeling (in Russian). In: Problems of salt accumulation, vol I. Nauka Press, Novosibirsk, pp 159–170

Kopnin VI, Zueva AS (1977) On the significance of transitional type of sections for study of paleogeographic conditions of sedimentation during Irenian time in the Solikamsk Basin (in Russian). In: Problems of salt accumulation, vol II. Nauka Press, Novosibirsk, pp 128–134

Kopnin VI, Oborin AA, Shishkin MA (1977) The role of terrigene spits and shall-anhydrite bars in the halogenesis of the Ufa-Solikamsk depression (in Russian). In: Problems of salt accumulation, vol II. Nauka Press, Novosibirsk, pp 134–138

Korenevsky SM (1973) The complex of useful minerals of halogenic formations (in Russian). Nedra Press, Moscow, 300 pp

Korzun VP (1977) The relationship of Upper Devonian volcanic and salt-bearing deposits in the Pripyat depression (in Russian). In: Problems of salt accumulation, vol II. Nauka Press, Novosibirsk, pp 178–181

Kozary MT, Dunlap JC, Humprey WE (1968) Incidence of saline deposits in geologic time. In: Saline deposits. Geol Soc Am Spec Pap 88: 43–57

Krasheninnikov GF (1960) Sedimentary Upper Paleozoic in Caledonides of Southern Siberia and North-Western Europe (in Russian). In: Caledonian Orogeny. Int Congr XXI Session. Dokl Sov Geol Probl 19. Izd Acad Nauk SSSR, Moscow, pp 89–104

Kravchinsky AJa (1970) Displacements of the Siberian Platform (according to paleomagnetic and paleoclimatic data) (in Russian). Geotektonica 6: 77–87

Kravchinsky AJa (1973) On the conjugation in paleomagnetic and paleogeographical rearrangements on the East-European Platform (in Russian). Geotektonica 6: 37–47

Krebs W, Wachendorf H (1973) Proterozoic evolution of Central Europe. Geol Soc Am Bull 84: 2611–2630

Kreidler WL (1957) Occurrence of Silurian salt in New York State. NY State Mus Bull 361: 1–56

Krichevsky GN, Kapustin IN, Milnichuk VS (1977) Distribution of Kungurian deposits in interior part of the Pre-Caspian depression (in Russian). In: Problems of salt accumulation, vol II. Nauka Press, Novosibirsk, pp 29–32

Krishnan MS (1966) Salt tectonics in the Punjab Salt Range, Pakistan. Geol Soc Am Bull 77: 115–122

Krishnan MS (1968) Geology of the salt deposits in the Punjab Salt Range, Pakistan. In: Saline deposits. Geol Soc Am Spec Pap 88: 410–411

Kroemmelbein K (1967) Devonian of the Amazonas Basin, Brazil. In: International Symposium on the Devonian System, vol II. Calgary, Alberta, Canada, pp 201–208

Kropotkin PN (1961) Paleomagnetism, paleoclimates, and problems of big horizontal movements of the Earth crust (in Russian). Sov Geol 5: 16–38

Kropotkin PN, Shakhvarstova KA (1965) Geological structure of the Pacific mobile belt (in Russian). Nauka Press, Moscow, 368 pp

Kropotkin PN, Valjaev BM (1970) Rock salt in deep grabens and depression containing oceans crust (in Russian). Byull Mosk Ova Ispyt Prir Otd Geol 45;27–42

Krylova AV, Levina VI, Chernova NI (1977) Regularity of distribution and formation conditions of salt sequences of Kungurian time in the Astrakhan-Kalmyk Pre-Caspian region (in Russian). In: Problems of salt accumulation, vol II. Nauka Press, Novosibirsk, pp 68–73

Kudryashov AI, Myagkov VF (1977) Spatial ratio of morphostructural and geochemical fields in sylvinite beds of Verkhnekama occurrence (in Russian). In: Problems of salt accumulation, vol II. Nauka Press, Novosibirsk, pp 120–126

Kudryavtsev NA (1966) Regularities in mineral salt accumulation (in relation with oil and salt paragenesis) (in Russian). Sov Geol 7: 17–35

Kulibakina IB, Zhuravlev VS, Goncharov ES (1972) On role of salt-bearing beds in the formation of the gas deposits of Permian sediments of the European Platform (in Russian). Geol Nefti Gaza 6: 55–59

Kumpan AS (1966) Upper Paleozoic of East Kazakhstan (Central and Southern Kazakhstan, Kolba, Altai) (in Russian). Nedra Press, Leningrad, pp 247

Kunin NJa (1968) Tectonics of Middle Syr-Darja and Chu-SarysuDepressions according geophysical data. Nedra Press, Moscow, 264 pp

Kurochka VP (1966a) Structure of the Upper salt sequence in the Pripyat Depression (in Russian). In: Geological description and perspective of the oil-and-gas potential of BSSR. Nedra Press, Moscow, 66–78

Kurochka VP (1966b) The comparison of principal sections of Lower salt sequence in the Pripyat Depression (in Russian). In: New data about geology and oil-and-gas content in Pripyat depression of the Byelorussian SSR. Nedra Press, Moscow, pp 103–114

Kuteinikov ES, Syagaev NA (1962) Tectonic structure and history of evolution of the Kyutingda transverse Trough (in Russian). In: Papers on the geology and oil-and-gas content in the Arctic Tr NIIGA 130, Leningrad, 83–90

Kutsyba AM (1954) Devonian in the Dnieper-Donets Depression and north-western margin of Donbass (in Russian). In: Geology of structure and oil-and-gas content of Dnieper-Donets Depression and north-western margins of Donbass. Naukova Dumka, Kiev, pp 122–126

Kutsyba AM (1959) Stratigraphy, facies, paleogeography and perspectives of oil-and-gas content of Devonian deposits of the Dnieper-Donets Depression (in Russian). In: Geologic structure and oil-and-gas potential of eastern regions of Ukraine. Kiev, pp 352–381

Kuznetsov VG (1972) On two types of salt-bearing beds (in Russian). Geol Geofiz 7: 22–30

Kuznetsov VG, Khenvin TI (1972) Lower Permian buried relief of the eastern slope of the Voronezh anteclise (in Russian). Byull Mosk Ova Ispyt Prir Otd Geol 47: 125–127

Landes KK (1945) The Saline and Bass Island rock in the Michigan Basin. US Geol Surv Oil and Gas Inv Prelim Map 40: 473–493
Landes KK (1951) Detroit River Group in the Michgan Basin. US Geol Surv Circ 133: 23 pp
Lane DM (1964) Sonris River Formation in Southern Saskatchewan. Sask Dep Miner Resour Rep 92
Langbein R (1973) Das Sedimentationsmodell des Plattendolomits (Ca 3) in Ostthüringen. Z Geol Wiss 1: 65–72
Langton JR, Chin GE (1968) Rainbow Member Facies and related reservoir properties, Rainbow Lake, Alberta. Am Assoc Petrol Geol Bull 52: 1925–1955
Latham JW (1973) Healdton oil and gas field (in Russian). In: Geology of gigant petroleum fields. Mir Press, Moscow, pp 219–241
Latskova VE (1961) Stratigraphy, lithology and facies of the Lower Permian deposits of the Saratov-Volgograd region of the Volga River Basin (in Russian). In: Mater. on Geol. and gas potential of the Lower Permian deposits of the Southern Russian Platform. Izd Kharkov Univ, Kharkov, pp 77–85
Latskova VE (1967) Permian system (in Russian). In: Geology and oil-and-gas potential of the Saratov regions of the Volga River Basin. Tr Nizhne-Volzhsk NIIGiG, Issue 10, Saratov, pp 24–27
Law J (1955) Geology of Northwestern Alberta and adjacent areas. Am Assoc Petrol Geol Bull 39: 1927–1975
Lecompte M (1967) Le Devonien de la Belgique et du Nord de la France. In: International Symposium on the Devonian System, vol I. Calgary, Alberta, Canada, pp 15–52
Lefond SJ (1969) Handbook of world salt resources. Plenum Press, New York, 384 pp
Legrand P (1967) Le Devonien du Sahare Algérien. In: International Symposium on the Devonian System, vol I. Calgary, Alberta, Canada, pp 245–284
Lepeshkov IN, Romasheva NN, Soloviev VK (1958) Potassium-bearing salt deposits of Tuva (in Russian). Dokl Akad Nauk USSR 119: 1156–1158
Leven ED (1975) Stage scale of Permian deposits of Tethys (in Russian). Byull Mosk Ova Ispyt Prir Otd Geol 1: 5–21
Levenko AI (1955) On age of salt-bearing strata of the Tuz-Tag Mountain (Tuva) (in Russian). Izv Akad Nauk USSR Ser Geol 3: 83–91
Levenko AI (1956) New data on the age of evaporites of Tuva (Devonian) (in Russian). Dokl Akad Nauk USSR,109: 1015–1018
Levenko AI (1960) Devonian of the Central and South Tuva (in Russian). Izd Akad Nauk USSR Moscow, 158 pp
Levenshtein ML, Gruba VI, Konashov VG (1977) Salt-bearing formation of Permian in the Donets Basin (in Russian). In: Problems of salt accumulation, vol II. Nauka Press, Novosibirsk, pp 105–108
Li AB, Mailibaev MM (1971) Devonian System, Chuya Depression (in Russian). In: Geology of the USSR, vol 40 (South Kazakhstan), Geological description, B 1. Nedra Press, Moscow, pp 201
Liepinsh PP (1963a) The stratigraphy of Frasnian deposits of Latvian SSR (in Russian). In: Frasnian deposits of Latvian SSR. Tr Inst Geol Akad Nauk Latv SSR, vol X. Izd Akad Nauk Latv SSR, Riga, pp 3–94
Liepinsh PP (1963b) Formation conditions of Frasnian deposits in the Baltic Basin (in Russian). In: Frasnian deposits of Latvian SSR. Tr Inst Geol Akad Nauk Latv SSR, vol X. Izd Akad Nauk Latv SSR, Riga, pp 311–337
Liepinsh PP (1973) Western part of major Devonian Area (in Russian). In: Stratigraphy of the USSR, Devonian System, B 1. Nedra Press, Moscow, pp 81–90
Lisitsyn AP (1970) Amorphous silica in the bottom sediments (in Russian). In: The Pacific Ocean, vol II. Sedimentation in the Pacific Ocean. Nauka Press, Moscow, pp 5–25
Lisitsyn AP (1974) Sedimentation in oceans (in Russian). Nauka Press, Moscow, 440 pp
Lisitsyn AP, Petelin VP (1970) Distribution of $CaCO_3$ in the Pacific bottom sediments (in Russian). In: The Pacific Ocean, vol II, Sedimentation in the Pacific Ocean. Nauka Press, Moscow, pp 26–68

Lithological and facial peculiarities of Devonian salt-bearing formations in Tungusska and Vilyui syneclises (1977) (in Russian). In: Problems of salt accumulation, vol II. (Authors: Fradkin GS, Matukhin RG, Menner VV, Sokolov PN, Kolodeznikov KE. Editors: Yanshin AL, Zharkov MA). Nauka Press, Novosibirsk, pp 197–202

Lithology and geochemistry of the Devonian deposits of the Pripyat trough in view of their oil content (1966) Nauka Tekhnika Press, Minsk, 316 pp

Litvin PA (1973) Central and East Turgai Trough (in Russian). In: Stratigraphy of the USSR, Devonian System, B 1. Nedra Press, Moscow, pp 391–392

Litvinovich NV (1972) Upper Paleozoic Tenis Depression (in Russian). In: Geology of the USSR, vol 20 (Central Kazakhstan), Geological description, B 1. Nedra Press, Moscow, pp 382–385

Llado NL, De Vilata JF, Cabanas R, Pruneda JRP, Vilas L (1967) Le Devonien de l'Espagne: In: International Symposium on the Devonian System, vol I, Calgary, Alberta, Canada, pp 171–188

Llewellyn PG, Stabbins R (1968) Core material from the anhydrite series, Carboniferous limestone, Hathern Borehole, Leicestershire. Proc Geol Soc 1650: 171–186

Lohmann HH (1970) Outline of tectonic history of the Bolivian Andes. Am Assoc Petrol Geol Bull 54: 735–757

Lotze F (1938) Steinsalz und Kalisalz, Geologie. In: Lagerstätten der Nichterze, vol III/1. Gebr. Borntraeger, Berlin, 830 pp

Lotze F (1957a) Steinsalz und Kalisalz, 1. Allgemein-geologischer Teil. Gebr. Borntraeger, Berlin, 465 pp

Lotze F (1957b) Vergleich der englischen Zechsteinentwicklung mit der deutschen (am Niederrhein). Z Dtsch Geol Ges 108: 259

Lotze F (1968) The distribution of evaporites in space and time (in Russian). In: Problems in Palaeoclimatology. Mir Press, Moscow, pp 321–336

Lovering TS, Mallory WW (1962) The Eagle Valley evaporite and its relation to the Minturn and Maroon Formations, Northwest Colorado. US Geol Surv Prof Pap 450-D: 45–48

Luchitskiy IV (1960) Middle Paleozoic history of the Siberian Early Caledonian folding (in Russian). In: Caledonian Orogeny. Izd Akad Nauk SSSR, Moscow, pp 105–111

Lukjanov VV (1973) Atmosphere perturbances of the Intertropical Convergence Zone in the Indian Ocean (in Russian). In: Tropical zone and related global processes. Nauka Press, Moscow, pp 48–55

Lupinivich YuI, Kislik VZ, Zelentsov II, Eroshina DM (1970) Some results of potassium research in the Pripyat Depression (in Russian). In: Geology and potassium-bearing of Siberian Platform and of other salt accumulation regions of the USSR. Nauka Press, Moscow, pp 229–239

Lyashenko AI (1962) Correlation of the Devonian deposits of the central and eastern regions of the Russian Platform and of the western slope of the Urals (in Russian). In: Stratigraphic schemes of Paleozoic deposits, Devonian System. Gostoptekhizdat, Moscow, pp 23–27

Lyashenko AI, Lyashenko GP (1967) Correlation of the Devonian deposits of the Russian Platform and the western part of the Urals (in Russian). In: International Symposium on the Devonian System, vol II. Calgary, Alberta, Canada, pp 511–525

Lyutkevich EM (1955) Permian and Triassic deposits of the north-western Russian Platform (in Russian). Tr VNIIGRI Nov Ser Issue 86, Leningrad, 236 pp

Lyutkevich EM (1963) Russian Platform between the Baltic shield and Timan Tectonics (in Russian). In: Geology of the USSR, vol 2 (Arkhangelsk and Vologda regions, Komi ASSR), part 1, Geological description. Gosgeoltekhizdat, Moscow, pp 762–791

Lyutkevich EM (1969) On the boundary of the Upper and Lower series of Permian System (in Russian). In: Permo-Trias of the Russian Platform in connection with oil and gas potential. Nedra Press, Moscow, pp 14–20

Lyutkevich EM (1975) Occurrences of salt tectonics in the western and southern margins of the Russian Platform (in Russian). Tr VNIIGRI, Issue 333. Nedra Press, Leningrad, 104 pp

Lyutkevich EM, Stepanov DL, Trizna VB (1953) Permian deposits of the Soviet Baltic Region (in Russian). Byull Mosk Ova Ispyt Prir Otd Geol 28: 3–14

Macauley G, Penner DG, Procter RM, Tisdall WH (1964) Carboniferous. In: Geological history of Western Canada. Alberta Soc Petrol Geol, Calgary, Alberta, pp 89–102

Magraw D (1975) Permian beds of the offshore and coastal areas of Durham and South-Eastern Northumberland. J Geol Soc 131: 397–414
Mahel M, Vozár J (1971) Prispevok k poznaniu permi a triasu v severogemeridnej synklinàle. Geol Pr Správy 56: 47–66
Makhlaev VG (1964) Sedimentation conditions in the Upper Famennian Basin of the Russian Platform (in Russian). Nauka Press, Moscow, 230 pp
Makhnach AS (1958) On the strike of the Narva horizon in Byelorussia (in Russian). Izv Vuzov Geol Razved 12: 38–45
Makhnach AS, Kurochka VP (1964) Some lithological peculiarities of the Devonian deposits of the Pripyat Depression (in Russian). In: Geology and oil content of Paleozoic deposits of the Pripyat Depression. Nauka Tekhnika Press, Minsk, pp 85–97
Makhnach AS, Britchenko AD, Feshchenko NI (1970) Comparative lithological characteristic of the Devonian deposits in North-Western Dnieper-Donets and South-Eastern Pripyat depressions (in Russian). In: Problems of geology in the Byelorussian SSR and some adjacent areas of Ukrainian SSR. Nauka Tekhnika Press, Minsk, pp 176–183
Makhnach AS, Pasyukevich VN, Semenyuk AD (1964) Narva horizon of Middle Devonian in the region of Polotsk (in Russian). Izv Akad Nauk BSSR Ser Fiz Tekh Nauk 1: 74–82
Mallory WW (1966) Cattle Creek anticline, a salt diapir near Glenwood Springs, Colorado. US Geol Surv Prof Pap 550-B: 12–15
Mallory WW (1971) The Eagle Valley evaporite, North-West Colorado, regional synthesis. Geol Surv Bull 1311-E, 37 pp
Mallory WW (1975) Middle and Southern Rocky Mountains, Northern Colorado Plateau, and Eastern Great Basin region. In: Paleotectonic investigations of the Pennsylvanian system in the United States, part I, US Geol Surv Prof Pap 853: 265–278
Martens IHC (1943) Rock salt deposits of West Virginia. W Va Geol Surv Bull 7: 67 pp
Matrosov PS (1954) On the finds of Upper Permian cordaite spores in salt-bearing sediments of Tuz-Tag (in Russian). Dokl Akad Nauk USSR 95: 623–624
Matthews RD (1970) The origin of Silurian potash in the Michigan Basin. Soc Am N Centr Sect 4th Annu Meet 2: 214
Mattox RB (1968) Paradox basin field area. In: Saline deposits. Geol Soc Am Spec Pap 88: 5–16
Matukhin RG, Menner VV (1974) Devonian and Lower Carboniferous of the North-East Siberian Platform (in Russian). Novosibirsk, 124 pp
Matukhin RG, Menner VV (1975) On the aims of lithological investigations with a view to working out stratigraphic charts of the Middle Paleozoic of the North-East Siberian Platform (in Russian). In: Lithological-geochemical investigations of the Paleozoic and Precambrian of Siberia. Rotaprint IGiG SO Acad Nauk USSR, Novosibirsk, pp 163–172
Maughan EK (1975) Montana, North Dakota, Northeastern Wyoming, and Northern South Dakota. In: Paleotectonic investigations of the Pennsylvanian system in the United States, Part I. US Geol Surv Prof Pap 853: 279–293
Maync V (1964) The Permian of Greenland (in Russian). In: Geology of the Arctic. Mir Press, Moscow, pp 138–149
Mayr U (1975) Correlation of Lower Paleozoic subsurface sections, Devon, Cornwallis, and Somerset Islands, District of Franklin. Rep Activ, Part B. Geol Surv Can Pap 75-1: 255–256
Mayr U (1976) Upper Paleozoic succession in the Yelverton Area, Northern Ellesmere Island, District of Franklin. Geol Surv Can Rep Act Pap 76-1A: 445–448
Mayrhofer H (1955) Beiträge zur Kenntnis des alpinen Salzgebirges. Z Dtsch Geol Ges 105:752–775
McCabe HR (1971) Stratigraphy of Manitoba. An introduction and review. Geol Assoc Can Spec Pap 9: 167–187
McCamis JG, Griffith LS (1968) Middle Devonian facies relations, Zama Area, Alberta. Am Assoc Petrol Geol Bull 52: 1899–1924
McKee, ED (1967) Arizona and Western New Mexico, Chap J. In: Paleotectonic investigations of the Permian System in the United States. US Geol Surv Prof Pap 515: 199–223
McKee ED (1968) Past hot and arid climates (in Russian). In: Problems in Palaeoclimatology. Mir Press, Moscow, pp 229–240
McKee ED, Oriel SS et al (1967) Paleotectonic maps of the Permian System in the United States. US Geol Surv Misc Geol Invest Map 1-450, 164 pp

McLaren DI (1963a) Devonian stratigraphy. In: Geology of the North-Central part of the Arctic Archipelago, Northwest Territories (Operation Franklin). Geol Surv Can Mem 320: 57–62

McLaren DI (1963b) Southwestern Ellesmere Island between Goose Fiord and Bjorne Peninsula. In: Geology of the North-Central part of the Arctic Archipelago, Northwest Territories (Operation Franklin). Geol Surv Can Mem 320: 310–337

McLaren DI (1963c) Meteorologist Peninsula Dome, Ellef Ringnes Island. In: Geology of the North-Central part of the Arctic Archipelago, Northwest Territories (Operation Franklin). Geol Surv Can Mem 320: 552–557

McMannis WJ (1965) Resume of depositional and structural history of Western Montana. Am Assoc Petrol Geol Bull 49: 1801–1823

McNaughton DA, Quinlan T, Hopkins RM, Wells AT (1968) Evolution of salt anticlines and salt domes in the Amadeus Basin, Central Australia. In: Saline deposits. Geol Soc Am Spec Pap 88: 229–247

Meier R (1975) Zu einigen Sedimentgefügen der Werra-Sulfate (Zechstein 1) am Osthang der Eichsfeld-Schwelle. Z Geol Wiss 3: 1333–1347

Meijer-Drees NC (1975) Geology of the Lower Paleozoic Formations in the subsurface of the Fort Simpson Area, District of Mackenzie, N.W.T. Geol Surv Can Pap 74-40: 65 pp

Meleshchenko VS, Predtechensky NN, Yanov EN (1973) Intermontane trough of the Altai-Sayan folded region (in Russian). In: Stratigraphy of the USSR, Devonian System, B 2. Nedra Press, Moscow, pp 95–118

Melnikov IV, Vozhov VI, Vorobiev VN, Kilina LI, Safronova IG (1977) The Cambrian salt-bearing deposits in the west of the Siberian Platform (in Russian). In: Problems of salt accumulation, vol II. Nauka Press, Novosibirsk, pp 231–234

Menner VV (1962) On the history of Devonian sedimentation in the north-western regions of the Siberian Platform (in Russian). Izv Vuzov Geol Razved 12: 3–15

Menner VV (1965) On the distribution of evaporites in Middle Paleozoic deposits of North Siberia (in Russian). Byull Mosk Ova Ispyt Prir Otd Geol 161: 666–669

Menner VV (1967a) Devonian of the Siberian Platform (in Russian). In: Paleozoic stratigraphy of Central Siberia. Nauka Press, Moscow, pp 121–125

Menner VV (1967b) On the facies and petrographic composition of Devonian deposits of North Siberia (in Russian). Byull Mosk Ova Ispyt Prir Otd Geol 4: 140

Menner VV, Fradkin GS (1969) Tectonic conditions of generation of the Devonian halogenic formation of the Siberian Platform (in Russian). In: Tectonics of Siberia, vol III. Nauka Press, Moscow, pp 61–67

Menner VV, Fradkin GS (1973) Middle Paleozoic salt-bearing formation (in Russian). In: Salt tectonics of the Siberian Platform. Nauka Press, Novosibirsk, pp 67–87

Menner VV, Mikhailov MV, Fradkin GS (1973a) Siberian Platform (in Russian). In: Stratigraphy of the USSR, Devonian System, B 2. Nedra Press, Moscow, pp 118–139

Menner VV, Nikolaev AA, Rzhonsnitskaya MA (1973b) North-East of the USSR (in Russian). In: Stratigraphy of the USSR, Devonian System, B 2. Nedra Press, Moscow, pp 157–181

Merzlyakov GA (1976) Permian salt accumulation in Eurasia (in Russian). In: Evolution of sedimentary rocks formation in the history of the Earth. Rotaprint IGiG SO Acad Nauk USSR, Novosibirsk, pp 31–38

Merzlyakov GA (1977) Volumes of halogenic rocks in the Permian salt-bearing formations of Eurasia (in Russian). In: Sedimentary and sedimentary-volcanogenic formations and methods for distinguishing them (in Russian). Nauka Press, Novosibirsk, pp 166–204

Merzlyakov VM (1971) Stratigraphy and tectonics of the Omulevskie Mountains (North-East of the USSR) (in Russian). Nauka Press, Moscow, 152 pp

Meyerhoff AA (1970a) Continental drift. I. Implications of Paleomagnetic studies, meteorology, physical oceanography and climatology. J Geol 78: 1–51

Meyerhoff AA (1970b) Continental drift, II. High-latitude evaporite deposits and geologic history of Arctic and North Atlantic oceans. J Geol 78: 406–444

Meyerhoff AA, Meyerhoff HA (1972) The new global tectonics: major inconsistencies. Am Assoc Petrol Geol Bull 56: 269–336

Meyerhoff AA, Teichert C (1971) Continental drift, III. Late Paleozoic centers, and Devonian-Eocene coal distribution. J Geol 79: 285–321

Miall AD (1974a) Stratigraphy of the Elf et al. Storkerson Bay A-15 well. Rep Act, April to October 1973. Geol Surv Can Pap 74-1: 335–336

Miall AD (1974b) Subsurface geology of Western Banks Island. Rep Act, November 1973 to March 1974. Geol Surv Can Pap 74-1: 178–281

Miall AD (1975) Stratigraphy of the Deminex CGDC Foc Amoca Orksut 1-44 well. Rep Act. Geol Surv Can Pap 75-1: 257–259

Mikhailov AE, Martynova MV, Yurina LL (1973) Western part of North-East Kazakhstan (in Russian). In: Stratigraphy of the USSR, Devonian System, B 1. Nedra Press, Moscow, pp 510–514

Milewicz J (1970a) The Sudetes. In: The Lower Permian (Rotliegendes), Geology of Poland, vol I, part 1. Publ House Wydawnictwa Geol, Warsaw, pp 511–516

Milewicz J (1970b) The Sudetes. In: The Upper Permian (Zechstein), Geology of Poland, vol 1, part 1. Publ House Wydawnictwa Geol, Warsaw, pp 561–566

Milewicz J, Pawłowska K (1970) Paleogeography. In: The Lower Permian (Rotliegendes), Geology of Poland, vol I, part 1. Publ House Wydawnictwa Geol, Warsaw, pp 528–530

Miljush P (1973) Geologic-tectonic structure and evolution of outher Dinarids and Adriatic Area. Am Assoc Petrol Geol Bull 57: 913–929

Mina P, Razaghnia MT, Paran Y (1967) Geological and geophysical studies and exploratory drilling of the Iranian continental shelf-Persian Gulf. 7th World Petrol Cong, Mexico 2: 870–903

Minko GM (1972) Potassium content potential of the Middle Devonian salt-bearing formation of Tuva (in Russian). In: Potassium content potential in Siberia. Nauka Press, Moscow, pp 40–45

Mirouse R (1967) Le Devonien des Pyrenées Occidentales et Centrales (France). In: International Symposium on the Devonian System, vol I. Calgary, Alberta, Canada, pp 153–170

Mitrofanova TV (1973) Upper Paleozoic red beds of Central Kazakhstan (in Russian). In: Problems of stratigraphy and tectonics of East Kazakhstan. Tr VSEGEI Nov Ser 160: 118–130

Morales LG (1959) General geology and oil possibilities of the Amazon Basin, Brazil. 5th World Petrol Congr Y: 925–942

Morales LJ (1961) General geological structure and oil content of the Amazon Basin in Brazil (in Russian). In: Trans 5th Int Cong Petrol, vol I. Gostoptekhizdat, Moscow, pp 196–200

Morgunov YuG, Rudakov SG (1972) Paleozoic Prepermian deposits of Iran and Afghanistan (in Russian). Izv Vuzov Geol Razved 2: 24–40

Morozov LN (1977) Paleogeographic conditions of formation of the potassium salts of the Upper Kama and Elton deposits (in Russian). In: Problems of salt accumulation, vol II. Nauka Press, Novosibirsk, pp 97–101

Morton DM (1959) The geology of Oman. World Petrol Congr S: 277–294

Mossakovsky AA (1970) On the Upper Paleozoic volcanic belt of Europe and Asia (in Russian). Geotektonika 4: 65–77

Mossop GD (1972a) Baumann Fiord Formation evaporites of Central Ellesmere Island. Rep Activ, Part A, April to October 1971. Geol Surv Can Pap 72-1: 216–217

Mossop GD (1972b) Baumann Fiord Formation evaporites of Central Ellesmere Island, Arctic Canada. Rep Activ, Part B, November 1971 to March 1972. Geol Surv Can Pap 72-1: 86–90

Mossop GD (1973) Lower Ordovician evaporites of the Baumann Fiord Formation, Ellesmere Island. Rep Activ, Part A, April to October 1972. Geol Surv Can Pap 73-1: 264–267

Mostler H (1972a) Die permoskytische Transgressions-Serie der Gailtaler Alpen. Verh Geol Bundesanst (Austria), pp 143–149

Mostler H (1972b) Zur Gliederung der Permoskyth-Schichtfolge im Raume zwischen Wörgl und Hochfilzen (Tirol). Verh Geol Bundesanst, Mitt Ges Geol Bergbaustud (Austria) 20: 155–162

Mostler H (1972c) Postvariscische Sedimente im Montafon (Vorarlberg). Verh Geol Bundesanst, Mitt Ges Geol Bergbaustud (Austria) 20: 171–174

Mueller KJ (1967) Devonian of Malaya and Burma. In: International Symposium on the Devonian System, vol I. Calgary, Alberta, Canada, pp 565–588

Muldakulov GG, Marchenko ON (1977) Some new data on the structure of the Kungurian deposits in the eastern marginal zone of the Pre-Caspian Depression (in Russian). In: Problems of salt accumulation, vol II. Nauka Press, Novosibirsk, pp 62–68

Müller A, Schwartz W (1955) Über das Vorkommen von Mikroorganismen in Salzlagerstätten (Geomikrobiologische Untersuchungen III). Z Dtsch Geol Ges 105: 789–802

Nakhabtsev YuS, Fradkin GS (1970) Carboniferous System, Kempendyai Trough (in Russian). In: Geology of the USSR, vol 18 (Western part of the Yakutian ASSR), part 1, Geological description, B 1. Nedra Press, Moscow, pp 272–274

Nalivkin DV (1963) Devonian system (in Russian). In: Geology of the USSR, vol 2 (Arckhangelsk, Vologda regions and Komi ASSR), part 1, Geological description. Gosgeoltekhizdat, Moscow, pp 255–345

Nalivkin VD (1967) The Devonian reefs of the Urals. In: International Symposium on the Devonian System, vol II. Calgary, Alberta, Canada, pp 331–333

Nalivkin VD, Sultanaev AA (1969) Carboniferous System, western slope of the Urals, Middle Division, Upper Division) (in Russian). In: Geology of the USSR, vol 12 (Perm, Sverdlovsk, Chelyabinsk and Kurgan regions), part 1, Geological description. Nedra Press, Moscow, pp 269–291

Naquib AIM (1963) Geology of the Arabian Peninsula, Southwestern Iraq. US Geol Surv Prof Pap 560-G: 54 pp

Narbutas VV (1959) Upper Devonian, Frasnian stage (in Russian). In: Brief account of the Lithuanian SSR. Vilnyus, pp 33–43

Narbutas VV (1961) Gypsums (in Russian). In: Geology of the USSR, vol 39 (Lithuanian SSR), Geological description. Gosgeoltekhizdat, Moscow, pp 186–190

Narbutas VV (1964) Stratigraphy and lithology of the early Frasnian deposits of the Poland-Litva syneclise in the light of facial zonality of the Frasnian Basin in the North-Western Russian Platform (in Russian). In: Problems of stratigraphy and paleogeography of the Devonian of the Baltic. Vilnyus, pp 89–103

Nassichuk WW (1975) Carboniferous ammonoids and stratigraphy in the Canadian Arctic Archipelago. Geol Surv Can Bull 237: 240 pp

Nelson SJ (1964) Ordovician stratigraphy of northern Hudson Bay Lowland, Manitoba. Geol Surv Can Bull 108: 36 pp

Neumann E, Schön K (1975) Bemerkungen zu einer paläotektonischen Analyse am Nordrand des Zechsteinbeckens. Z Geol Wiss 2: 211–224

Newell ND, Chronic I, Roberts TG (1953) Upper Paleozoic of Perm. Geol Soc Am Mem 58: 276

Nikolaev VA (1933) On the most important structural line of Tien-Shan (in Russian). Zap Vses Mineral Ova Ser 2, part 62, Issue 2, pp 354–355

Norford BS, Macqueen RW (1975) Lower Paleozoic Franklin Mountain and Mount Kindle Formations, District of Mackenzie, their type sections and regional development. Geol Surv Can Pap 74-34: 37 pp

Norris AW (1963) Amund Ringnes Piercement Dome, Amund Ringnes Island. In: Geology of the North-Central part of the Arctic Archipelago, Northwest Territories (Operation Franklin). Geol Surv Can Mem 320: 537–544

Norris AW (1965) Stratigraphy of Middle Devonian and older Paleozoic rocks of the Great Slave Lake Region, Northwest Territories. Can Geol Surv Mem 322: 180 pp

Norris AW (1967) Devonian of Northern Yukon Territory and adjacent district of Mackenzie. In: International Symposium on the Devonian System, vol I. Calgary, Alberta, Canada, pp 753–780

Norris AW, Sanford BV (1969) Paleozoic and Mesozoic geology of the Hudson Bay Lowlands. In: Earth Science Symposium on Hudson Bay. Geol Surv Can Pap 68-53: 169–205

North-East Africa and Arabia (in Russian). In: Geology and mineral resources of Africa (1973) (Authors: Ponikarov VP, Uflyand AK, Dolginov EA, Sulidi-Kondratiev ED, Kozlov VV) Nedra Press, Moscow, pp 313–359

Norwood EE (1965) Geologic history of Central and South-Central Montana. Am Assoc Petrol Geol Bull 49: 1824–1832

Novik EO (1952) Devonian deposits of the Dnieper-Donets Depression and its geological history in the Devonian period (in Russian). Geol J 12: 10–25

Novik EO (1954) Stratigraphy of Devonian deposits of the Dnieper-Donets Depression (in Russian). Izv Akad Nauk SSSR, Ser Geol 2: 44–54

O'Brien CAE (1957) Salt diapirism in South Persia. Geol Mijnbouw 19: 357–376

Ohlen HR, McIntyre LB (1965) Stratigraphy and tectonic features of Paradox Basin, Four Corners Area. Am Assoc Petrol Geol Bull 49: 2020–2040

Oliveira AI (1959) Brazil (in Russian). In: Accounts of the geology of South America. In Lit Press, Moscow, pp 23–82
Oliver WA, De Witt W, Dennisok JM, Hoskins DM, Huddle JW (1967) Devonian of the Appalachian Basin, United States. In: International Symposium on the Devonian System, vol I. Calgary, Alberta, Canada, pp 1001–1040
On the potassium salt shows in Morsovsky salt-bearing formation, Central part of the Russian Platform (1970) (in Russian) (Authors: Rozov BN et al). Dokl Akad Nauk USSR 191: 211–213
Opdyke ND (1966) Paleoclimatology and continental drift (in Russian). In: Continental drift. Mir Press, Moscow, pp 159–162
Ordovician stratigraphy of the Siberian Platform (1975) (in Russian). Sokolov BS, Tesakov JuI (eds). Nauka Press, Novosibirsk, 256 pp
Oriel SS (1967a) Base of the Permian System. In: Paleotectonic maps of the Permian System in the United States. US Geol Surv Misc Geol Invest Map 1-450: 26–28
Oriel SS (1967b) Divisions of the Permian System. In: Paleotectonic maps of the Permian System in the United States. US Geol Surv Misc Geol Invest Map 1-450: 28–33
Orlov IV, Gabai NL, Stukalova NI (1969) Salt content and potentials of oil-and-gas presence in the Dzhezkazgan Trough (in Russian). In: Geology of Central Kazakhstan. Nauka Press, Alma-Ata, pp 144–155
Oswald DH (1955) The Carboniferous rocks between the Ox Mountains and Danegal Bay. Geol Soc London Q J III: 167–186
Oxburgh ER (1968) An outline of the geology of the Central Eastern Alps. The Eastern Alps, a geological excursion guide. Proc Geol Assoc 79: 124 pp
Padget P (1953) The stratigraphy of Cuilcagh, Ireland. Geol Mag 90: 17–24
Padula EL, Rolleri ED, Mingramm ARG, Reque PC, Tlores MA, Baldis BA (1967) Devonian of Argentina. In: International Symposium on the Devonian System, vol II. Calgary, Alberta, Canada, pp 165–199
Pajchlowa M (1967) Le Devonien de la Pologne. In: International Symposium on the Devonian System, vol I. Calgary, Alberta, Canada, pp 311–330
Pakhtusova NA (1957) On the problem of salt stone prospecting in Northern Russian Platform (in Russian). In: Mater. on geology and mineral resources of the North-West of the USSR, vol I. Gostoptekhizdat Press, Moscow, pp 18–27
Pakhtusova NA (1963a) Permian System, Lower Division, North of the Russian Platform and South-West Pretiman Region (in Russian). In: Geology of the USSR, vol 2 (Arkhangelsk, Vologda regions and Komi ASSR), part 1, Geological description. Gosgeoltekhizdat, Moscow, pp 463–478
Pakhtusova NA (1963b) Permian System, North of the Russian Platform and South-West Pretiman Region, Upper Division, Ufimian stage (in Russian). In: Geology of the USSR, vol 2 (Arkhangelsk, Vologda Regions and Komi ASSR), part 1, Geological description. Gosgeoltekhizdat Press, Moscow, pp 478–488
Paleogeographic atlas of China (1962) (in Russian). In Lit Press
Paleomagnetic directions and paleomagnetic poles (reference data on the USSR) (1971) (in Russian). Khramov AN (ed). Izd VNIGRI, Leningrad, 124 pp
Paleotectonic investigations of the Pennsylvanian system in the United States, part I, Introduction and regional analyses of the Pennsylvanian system; part II, Interpretative summary and special features of the Pennsylvanian system; part III, Plates (1975). US Geol Surv Prof Pap 853, part I: 349 pp, part II: 192 pp, part III: 14 pp
Papa A (1970) Conceptions nouvelles sur la structure des Albanides (présentation de la Carte tectonique de l'Albanie an 500 000e). Bull Soc Géol Fr Sér 7 XII: 1096–1109
Parker M (1974) Iowa and Minnesota (in Russian). In: Future petroleum provinces of the United States – Their geology and potential. Nedra Press, Moscow, pp 531–536
Pashkevich EI, Pistrak RM, Shulga PL, Krylova AK (1973) Dnieper-Donets Depression (in Russian). In: Stratigraphy of the USSR, Devonian System, B 1. Nedra Press, Moscow, pp 190–201
Pashkevichus I (1961) Silurian System (in Russian). In: Geology of the USSR, vol 39 (Lithuanian SSR), part 1, Geological description. Gosgeoltekhizdat, Moscow, pp 45–62

Pastukhova MV (1960) Mineral composition of the salt rocks of the salt-bearing Middle Devonian sequence in the Tuva Autonomous Region (in Russian). In: Petrography of halogenic rocks. Tr VNII Galurgii, Issue 40, Leningrad, pp 245–302

Pastukhova MV (1965) Authigenous minerals in chemogene-terrigenous rocks of Tuz-Tag salt-bearing strata (in Russian). Lothol Polezn Iskop 1: 31–52

Patrulius D, Iorden M, Mirauta E (1967) Devonian of Romania. In: International Symposium on the Devonian System, vol I. Calgary, Alberta, Canada, pp 127–134

Patterson JR (1961) Ordovician stratigraphy and correlations in North America. Am Assoc Petrol Geol Bull 45: 1364–1378

Pattison J, Smith DB, Warrington G (1973) A review of Late Permian and Early Triassic biostratigraphy in the British Isles. Mem Can Soc Petrol Geol 2: 220–260

Pawłowska K (1970a) The Polish Lowlands, the Góry Świetokrzyskie and the Carpathian Foreland, the Lower Permian (Rotliegendes). In: Geology of Poland, vol I, part 1. Publ House Wydawnictwa Geol, Warsaw, pp 525–527

Pawłowska K (1970b) The extra-salinary facies. The Polish Lowlands, The Upper Permian (Zechstein). In: Geology of Poland, vol I, part 1. Publ House Wydawnictwa Geol, Warsaw, pp 539–546

Pawłowska K (1970c) The Góry Swietokrzyskie and the Carpathian Foreland, The Upper Permian (Zechstein). In Geology of Poland, vol I, part 1. Publ House Wydawnictwa Geol, Warsaw, pp 546–561

Pawłowska K, Poborski J (1970) Paleogeography, The Upper Permian (Zechstein). In: Geology of Poland, vol I, part 1. Publ House Wydawnictwa Geol, Warsaw, pp 569–571

Pearson WJ (1960) Developments in potash in Saskatchewan. Can Mines Met Bull Oct: 759–764

Pearson WJ (1963) Salt deposits of Canada. In: Symposium on salt, Cleveland, Ohio. Ohio Geol Soc, pp 197–239

Peirce HW, Gerrard TA (1966) Evaporite deposits of the Permian Holbrook Basin, Arizona. In: 2nd Symposium on salt, vol I, Cleveland, Ohio. Northern Ohio Geol Soc, pp 1–10

Peive AV, Shtreis NA, Mossakovskiy AA et al (1972) Paleozoides of Eurasia and some questions of geosynclinal process evolution (in Russian). Sov Geol 12: 7–25

Pepper IF (1947) Areal extent and thickness of the salt deposits of Ohio. Ohio J Sci 47: 225–239

Permian System (1966) (in Russian). Nedra Press, Moscow, 536 pp

Permian System of the Baltic Region (fauna and stratigraphy) (1975) (in Russian). Suveizdis (ed). Mintis Press, Vilnyus, 218 pp

Pershina AI (1962) Silurian and Devonian deposits of the Chernyshev Ridge (in Russian). Izd Akad Nauk USSR, Moscow Leningrad, 122 pp

Pershina AI (1966) Silurian and Devonian deposits on the western slope of the Urals (in Russian). In: Paleozoic deposits of the North Urals Region. Nauka Press, Moscow Leningrad, pp 3–34

Pershina AI (1972) Paleogeography of the Early Silurian epoch (in Russian). In: History of geological evolution of the North Region in the Paleozoic and Mesozoic. Nauka Press, Leningrad, pp 11–12

Peterson JA (1968) Regional stratigraphy of the Paradox Basin, Utah and Colorado. In: Saline deposits. Geol Soc Am Spec Pap 88: 422–423

Peterson JA, Hite RJ (1969) Pennsylvanian evaporite-carbonate cycles and their relation to petroleum occurrence, Southern Rocky Mountains. Am Assoc Petrol Geol Bull 53: 884–908

Petrushevsky BA (1938a) A note on the domes of the Dzhezkazgan Region (in Russian). Dokl Akad Nauk USSR, Nov Ser XIX: 69–71

Petrushevsky BA (1938b) Sarysu Domes (in Russian). Izv Akad Nauk USSR, Ser Geol 3: 409–433

Pierce WG, Rich EI (1962) Summary of rock salt deposits in the United States as possible storage sites for radioactive waste materials. US Geol Surv Bull 1148: 91 pp

Pisarchik YaK, Minaeva MA, Rusetskaya GA (1975) Paleogeography of the Siberian Platform in the Cambrian (in Russian). Nedra Press, Leningrad, 195 pp

Pisarenko YuA, Belonozhko VS, Burunkov VA, Fainitsky SB, Chudin AV (1977) Results of correlation of the salt-bearing sequence in the north-west of the marginal zone of the Pre-Caspian Depression (in Russian). In: Problems of salt accumulation, vol II. Nauka Press, Novosibirsk, pp 36–40

Poborski J (1970) The salinary facies, The Upper Permian (Zechstein). In: Geology of Poland, vol I, part 1. Publ House Wydawnictwa Geol, Warsaw, pp 533–539

Poborski J, Pawlowska K (1970) General geological description. The Upper Permian (Zechstein). In: Geology of Poland, vol I, part 1. Publ House Wydawnictwa Geol, Warsaw, pp 530–532

Polster LA, Janev S, Schustova LG, Gnoevaya N, Ulizlo BM (1976) Geological development of the North Bulgaria territory in the Paleozic (in Russian). Izv Akad Nauk SSSR, Ser Geol 5: 113–128

Pomyanovskaya GM (1964) Comparative characteristic of the Middle Devonian salt deposits of the Volyn-Podolia marginal zone of the Russian Platform and the Pripyat Depression (in Russian). In: Geology and oil-and-gas content in the Volyn-Podolia marginal zone of the Russian Platform. Tr Ukr NIGRI, Issue IX. Nedra Press, Leningrad, pp 33–39

Ponikarov VP, Uflyand AK, Dolginov EA, Sul'di-Kondratiev ED, Kozlov VV (1973) North-East Africa and Arabia (in Russian). In: Geology and mineral resources of Africa. Nedra Press, Moscow, pp 313–359

Poole FG, Baars DL, Dreves H, Hayes PT, Ketner KB, McKee ED, Teichert C, Williams JS (1967) The Southwestern United States. In: International Symposium on the Devonian system, vol I. Calgary, Alberta, Canada, pp 879–912

Porter IW et al (1964) Ordovician-Silurian, Chap 4. Geological history of Western Canada. Alberta Soc Petrol Geol, Calgary, Alberta, pp 34–48

Potassium salt deposits of the USSR. Methods of their exploration and prospecting (1973) (in Russian) (Authors: Raevsky VI, Fiveg MP, Gerasimova VV, Gorbov AF et al. Editors: Raevsky VI, Fiveg MP). Nedra Press, Leningrad, 344 pp

Poyarkov BV (1972) Carboniferous System, Lower Division (in Russian). In: Geology of the USSR, vol 25 (Kirghiz SSR), Geological description, B 1. Nedra Press, Moscow, pp 149–170

Praehauser-Enzenberg (1972) Das Permoskyth vom Penken (Zillertal). Verh Geol Bundesanst, Mitt Ges Geol Bergbaustud (Austria) 20: 167–170

Prather RW, McCourt GB (1968) Geology of gas accumulations in Paleozoic rocks of Alberta Plains. In: Natural gases of North America, vol II. Tulsa, Oklahoma USA, pp 1238–1284

Price LL, Ball NL (1971) Stratigraphy of Duval Corporation potash, Shaft no. 1, Saskatoon, Saskatchewan. Geol Surv Can Pap 70-71: 107 pp

Prichard GE (1975) Nebraska and adjoining parts of South Dakota and Wyoming. In: Paleotectonic investigations of the Pennsylvanian system in the United States, Part I. US Geol Surv Prof Pap 853: 115–126

Privalova LA, Antipova AS, Savitskaya VN (1968) Salt deposits and salt manifestations in the European part of the USSR and Caucasus (in Russian). Tr VNIIsol, Issue 13 (21). Nedra Press, Leningrad, 182 pp

Problems of salt accumulation (1977) (in Russian), vol I–II. Ynashin AL, Zharkov MA (eds). Nauka Press, Novosibirsk, vol I: 305 pp, vol II: 311 pp

Proctor RM, Macauley G (1968) Mississippian of Western Canada and Williston Basin (1968) Am Assoc Petrol Geol Bull 52: 1956–1968

Pronin AA (1969) Carboniferous System. Eastern slope of the Middle Urals (in Russian). In: Geology of the USSR, vol 12 (Perm, Sverdlovsk, Chelyabinsk and Kurgan regions), part 1, Geological description, B 1. Nedra Press, Moscow, pp 291–306

Quigley MD (1965) Geologic history of Piceance Creek-Eagle Basins. Am Assoc Petrol Geol Bull 49: 1974–1996

Ramovs A (1972) Mittelpermische Klastite und deren marine Altersäquivalente in Slowenien, NW Jugoslavien. Verh Geol Bundesanst, Mitt Ges Geol Bergbaustud (Austria) 20: 35–45

Rampnoux JP (1970) Regards sur les Dinarides internes yougoslaves (Serbie-Monténégro oriental): stratigraphie, évolution paléogéographique, magmatisme. Bull Soc Géol Fr Sér 7 XII: 948–966

Ranford LC, Cook PJ, Wells AT (1965) The geology of the Central part of the Amadeus Basin, Northern Territory. Bur Miner Resour Aust Rep 86: 48 pp

Rauzer-Chernousova DM, Khachatryan RO, Yarikov GM (1967) The Carboniferous System. In: Geology of the USSR, vol XI. Volga Region and Kama Region (in Russian), part 1. Nedra Press, Moscow, pp 226–331

Reedy HJ (1968) Carter-Knox gas field, Oklahoma. In: Natural gases of North America, vol II. Am Assoc Petrol Geol Mem 9. Tulsa, Oklahoma, USA, pp 1476–1491
Reefs of the Ural-Volga Region, their role in the distribution of the oil and gas fields and prospecting methods (1974) (in Russian) (Authors: Mirchink MF, Mkrtchyan OM, Khatyanov FI et al). Nedra Press, Moscow, 153 pp
Reference section of the Silurian and Lower Devonian of Podolia (1972) (in Russian) (Authors: Nikiforova OI, Predtechensky NN, Abushik AF, Ignatovich MM, Modzalevskaya TZ, Burkov YuK). Nauka Press, Leningrad, 262 pp
Regional stratigraphy of China (1960) (in Russian). Issue 1. In Lit Press, Moscow, 349 pp
Regional stratigraphy of China (1963) (in Russian). Issue 2. In Lit Press, Moscow, 274 pp
Renaud A, Poncet J, Cavet P, Lardeux H, Babin C (1967) Le Devonien du Massif Armorican. In: International Symposium on the Devonian system, vol I. Calgary, Alberta, Canada, pp 135–152
Repina LN (1968) Biogeography of the Early Cambrian of Siberia based on trilobites (in Russian). In: Problems of paleontology. Papers presented by the Soviet geologists at the XXIIIrd Session of the International Geological Congress. Nauka Press, Moscow, pp 46–56
Richter DK, Zinkernagel U (1975) Petrographie des „Permoskyth" der Jaggl-Plawen-Einheit (Südtirol) und Diskussion der Detritusherkunft mit Hilfe von Kathoden-Lumineszenz-Untersuchungen. Geol Rundsch 64: 783–807
Richter-Bernburg G (1955) Über salinare Sedimentation. Z Dtsch Geol Ges 105: 593–645
Richter-Bernburg G (1957) Zur Paläogeographie des Zechsteins. Acti del Convegno di Milano su i Giacimenti Gassiferi dell Europa Occidentale, Mailand
Richter-Bernburg G (1959) Zur Paläogeographie des Zechsteins. In: Accad Naz Lincei, Giacimenti gassiferi dell Europa occidentale (Rome) 1: 87–99
Richter-Bernburg G (1972) Saline deposits in Germany, a review and general introduction to the excursions, geology of saline deposits, Proc Hannover Symposium 1968 (Earth Sci 7). Unesco, Paris, pp 275–287
Riehl-Herwirsch G (1972) Vorstellung zur Paläogeographie – Verrucano. Verh Geol Bundesanst (Austria), Mitt Ges Geol Bergbaustud 20: 97–106
Riehl-Herwisch G, Wascher W (1972) Die postvariscische Transgressionsserie im Bergland vom Magdalensberg (Basis der Krappfeldtrias, Kärnten). Verh Geol Bundesanst, Mitt Ges Geol Bergbaustud (Austria) 20: 127–138
Rivas S (1968) Geologiá de la regioń norte del Lago Titicaca. Servi Geol Bolivia 1: 88 pp
Robinson PL (1973) Palaeoclimatology and continental drift. In: Implications of continental drift in the Earth Sciences, vol I. Academic Press, London New York, pp 451–476
Rodgers J (1970) The Pulaski fault and the extent of Cambrian evaporites in the Central and Southern Appalachians. In: Studies of Appalachian geology: Central and Southern. Wiley, New York, pp 12–18
Roehl PO (1967) Stony Mountain (Ordovician) and Interlake (Silurian) facies analogs of recent low-energy marine and subaerial carbonates. Am Assoc Petrol Geol Bull 51: 1979–2032
Rogozov YuG (1970) Novaya Zemlya and Vaigach, Carboniferous System (in Russian). In: Geology of the USSR, vol 24 (The Soviet Arctic Islands), Geological description. Nedra Press, Moscow, pp 138–148
Ronov AB (1959) Contribution to the post-Precambrian geochemical history of the atmosphere and hydrosphere (in Russian). Geokhimia 5: 397–409
Ronov AB (1964) Common tendencies in the evolution of the Earth crust, the ocean and the atmosphere (in Russian). Geokhimia 8: 715–743
Ronov AB (1976) Volcanism, carbonate accumulation, life (the regularities of the global geochemistry of carbon) (in Russian). Geokhimia 8: 1252–1277
Ronov AB, Khain VE (1954) Devonian lithological formations of the World (in Russian). Sov Geol 41: 46–76
Ronov AB, Khain VE (1955) Carboniferous lithological formations of the World (in Russian). Sov Geol 48: 92–117
Ronov AB, Khain VE (1956) Permian lithological formations of the World (in Russian). Sov Geol 54: 20–36

Ronov AB, Khain VE (1957) History of sedimentation in Middle and Upper Paleozoic in connection with hercynian stage of tectonic evolution of the Earth's crust (in Russian). Sov Geol 58: 8–24

Ronov AB, Magdisov AA, Barskaya NB (1969) Regularities of evolution of the sedimentary rocks and the paleogeographic sedimentation conditions on the Russian Platform (an attempt at a quantitative investigation) (in Russian). Lithol Polezn Iskop 6: 3–36

Ronov AB, Seslavinsky KB, Khain VE (1974) Cambrian lithological formations of the world (in Russian). Sov Geol 12: 10–33

Ronov AB, Khain VE, Seslavinsky KB (1976) Ordovician lithological formations of the world (in Russian). Sov Geol 1: 7–27

Roots EF (1963a) Axel Heiberg and Stor Islands, Cape Level gypsum diapir. In: Geology of the North-Central part of the Arctic Archipelago, Northwest Territories (Operation Franklin). Geol Surv Can Mem 320: 462–465

Roots EF (1963b) Axel Heiberg and Stor Islands, gypsum body west of Mökka Fiord. In: Geology of the North-Central part of the Arctic Archipelago, Northwest Territories (Operation Franklin). Geol Surv Can Mem 320: 482–295

Roots EF (1963c) Axel Heiberg and Stor Islands, gypsum body west of Flat Sound. In: Geology of the North-Central part of the Arctic Archipelago, Northwest Territories (Operation Franklin). Geol Surv Can Mem 320: 496–498

Rudich EM, Shapiro MN (1976) Plate tectonics and correlation of Eurasia and North America (in Russian). In: Continental drift. Nauka Press, Moscow, pp 3–27

Rutten MG (1972) Geology of Western Europe (in Russian). Mir Press, Moscow, 445 pp

Ruttner A, Nabavi M, Hajian I (1968) Geology of theShirhesth Area, Tabas Area, East Iran. Iran Geol Surv Rep 4: 28 pp

Rzhonsnitskaya MA (1967) Devonian of the USSR. In: International Symposium on the Devonian System, vol I. Calgary, Alberta, Canada, pp 331–348

Rzhonsnitskaya MA (1973) The Kuznetsk Trough and the adjacent regions of Western Siberia (in Russian). In: Stratigraphy of the USSR, Devonian System, B 2. Nedra Press, Moscow, pp 15–55

Said R (1962) The geology of Egypt. Elsevier, Amsterdam New York, 378 pp

Salamuni R, Bigarella JJ (1967) Some palaeogeographical features of the Brazilian Devonian. In: International Symposium on the Devonian System, vol II. Calgary, Alberta, Canada, pp 1313–1326

Salas GP (1968a) Saline deposits of Northern Mexico (abs.). In: Saline deposits. Geol Soc Am Spec Pap 88: 414

Salas GP (1968b) Petroleum evaluation of North-Central Mexico. Am Assoc Petrol Geol Bull 52: 665–674

Salt potential of the Devonian in Siberian Platform (1968) (in Russian). In: Mater.on regional geology of Siberia (Authors: Bgatov VI, Matukhin RG, Menner VV, Fradkin GS). Nauka Press, Novosibirsk, pp 169–173

Samoilenko VS (1973) Some results and further aims to the meteorologic investigations in the tropical zone of the ocean (in Russian). In: Tropical zone of the World Ocean and the associated global processes. Nauka Press, Moscow, pp 24–35

Sandberg CA (1962) Geology of the Williston Basin, North Dakota, Montana, and South Dakota, with reference to subsurface disposal of radioactive waste. US Geol Surv TEI 809: 148 pp

Sandberg CA, Mapel WJ (1967) Devonian of the Northern Rocky Mountains and Plains. In: International Symposium on the Devonian System, vol I. Calgary, Alberta, Canada, pp 843–877

Sanford BV (1965) Salina salt beds Southwestern Ontario. Geol Surv Can Pap 65-9: 7 pp

Sanford BV (1974) Paleozoic geology of Hudson Bay Region. Rep Act, Part B, November 1973 to March 1974. Geol Surv Can Pap 74-1: 144–146

Sanford BV, Norris AW (1975) Devonian stratigraphy of the Hudson Platform. Geol Surv Can Mem 379: Part I – 124 pp, Part II – 248 pp

Sanford BV, Norris AW, Bostock HH (1968) Geology of the Hudson Bay Lowlands (Operation Winisk). Geol Surv Can Pap 67-60: 118 pp

Sapegin BM (1977) Distribution of the potassium chloride in the sylvinite layers of the Upper Kama deposit (in Russian). In: Problems of salt accumulation, vol II. Nauka Press, Novosibirsk, pp 118–120

Sarkisyan SG, Teodorovich GI (1955) Main features of paleogeography of the Devonian epoch in the Ural-Volga Region (in Russian). Izd Akad Nauk USSR, Moscow, 256 pp

Schauberger O (1955) Zur Genese des alpinen Haselgebirges. Z Dtsch Geol Ges 105: 736–751

Schauberger O, Zankl H, Kühn R, Klaus W (1976) Die geologischen Ergebnisse der Salzbohrungen im Talbecken von Bad Reichenhall. Geol Rundsch 65: 558–579

Schenk PE (1969) Carbonate-sulfate red bed facies and cyclic sedimentation of the Windsorian Stage (Middle Carboniferous). Maritime Provinces. Can J Earth Sci 6: 1037–1066

Schindewolf OM (1954) Über die Faunenwende vom Palaeozoikum zum Mesozoikum. Z Dtsch Geol Ges 105: 153–182

Schlatter LE, Nederlof MH (1969) Bosquejo de la geologiá y paleogeografiá de Bolivia. Serv Geol Bolivia 8: 49 pp

Schmalz RF (1969) Deep-water evaporite deposition. A genetic model. Am Assoc Petrol Geol Bull 53: 798–823

Schwarzbach M (1963) Climates of the Past. Van Nostrand, London, 328 pp

Schwerdtner WM (1964) Genesis of potash rocks in Middle Devonian Prairie Evaporite Formation of Saskatchewan. Am Assoc Petrol Geol Bull 48: 1108–1115

Sedimentary and volcanogenic-sedimentary formations of Verkhoyanye, the Paleozoic and Lower Mesozoic (1976) (in Russian) (Authors: Bulgakova MD, Korobitsin AV, Semyonov VP, Ivensen VYu). Nauka Press, Novosibirsk, 134 pp

Seifert J (1972) Das Perm am Südostrand des Thüringer Beckens. Jahrb Geol, vol IV. Akademie-Verlag, Berlin, pp 97–180

Selivanova VA (1971) Ordovician System (in Russian). In: Geology of the USSR, vol 1 (Leningrad, Pskov, Novgorod regions), Geological description. Nedra Press, Moscow, pp 127–174

Shafiro JSh (1972) Correlation of Permian halogene deposits of the Volgograd Volga region and tectonic conditions of their formation (in Russian). Byull Mosk Ova Ispyt Prir Otd Geol 47: 45–59

Shafiro JSh (1975) Lower Permian halogene formation of Pre-Caspian syneclise and its northwestern regions (in Russian). Byull Mosk Ova Ispyt Prir Geol 1: 22–39

Shafiro JSh (1977) Structure and formation conditions of the Lower Permian halogenic deposits of north-western margin of the Pre-Caspian Depression and the adjacent areas (in Russian). In: Problems of salt accumulation, vol II. Nauka Press, Novosibirsk, pp 32–36

Shafiro JSh, Sipko TA (1973) On the volume of Verkhneartinsk substage in Saratov-Volgograd Povolzhiie (in Russian). Dokl Akad Nauk USSR 213: 685–687

Shakhov RA (1965) Major features of the geological structure and oil-and-gas presence in the Chu-Sarysu Syneclise (in Russian). In: Data on the geology and mineral resources of South Kazakhstan, Issue 3 (28). Nauka Press, Alma-Ata, pp 192–202

Shakhov RA (1965b) Geological structure and oil-and-gas presence potentials of the Chu-Sarysu Syneclise (in Russian). Geol Nefti Gaza 5: 14–18

Shakhov RA (1968) Major stages of geological evolution of the Chu Depression in the Middle and Upper Paleozoic (in Russian). In: Geological structure and potentials of oil-and-gas content in the depressions of South Kazahkhstan. Kazakhstan Press, Alma-Ata, pp 145–154

Shaw WS, Blanchard JE (1968) Salt deposits of the Maritime provinces of Canada (abs.). In: Saline deposits. Geol Soc Am Spec Pap 88: 414–415

Shchedrovitskaya ES (1973) Pripyat potassium Basin (in Russian). In: Potassium salt deposits of the USSR. Nedra Press, Leningrad, pp 16–59

Shcherbina VN (1945) Generic and age types of the gypsum deposits of the Kazakhstan (in Russian). Izv Kaz Fil Acad Nauk USSR Ser Geol 4–5: 99–116

Shcherbina VN (1954) Fossil soda gypsum-bearing deposits (in Russian). Tr Inst Geol Kirg Fil Acad Nauk USSR 5: 123–129

Shik EM (1971) Carboniferous System, Middle Division (in Russian), Geological description. In: Geology of the USSR, vol 4 (Centre of the European regions of the USSR). Nedra Press, Moscow, pp 258–291

Shirley J (1968) Range of the Early Devonian fauna (in Russian). In: Problems of Palaeoclimatology. Mir Press, Moscow, pp 154–162
Shmalz RF (1972) Deep-water evaporite deposition: A genetic model (in Russian). In: Salt accumulation and salt-bearing deposits of the sedimentary basins. Nedra Press, Moscow, p 5–45
Siedlecka A (1970) The Silesia-Cracow Upland, the Lower Permian (Rotliegendes). In: Geology of Poland, vol I, Part 1. Publ House Wydawnictwa Geol, Warsaw, pp 519–525
Silurian of Estonia (1970) (in Russian). Kalio DL (ed). Valgus Press, Tallin, 343 pp
Sinitsyn VM (1970) Ancient climates of Eurasia, Part 3 (in Russian). In: The upper half of the Paleozoic (Devonian, Carboniferous, Permian). Izd LGU, Leningrad, 134 pp
Sinitsyn FE, Filipiev GP, Khromova NP (1977) Salt-bearing formations of the epigeosynclinal Paleozoic of the Chu-Sarysu Syneclise and problems of their oil-and-gas content (in Russian). In: Problems of salt accumulation, vol II. Nauka Press, Novosibirsk, pp 141–146
Sitter LU, Sitter-Koomans CM (1949) The geology of the Bergamasc Alps, Lombardia, Italy. Leid Geol Med 14: 257 pp
Sloss LL, Dapples ES, Krumbein WC (1960) Lithofacies maps: an atlas of the United States and Southern Canada. John Wiley and Sons, New York, 108 pp
Smith A, Hallam AH (1970) The fit of the southern continent. Nature (London) 225: 139–144
Sokolov BS, Polenova EN (1968) Silurian-Devonian boundary (in Russian). In: Biostratigraphy of the Silurian-Devonian boundary deposits. Nauka Press, Moscow, pp 3–24
Sokolov PN (1977) Distribution regularities of some trace elements in the rocks of the Cambrian halogenic formation in the South-West of the Siberian Platform (in Russian). In: Problems of salt accumulation, vol II. Nauka Press, Novosibirsk, pp 240–244
Sokolov PN, Matukhin RG (1975) To the lithological characteristic of the Middle Devonian salt-bearing deposits of the Imangdy River Basin, North-West of the Siberian Platform (in Russian). In: Lithological-geochemical investigations of the Paleozoic and Precambrian in Siberia. Rotaprint IGiG SO Acad Nauk SSSR, Novosibirsk, pp 145–152
Sokolov PN, Mathukhin RG, Divina TA, Mokrousov YuN, Cherevkov EA (1977) Salinity of Lower Devonian and Upper Silurian rocks of the North-West Siberian Platform (in Russian). Sov Geol 1: 160–164
Sokołowski S (1970) The Carpathians, the Upper Permian (Zechstein). In: Geology of Poland, vol I, part 1. Publ House Wydawnictwa Geol, Warsaw, pp 567–569
Some aspects of the Paleozoic of the Tarim Platform (in China) (1965) (Authors: Khu Bin et al). Geol J China 45: 131–141
Sommer D (1972) Die Prebichlschichten als permotriadische Basis der nördlichen Kalkalpen in der östlichen Grauwackenzone (Steiermark, Österreich). Verh Geol Bundesanst, Mitt Ges Geol Bergbaustud (Austria) 20: 119–122
Sosipatrova GP (1967) Foraminifer assemblages of the Upper Paleozoic of Spitsbergen. Dokl Akad Nauk SSSR 176: 182–185
Souther JG (1963) Axel Heiberg and Stor Islands, Buchanan Lake to Strand Fiord. In: Geology of the North-Central part of the Arctic Archipelago, Northwest Territories (Operation Franklin). Geol Surv Can Mem 320: 426–447
Sozansky VI (1973) Geology and genesis of the salt-bearing formations (in Russian). Naukova Dumka Press, Kiev, 200 pp
Spasov Kh, Yanev S (1966) Stratigraphy of Paleozoic sediments of North-East Bulgaria (in Bulgarian). Bull Geol Inst Bulg Acad Sci Ser Stratigr Lithol 15: 93–101
Spizharsky TN (1973) Sketched tectonic maps of the USSR (in Russian). Nedra Press, Leningrad, 240 pp
Stanoiu I (1971) Nota, preliminara asupra prezentei silurianuluc fosilifer in Carpatii meridionali-Dari de seama all sedintelor, vol 57 (1969–1970), pp 5–15
Stehli F (1963) Probable climatic zoning in the Permian and its implication (in Russian). In: Problems of the continent displacements. In Lit Press, Moscow, pp 212–219
Stewart FH (1963a) The Permian Lower evaporites of Fordon in Yorkshire. Proc Yorks Geol Soc 34: 1
Stewart FH (1963b) Marine evaporites, Chap Y. In: Data of geochemistry, 6th edn. US Geol Surv Prof Pap 440-Y: 1–52

Stewart JS (1945) Recent exploratory deep-well drilling in the Mackenzie River Valley, Northwest Territories. Can Geol Surv Pap 45-29: 24
Stibane FR (1967) Devonian of the Cordillera of Colombia. In: International Symposium on the Devonian System, vol II. Calgary, Alberta, Canada, pp 209–213
Stöcklin J (1962) Salt deposits of the Middle East. Geol Soc Am Prog Annu Meet, Houston, 185A
Stöcklin J (1968a) Salt deposits of the Middle East. In: Saline deposits. Geol Soc Am Spec Pap 88: 157–181
Stöcklin J (1968b) Structural history and tectonics of Iran, a review. Am Assoc Petrol Geol Bull 52: 1229–1258
Stöcklin J, Ruttner A, Nabavi M (1964) New data on the Lower Paleozoic and Precambrian of North Iran. Iran Geol Surv Rep 1: 29 pp
Stołarczyk F (1972) Nowe dane o permie wshodniej czesci syneklizy perybaltyckiej. Kwart Geol 16: 113–130
Stolle E, Döhner C (1976) Die Erforschung der Kalilagerstätten der DDR in den vergangenen 20 Jahren. Z Geol Wiss 4: 577–590
Stolle E, Reichenbach W, Bohm G, Liebisch K, Lorenz S (1975) Genetische Aspekte der vertikalen und horizontalen Qualitätsgliederung des Flözes Ronnenberg (Zechstein 3) auf der Scholle von Calvorde und ihre Bedeutung für die Erkundung. Z Geol Wiss 2: 143–156
Strakhov NM (1960) Foundations of the lithogenesis theory, vol I, Types of lithogenesis and their distribution on the surface of the Earth (in Russian). Izd Akad Nauk USSR, Moscow, 212 pp
Strakhov NM (1962) Foundations of the lithogenesis theory, vol III (in Russian). Izd Akad Nauk USSR, Moscow, 550 pp
Strakhov NM (1963) Types of lithogenesis and their evolution in the history of the Earth (in Russian). Gosgeoltekhizdat, Moscow, 535 pp
Strakhov NM (1971) Evolution of lithogenetic notions in Russia and the USSR. Critical review (in Russian). Nauka Press, Moscow, 622 pp
Strakhov NM (1976) Problems of geochemistry of the Recent oceanic lithogenesis (in Russian). Nauka Press, Moscow, 300 pp
Stratigraphy of the sedimentary formations of Moldavia (1964) (in Russian) (Authors: Bobrinskaya OG, Bobrinsky VM, Bukatchuk PO et al). Map of Moldavia Press, Kishinev, 132 pp
Streltsova TV (1972) To the question about the Paleozoic history of the Mizian median mass (in Russian). Vest Mosk Gos Univ Ser Geol 4: 85–88
Structure and formation conditions of the Cambrian salt-bearing deposits within the territory of the USSR (1977) (in Russian). In: Problems of salt accumulation, vol II (Authors: Britain IV, Zharkov MA, Kavitsky ML, Kolosov AS, Mashovich YaG, Chechel EI). Nauka Press, Novosibirsk, pp 203–226
Structure and formation conditions of the Permian salt-bearing deposits on the territory of the Soviet Union (1977) (in Russian). In: Problems of salt accumulation, vol II (Authors: Tikhvinsky IN, Merzlyakov GA, Sementovsky YuV, Suveizdis PI). Nauka Press, Novosibirsk, pp 3–16
Summet AYu (1971) Devonian system (in Russian). In: Geology of the USSR, vol 1 (Leningrad, Pskov and Novgorod regions), Geological description. Nedra Press, Moscow, pp 174–245
Summet AYu (1973) Eastern main Devonian Area (in Russian). In: Stratigraphy of the USSR, Devonian system, B 1. Nedra Press, Moscow, pp 90–106
Suveizdis PI (1963) Upper Permian of the Polish-Lithuanian Syneclise (in Russian). In: Problems of the geology of Lithuania. Vilnyus, pp 225–372
Suveizdis PI, Smilgis II (1975) To the probem of evolution and the reservoir properties of the reefogenic formations in the Zechstein of the Baltic Region (in Russian). In: Lithology and paleogeography of the Biogerm Massifs. Nauka Press, Moscow, pp 139–145
Svidzinsky SA, Kovalsky FI, Morozov LN, Anoshin LV, Muzalevsky MM, Bordyugov VP (1977) Composite section of the halogenic formations of the Elton structure and principles of its correlation (in Russian). In: Problems of salt accumulation, vol II. Nauka Press, Novosibirsk, pp 49–54
Swann DH, Willman HB (1961) Megagroups in Illinois. Am Assoc Petrol Geol Bull 45: 471–483

Tanner JJ (1967) Devonian of the Adavale Basin, Queensland, Australia. In: International Symposium on the Devonian System, vol II. Calgary, Alberta, Canada, pp 111–116
Tassonyi EI (1969) Subsurface geology, Lower Macken River and Anderson River Area, District of Mackenzie. Geol Surv Can Pap 68-25: 207 pp
Taylor FC (1969) Geology of the Annapolis-St. Marys Bay map-area, N S (21A, 21B, East Half). Geol Surv Can Mem 358: 65 pp
Tectonic map of Eurasia, 1:5 000 000 (1966) Yanshin AL (ed). Izd GUGKGGK SSSR, Moscow
Tectonics of Africa (1973). Mir Press, Moscow, 543 pp
Tectonics of Byelorussia (1976) (in Russian) Garetsky RG (ed). Nauka Tekhnika Press, Minsk, 200 pp
Tectonics of Eurasia: Explanatory note to the Tectonic Map of Eurasia on the scale 1:5 000 000 (1966) (in Russian). Nauka Press, Moscow, 488 pp
Tectonics of Europe: Explanatory note to the International Tectonic Map on the scale 1:2 500 000 (1964) (in Russian). Nauka Press, Moscow, 364 pp
Teixeira C, Thadeau D (1967) Le Devonien du Portugal. In: International Symposium on the Devonian system, vol I. Calgary, Alberta, Canada, pp 189–199
Termier H, Termier G (1964) Les temps fossilifères. 1. Paleozoique Inferieur, 690 pp
Teruggi ME, Andreis RA, Inigues AM, Abait JP, Mazzoni MM, Spalletti LA (1969) Sedimentology of the Paganzo beds at Cerro Guandacol, Province of la Rioja. In: Gondwana stratigraphy. Unesco, Paris, pp 857–880
Tesakov YuI (1971) Favositids of Podolia (in Russian). Nauka Press, Moscow, 121 pp
The Carboniferous System, Lower Section (1971) (in Russian). In: Geology of the USSR, vol 4 (Centre of the European part of the USSR), Geological description. (Authors: Birina LM, Sorskaja LS et al). Nedra Press, Moscow, pp 194–258
The Devonian of the Russian Platform (1967). In: International Symposium on the Devonian System, vol I (Authors: Aronova SM, Gassanova IG et al). Calgary, Alberta, Canada, pp 379–396
The Devonian System (1973) (in Russian). Nedra Press, Moscow, B 1: 520 pp, B 2: 376 pp
The geologic evolution of the Japanese Islands (1968) (in Russian). Mir Press, Moscow, 720 pp
The Paleozoic salt-bearing formations of the North-Western Dnieper-Donets-Depression (1977) (in Russian). In: Problems of salt accumulation, vol II (Authors: Britchenko AD, Vakarchuk GI, Tkachuk SV, Khmel FF). Nauka Press, Novosibirsk, pp 170–175
Thomas GA (1959) The Lower Carboniferous Lanrel Formation of the Fitzroy Basin. In: Papers on Western Australian stratigraphy and paleontology. Bur Miner Resour Aust Rep 38: 21–36
Thorsteinsson R (1963) Axel Heiberg and Stor Islands North Side of Strand Fiord. In: Geology of the North-Central part of the Arctic Archipelago, Northwest Territories (Operation Franklin). Geol Surv Can Mem 320: 457–461
Thorsteinsson R (1974) Carboniferous and Permian stratigraphy of Axel Heiberg Island and Western Ellesmere Island, Canadian Arctic Archipelago. Geol Surv Can Bull 224: 115 pp
Thorsteinsson R, Kerr JW (1968) Cornwallis Island and adjacent Smaller Island, Canadian Arctic Archipelago. Geol Surv Can Pap 67-64: 16 pp
Thorsteinsson R, Tozer ET (1960) Summary account of structural history of the Canadian Arctic Archipelago since Precambrian time. Geol Surv Can Pap 60-7: 122
Thorsteinsson R, Tozer ET (1961) Banks, Victoria and Stefansson Islands, District of Franklin, Northwest Territories. Geol Surv Can Pap 61-12: 7 pp
Tikhomirov SV (1967) Devonian sedimentation stage on the Russian Platform (in Russian). Nedra Press, Moscow, 268 pp
Tikhvinsky IN (1971) Lower Permian of the Middle Volga Region (stratigraphy and paleogeography (in Russian). Avtoref Dokt Diss, Leningrad, 42 pp
Tikhvinsky IN (1973) Stratigraphic position of potassium-bearing series of the Central and Western Parts of the Pre-Caspian Depression (in Russian). Dokl Akad Nauk USSR 211: 668–671
Tikhvinsky IN (1974) The stratigraphy and potassium-bearing horizons of Kungur stage in the Fore-Caspian Depression (in Russian). Sov Geol 5: 44–54
Tikhvinsky IN (1976a) Regularities of accumulation of chloride-potash salts in the Preduralian-Precaspian Basin (in Russian). Lithol Polezn Iskop 1: 163–168
Tikhvinsky IN (1976b) Regular features of potassium salt deposits distribution in the Fore-Urals-Fore-Caspian Basin (in Russian). Sov Geol 2: 102–113

Tikhvinsky IN, Blizeev BI (1972) Major regularities of structure and accumulation conditions of the Kungurian deposits in the Ural-Volga Region (with a view to potassium content potentials) (in Russian). In: Problems of lithology and stratigraphy of the Pre-Paleozoic, Paleozoic and Mesozoic sediments of the Russian and the Scythian Platforms). Nedra Press, Moscow, pp 79–89

Tikhvinsky IN, Merzlyakov GA, Sementovsky YuV, Suveizdis PI (1977) Structure and formation environment of Permian salt-bearing deposits of the USSR (in Russian). In: Problems of salt accumulation, vol II. Nauka Press, Novosibirsk, pp 3–16

Tikhy VN (1964) Devonian period (in Russian). In: History of geological evolution of the Russian Platform and the adjacent areas. Nedra Press, Moscow, pp 52–85

Tikhy VN (1967) Devonian history of USSR. In: International Symposium on the Devonian system, vol I. Calgary, Alberta, Canada, pp 349–359

Tikhy VN (1973) Volga-Uralian oil-bearing Region (in Russian). In: Stratigraphy of the USSR, Devonian System, B 1. Nedra Press, Moscow, pp 128–145

Tilman SM, Bogdanov NA, Byalobzhensky SG, Chekhov AD (1970) Vrangel Island (in Russian). In: Geology of the USSR, vol XXVI, Soviet Arctic Islands. Nedra Press, Moscow, pp 377–404

Tolstikhina MM (1952) Devonian sediments in the center of the Russian Platform and the evolution of its basement in the Paleozoic (in Russian). Gosgeoltekhizdat, Moscow, 142 pp

Tozer ET (1963) Axel Heiberg and Stor Islands, South Side of Strand Fiord. In: Geology of the North-Central Part of the Arctic Archipelago, Northwest Territories (Operation Franklin). Geol Surv Can Mem 320: 448–456

Tozer ET, Thorsteinsson R (1964) Western Queen Elisabeth Islands, Arctic Archipelago. Geol Surv Can Mem 332: 242 pp

Tretjakov YuA (1977) Lithological control of the spatial distribution of the depletion zones of the Upper Kame potassium deposit (in Russian). In: Problems of salt accumulation, vol II. Nauka Press, Novosibirsk, pp 127–128

Trettin HP (1967) Devonian of the Franklinian Eugeosyncline. In: International Symposium on the Devonian System, vol I. Calgary, Alberta, Canada, pp 693–702

Trettin HP (1969) Geology of Ordovician to Pennsylvanian rocks, M'Clintock Inlet, north coast of Ellesmere Island, Arctic Archipelago. Geol Surv Can Bull 183: 93 pp

Trettin HP (1971) Geology of Lower Paleozoic Formations Hazen Plateau and Southern Grant Land Mountains, Ellesmere Island, Arctic Archipelago. Geol Surv Can Bull 203: 134 pp

Trevisan L (1972) Ähnlichkeiten und Unterschiede zwischen dem Verrucano der Toscana und dem der Alpen und eine Definition des Begriffes Verrucano. Verh Geol Bundesanst (Austria), pp 7–10

Trofimova AN, Efremov PE (1977) Structure of the Lower Permian salt-bearing sequence in review of the potassium salt prospecting. In: Problems of salt accumulation, vol II. Nauka Press, Novosibirsk, pp 105–108

Trunkó L (1969) Geologie von Ungarn. Gebr Borntraeger, Berlin, 257 pp

Tschopp RH (1967) The general geology of Oman. 7th World Petrol Congr Mexico 2: 870–903

Tsegelnyuk PD (1976) Brachipoda and stratigraphy of the Lower Paleozoic of the Volyn-Podolia region (in Russian). Naukova Dumka Press, Kiev, 156 pp

Tsegelnyuk PD, Bukatchuk PD (1974) South-Western Part of the Ukraina shield (in Ukrainian). In: Stratigraphy of the UkrSSR, IV. Silurian, vol I. Naukova Dumka Press, Kiev, pp 275–296

Tszu ZI (1964) Principal features of the tectonic development of Timan-Pechora Provinces. In: Oil-and-gas geology of the USSR North-Eastern Europe. Nedra Press, Moscow, pp 3–25

Tszu ZI, Kossovoy LS (1973) Timan-Pechora region (in Russian). In: Stratigraphy of the USSR, Devonian system, B 1. Nedra Press, Moscow, pp 145–166

Ulst RZ (1963) Frasnian carbonate rocks in the Latvian SSR (in Russian). In: Frasnian deposits in the Latvian SSR. Tr Inst Geol Latv SSR, vol X. Izd Acad Nauk Latv SSR, Riga, pp 143–200

Urazov GG (1932) On the deposition succession of the salts of the Solikamsk potassium deposit (in Russian). Tr Gl Geol Razved Upravleniya, Issue 43. Leningrad Moscow, pp 28–58

Ustritsky VI (1967) On situation of the North Pole in Late Paleozoic on the base of paleontological data (in Russian). Geol Geofiz 1: 25–32

Ustritsky VI (1972) The Permian climate (in Russian). Dokl Akad Nauk USSR Ser Geol 4: 3–12

Ustritsky VI (1975) History of evolution of the North-East USSR in the Late Paleozoic (in Russian). In: Upper Paleozoic of the North-East USSR. Rotaprint NIIGA, Leningrad, pp 54–75

Utekhin DN (1971) Devonian System, Lower Division, Middle Devision (in Russian), Geological description. In: Geology of the USSR, vol 4 (Centre of the European regions of the USSR). Nedra Press, Moscow, pp 127–154

Utekhin DN, Sorskaya LS (1971) Devonian System, Upper division (in Russian). In: Geology of the USSR, vol 4 (Centre of the European Regions of the USSR), Geological description. Nedra Press, Moscow, pp 154–188

Vadas E (1964) Geology of Hungary (in Russian). Mir Pres, Moscow, 530 pp

Valyashko MG (1962) Geochemical regularities of formation of the potassium salt deposits (in Russian). Izd MGU, Moscow, 395 pp

Valyashko MG (1963) On the constancy of World Ocean waters (in Russian). Vestn Mosk Gos Univ Ser Geol 1: 18–27

Valyashko MG (1967) Principles of the geochemistry of natural waters (in Russian). Geokhimia 11: 1395–1405

Valyashko MG, Vlasova NK (1975) Stability of magnesium salts in marine solutions and its geochemical importance (in Russian). Vestn Mosk Gos Univ 4: 16–27

Varentsov MI, Ditmar VI, Shmakova EI (1963) The salt domes of Sarysu (in Russian). Dokl Acad Nauk SSSR 151: 396–398

Vary JA, Elenbaas JR, Johnson MA (1968) Gas in Michigan Basin. In: Natural gases of North America. Tulsa, Oklahoma, USA, Published by the AAPG, pp 1761–1797

Vasiljev YuM (1968) Geological structure of the Pre-Caspian Depression and the regularities of oil and gas distribution in its interior (in Russian). Nedra Press, Moscow, 178 pp

Veevers JJ (1967) The Phanerozoic geological history of Northwest Australia. J Geol Soc Aust 14: 235–272

Veevers JJ, Wells AT (1959) Probable salt dome at Woolnaugh Hills, Canning Basin, Western Australia. In: Papers on Western Australia stratigraphy and paleontology. Bur Miner Resour Rep 38: 97–113

Veevers JJ, Wells AT (1961) The geology of the Canning Basin, Western Australia. Bur Miner Resour Aust Bull 60: 323 pp

Vertunov LN (1968) Lithology, paleogeography and the problem of oil-and-gas content in the Cenozoic molasse formations of the Issyk-Kul Depression (in Russian). Avtoref Dokt Diss, Frunze, 65 pp

Vicente JC (1970) Étude géologique de l'Ile de Gavdos (Grèce), la plus méridionale de l'Europe. Bull Soc Géol Fr Sér 7, 12: 481–495

Vicente JC (1975) Essai d'organisation paléogéographique et structural du Paléozoique des Andes Méridionales. Geol Rundsch 64: 343–394

Vinogradov AP (1959) Chemical evolution of the Earth (in Russian). Izd Akad Nauk SSSR, Moscow, 44 pp

Vinogradov PD (1959) Permian deposits of South-East Tadjikistan (in Russian). In: Geology of the USSR, vol 24 (Tadjik SSR), part 1, Geological description. Gosgeoltekhizdat, Moscow, pp 149–163

Vinogradov VI, Pustylnikov AM (1977) Isotopic composition of sulfur in the Cambrian salt-bearing deposits of the Siberian Platform (in Russian). In: Problems of salt accumulation, vol II. Nauka Press, Novosibirsk, pp 237–240

Visloguzova AB, Karagodin PF, Kozlovsky GM (1968) Tectonic structure of the Permian oil-bearing region of the Bestyube group of salt domes (in Russian). In: Geological structure and oil-and-gas presence potentials of the depressions of Southern Kazakhstan. Alma-Ata, pp 191–197

Vodzinskas EV, Kadunas VB (1969) Carbonate resources of the Lithuanian SSR, dolomites and limestones (in Russian). Tr Inst Geol, Issue 13. Mintis Press, Vilnyus, 199 pp

Voinovsky-Kriger KG (1963) Carboniferous deposits of the Lemva facies-structural zone, western slope of the Polar Urals (in Russian). Byull Mosk Ova Ispyt Prir Otd Geol 38: 56–77

Volkheimer W (1969) Paleoclimatic evolution in Argentina and relations with other regions of Gondwana. In: Gondwana stratigraphy. Unesco, Paris, pp 551–588

Vorobiev VN, Safronova IG (1977) Peculiarities of the structure of the marginal (south-western) zone of the Cambrian halogenic formation of the Siberian Platform (in Russian). In: Problems of salt accumulation, vol II. Nauka Press, Novosibirsk, pp 234–237

Vozárová A, Vozár J (1975) Zàkladué crty paleogeographie mladsieho paleozoika Západných Karpát. Geol Pr Spravy 64: 81–94

Vyalov OS (1966) Riphean and Paleozoic formation of the Cis-Carpathian Trough (in Russian). In: Geology of the USSR, vol 48 (Carpathians), part 1, Geological description. Nedra Press, Moscow, pp 53–58

Walter R (1972) Palaeogeographie des Siluriums in Nord-, Mittel- und Westeuropa. Geotektonische Forsch 41: 180 pp

Wanless HR (1975a) Missouri and Iowa. In: Paleotectonic investigations of the Pennsylvanian system in the United States, Part I. US Geol Surv Prof Pap 853: 97–114

Wanless HR (1975b) Illinois Basin Region. In: Paleotectonic investigations of the Pennsylvanian system in the United States, Part I. US Geol Surv Prof Pap 853: 71–95

Wanless HR, Shideler GL (1975) Michigan Basin Region. In: Paleotectonic investigations of the Pennsylvanian system in the United States, Part I. US Geol Surv Prof Pap 853: 63–70

Wardlaw NC (1968) Carnallite-sylvite relationships in the Middle Devonian Prairie evaporite formation, Saskatchevan. Bull Geol Soc Am 79: 1273–1294

Waterhouse JB (1976) World correlations for Permian marine faunas. Univ Queensl Pap Dep Geol 7: 232 pp

Watson JM, Swanson CA (1975) North Sea – major petroleum province. Am Assoc Petrol Geol Bull 57: 1098–1112

Weisbord NE (1967) The Devonian System in Western Venezuela. In: International Symposium on the Devonian System, vol II. Calgary, Alberta, Canada, pp 215–226

Wells AT (1969) 1:250 000 Geological series – explanatory note. Alice Springs, Northern Territory, Sheet SS/53-14. Bur Miner Resour Aust, Canberra, 24 pp

Wells AT, Forman DJ, Ranford LC (1964) Geological reconnaissance of the Rawlinson and Macdonald 1:250 000 Sheet areas, Western Australia. Bur Miner Resour Aust Rep 65: 35 pp

Wells AT, Forman DJ, Ranford LC (1965) The geology of the North-Western part of the Amadeus Basin, Northern Territory. Bur Miner Resour Aust Rep 85: 45 pp

Wells AT, Stewart AJ, Skwarko SK (1966) Geology of the South-Eastern part of the Amadeus Basin, Northern Territory. Bur Miner Resour Aust Rep 88: 59 pp

Wells AT, Ranford LC, Stewart AJ, Cook PI, Shaw RD (1967) Geology of the North-Eastern part of the Amadeus Basin, Northern Territory. Bur Miner Resour Aust Rep 113: 93 pp

Wells AT, Forman DJ, Ranford LC, Cook PI (1970) Geology of the Amadeus Basin, Central Australia. Bur Miner Resour Bull 100: 222 pp

Wells JW (1957) Coral reefs. In: Treatise on marine ecology and paleoecology, vol I. Geol Soc Am 67: 609–631

West IM, Brandon A, Smith M (1968) A tidal flat evaporitc facies in the Visean of Ireland. J Sediment Petrol 38: 1079–1093

Whittington HB, Williams A (1964) The Ordovician period. In: The Phanerozoic time scale. A Symposium. Q J Geol Soc London 120 S: 241–254

Williams MY (1923) Reconnaissance across Northeastern British Columbia and the geology of the northern extension of Franklin Mountains. NWT Geol Surv Can Sum Rep Part B, pp 97–104

Willis RP (1959) Upper Mississippian – Lower Pennsylvanian stratigraphy of Central Montana and Williston Basin. Am Assoc Petrol Geol Bull 43: 1940–1966

Willis RP (1967) Geology of the Arabian Peninsula, Bahrain. US Geol Surv Prof Pap 560-E: 4 pp

Wilson ME (1968) Precambrian of Canada, Canadian Shield (in Russian). In: Precambrian of Canada, Greenland, British Isles and Spitsbergen. Mir Press, Moscow, pp 236–369

Withington CF (1965) Suggestions for prospecting for evaporite deposits in South-Western Virginia. US Geol Surv Prof Pap 525-B: 29–33

Wolfort R (1967) Zur Entwicklung der paläozoischen Tethys in Vorderasien. Erdöl Kohle Erdgas Petrochem, Jahrg 20, 3: 168–180

Workum RH (1965) Lower Paleozoic salt, Canadian Arctic Islands. Bull Can Petrol Geol 13: 181–191

Wyźykowski J (1970a) The Sudetic Foreland, the Lower Permian (Rotliegendes). In: Geology of Poland, vol I, part 1. Publ House Wydawnictwa Geol, Warsaw, pp 516–519
Wyźykowski J (1970b) The Sudetic Foreland, the Upper Permian (Zechstein). In: Geology of Poland, vol I, Part 1. Publ House Wydawnictwa Geol, Warsaw, pp 566–567
Yanev S (1969) Facies and lithogenic types of Early Paleozoic sediments in North-West Bulgaria (in Bulgarian). Bull Geol Inst Bulg Acad Sci Ser Stratigr Lithol 18: 2–12
Yanev S (1970a) Type rocks in the Early Paleozoic sediments of North-West Bulgaria (in Bulgarian). Bull Geol Inst Bulg Acad Sci Ser Stratigr Lithol 19: 26–40
Yanev S (1970b) Paleogeography of North-West Bulgaria during the Paleozoic (in Bulgarian). Rev Bulg Geol Soc 31: 197–208
Yanev S (1971) Structure of the Early Paleozoic complexes in North-West Bulgaria (in Bulgarian). Bull Geol Inst Bulg Acad Sci Ser Stratigr Lithol 20: 111–132
Yanev S (1976) The lithologic facies criteria of the oil-and-gas presence potentials of the Devonian-Carboniferous sediments in North-East Bulgaria (in Bulgarian). Rev Bulg Geol Soc 37: 149–159
Yanshin AL (1961) On the depth of salt-bearing basins and some problems of thick salt series forming (in Russian). Geol Geofiz 1: 3–15
Yanshin AL (1963) Principle of actualism and problems of evolution of the geological processes (in Russian). In: Ways and methods of study of the regularities of Earth evolution (abstracts of the papers presented at the Joint Theoretical Conference of the Philosophic Seminars of the Scientific Institutions of Acad Sci USSR). Izd Akad Nauk USSR, Moscow, pp 1–8
Yanshin AL (1965) The tectonic structure of Eurasia (in Russian). Geotektonika 5: 7–34
Yanshin AL (1973) On so-called world transgressions and regressions (in Russian). Byull Mosk Ova Ispyt Prir Otd Geol 48: 9–44
Zadorozhnaya NM, Osadchaya DV, Zhuravleva IT, Luchinina VA (1973) Early Cambrian organogenic structures on the territory of Tuva, Sayany-Altaian folded region (in Russian). In: Environment and life in the geological past, Late Cambrian and Paleozoic of Siberia. Nauka Press, Novosibirsk, pp 53–65
Zaikov VV, Kolosov AS, Onufrieva EV (1967) New data on the salt presence in the Devonian rocks of Tuva (in Russian). Geol Geofiz 8: 21–27
Zaitsev NS (1963) Peculiarities of evolution of the Siberian Caledonides in the Middle and Upper Paleozoic (in Russian). In: Problems of regional tectonics of Eurasia. Izd Akad Nauk USSR, Moscow, pp 90–129
Zaitsev NS, Pokrovskaya NV (1948) Geological structure of the Tes-Bulak Region in Bet-Pak-Dala (in Russian). Tr GIN Acad Nauk SSSR, Issue 102, Ser Geol No 33. Izd Akad Nauk USSR, Moscow, pp 36–84
Zamaraev SM, Ryazanov GV (1972) Tectonics of the Paleogene formation of the southern Siberian Platform and the structural pecularities of some fold zones in view of the potasssium content potentials (in Russian). In: Potassium content potentials of Siberia. Nauka Press, Moscow, pp 76–81
Zamarenov AK, Kukhtinov DA, Bulekbaev ZY, Taran LV, Zamarenova EN (1972) The Upper Permian sediments of the eastern side part of the Fore-Caspian Depression (in Russian). Sov Geol 6: 59–69
Zavidonova AG (1956) Pre-Paleozoic and Paleozoic sediments of the Moldavian SSR (in Russian). Byull Mosk Ova Ispyt Prir Otd Geol 61:31–50
Zelentsov II, Zingerman AYa (1977) On the zoning of the Petrikov potassium salt deposits, Pripyat Depression (in Russian). In: Problems of salt accumulation, vol II. Nauka Press, Novosibirsk, pp 192–193
Zhaiba S, Narbutas V, Karatayute V (1961) Devonian system (in Russian). In: Geology of the USSR, vol 39 (Lithuanian SSR), Geological description. Gosgeoltekhizdat, Moscow, pp 62–85
Zhalybin FI, Bakirov KKh, Tsvetkov DV (1974) The Upper Permian salt-bearing deposits of the central and south-eastern parts of the Fore-Caspian Depression (in Russian). Sov Geol 6: 54–64
Zharkov MA (1966) The Siberian Platform salt-bearing formation of Cambrian age (in Russian). Sov Geol 2: 32–45

Zharkov MA (1969) On the scale of salt accumulation at the Cambrian period (in Russian). Dokl Akad Nauk SSSR 184: 913–914

Zharkov MA (1970) Conditions of formation of the salt-bearing deposits in Siberia and their potassium content potentials (in Russian). In: State and aims of Soviet Lithology, vol III. Nauka Press, Moscow, pp 40–48

Zharkov MA (1971a) Evolution of salt accumulation in geological history (in Russian). In: Problems of general and regional geology. Nauka Press, Novosibirsk, pp 260–299

Zharkov MA (1971b) On the transitional basins of the salt accumulation epoch and the paragenetic association with the salt-bearing series of oil-and-gas fields (in Russian). In: Problems of oil presence in Siberia. Nauka Press, Novosibirsk, pp 163–185

Zharkov MA (1974a) Paleozoic salt formations of the world (in Russian). Nedra Press, Moscow, 392 pp

Zharkov MA (1974b) Regularities of the spatial and age distribution of the Paleozoic salt-bearing series (in Russian). In: Physico-chemical and paleogeographic problems of salt accumulation and the formation of the potassium salt deposits. Rotaprint IGiG SO Akad Nauk SSSR, Novosibirsk, pp 21–24

Zharkov MA (1975) Oil-and-gas content of the evaporite formations (in Russian). In: Sedimentary basins and their oil-and-gas content. Seminar held at the Moscow State University, December 25–27, 1975, pp 102–110

Zharkov MA (1976) Evolution of the Paleozoic salt accumulation (in Russian). In: Evolution of sedimentary rock formation in the history of the Earth. Rotaprint IGiG SO Akad Nauk USSR, Novosibirsk, pp 5–19

Zharkov MA (1977) Stages, volume, and area of the Paleozoic salt accumulation (in Russian). In: Problems of salt accumulation, vol I. Nauka Press, Novosibirsk, pp 57–88

Zharkov MA, Chechel EI (1973) Cambrian sedimentary formations of the Angara-Lena Depression, 1. General characteristic of the Cambrian deposits and the carbonate formation of the Angara-Lena Depression (in Russian). Nauka Press, Novosibirsk, 240 pp

Zharkov MA, Khomentovsky VV (1965) Principal problem of stratigraphy of the Cambrian and Vendian in the South of the Siberian Platform in connection with salt content (in Russian). Byull Mosk Ova Ispyt Prir Otd Geol 19: 100–118

Zharkov MA, Skripin AI (1971) Upper Cambrian deposits of the South of the Siberian Platform (in Russian). Nauka Press, Novosibirsk, 100 pp

Zharkova TM (1976a) Rock types of the Cambrian salt-bearing formation of the Siberian Platform (in Russian). Nauka Press, Novosibirsk, 304 pp

Zharkova TM (1976b) Classification and nomenclature of rocks of the Cambrian and Devonian salt-bearing formations of Eurasia (in Russian). In: Evolution of sedimentary rock formation in the history of the Earth. Rotaprint IGiG SO Akad Nauk USSR, Novosibirsk, pp 20–30

Zhukov FI (1963) Jedna z variant rozdĕleni svrhcuepermskych sedimentú Spišsko-gemerského rudohoři. Geol Pr Správy 30: 39–46

Zhukov FI (1965b) Paleograficeskije osobenosti razvitija Spišsko-gemerského rudogoria v verchnepermskij period. Geol Pr Správy 35: 5–22

Zhukov FI (1965b) Possible dissection of red beds of the Rahov massive (in Russian). Geol J 25: 98–102

Zhukov FI, Yanev SN (1971) Carpathian-Balkan Region in the Permian period. Acta Geol Acad Sci Hung 15: 323–331

Zhukov FI, Vozar II, Yanev SN (1976) Permian sedimentary-volcanogenic formations and ore deposits of the Carpathian-Balkan Region (in Russian). Naukova Dumka Press, Kiev, 182 pp

Zhukov FI, Sergeeva LO, Pasechnik YaV (1964) Some characteristic features of Carpathians Paleozoic in the Rachov massive (in Russian). Geol J 24: 31–37

Zhuravleva IT (1966) Early Cambrian organogenic structures on the territory of the Siberian Platform (in Russian). In: Organism and environment in the geological past. Nauka Press, Moscow, pp 61–84

Zhuravleva IT (1968) Biogeography and geochronology of the Early Cambrian based on archaeocyatids (in Russian). In: Problems of paleontology. Papers presented by the Soviet geologists at the XXIIIrd Session of the International Geological Congress. Nauka Press, Moscow, pp 33–45

Ziegler PA (1969) The development of sedimentary basins in Western and Arctic Canada. Alberta Soc Petrol Geol, Calgary, Alberta, Canada, 89 pp

Ziegler PA (1973) Geologic evolution of North Sea and its tectonic framework. Am Assoc Petrol Geol Bull 57: 1073–1097

Zonenshein ZP (1976) Reconstruction of the Paleozoic oceans (in Russian). In: Continental drift. Nauka Press, Moscow, pp 28–71

Zoricheva AI (1966) North-west of the Russian Platform (in Russian). In: Stratigraphy of the USSR, Permian System. Nedra Press, Moscow, pp 117–132

Zoricheva AI (1973) Northern regions of the Russian Platform (in Russian). In: Stratigraphy of the USSR, Devonian System, B 1. Nedra Press, Moscow, pp 106–113

Subject Index

H.-E. Reineck, I. B. Singh

Depositional Sedimentary Environments

With Reference to Terrigenous Clastics

Springer Study Edition

2nd revised and updated edition. 1980.
683 figures, 38 tables. XIX, 549 pages
ISBN 3-540-10189-6

From the reviews of the 1st edition:

"...In conclusion, both the authors and the publisher are to be congratulated on the production of an excellent textbook. The text is beautifully illustrated with 597 figures and photographs printed on glossy paper, and it provides an up-to-date, concise, and wellorganized review of the voluminous literature written on the subject of depositional environments. It represents a valuable addition not only to the literature of sedimentology, but also to that of earth science in general."

Palaeogeography, -climatology, -ecology

Springer-Verlag
Berlin
Heidelberg
New York